The Age of Johnson

A Scholarly Annual

8

THE AGE OF JOHNSON

A Scholarly Annual

8

Edited by
Paul J. Korshin

AMS PRESS, Inc.
New York

The Age of Johnson welcomes contributions on all aspects of the literature, history, and culture of the period of Samuel Johnson's literary career and primary influence, roughly the years from 1730 to 1810. Contribution should not exceed 12,000 words; the style of documentation should follow the latest edition of *The Chicago Manual of Style*. Authors who would like to contribute or to suggest books for review should write to Paul J. Korshin, Editor; *The Age of Johnson*; Department of English; University of Pennsylvania; Philadelphia, Pennsylvania 19104.

ISSN 0884-5816
Set ISBN 0-404-62750-1
Volume 8 ISBN 0-404-62758-7
Library of Congress Catalog Card Number 85-48058

AMS PRESS, INC.
56 East 13th Street
New York, NY 10003-4686

Manufactured in the United States of America

CONTENTS

Review Essay

Reviews

List of Illustrations

PREFACE

Volume 8 of *The Age of Johnson* continues the debate on Samuel Johnson's ties to Jacobitism, which are substantial or insubstantial, depending on one's view. The disputants who introduced this topic in Volume 7, J. C. D. Clark and Howard Erskine-Hill, write again, this time in response to the essays by Donald Greene and Thomas Curley and to Howard Weinbrot's lengthy review of Professor Clark's book on Johnson's politics, all of which I published in Volume 7. Professors Greene, Curley, and Weinbrot then respond to Professors Clark and Erskine-Hill. Further essays in this volume by Michael Caldwell and Lawerence Lipking pursue the subjects of Johnson's political thought and his relationship to Jacobitism. This volume continues with our practice of publishing original documents from eighteenth-century archives, in this case with William McCarthy's history of Anna Letitia and Rochemont Barbauld's famous Palgrave School. Volume 8 contains one further response, James Basker's essay on the myth of Johnson's misogyny, which is a comment on a book review in *The Age of Johnson*, Volume 6 (1994). We also include several essays and reviews which the length of the debate on Jacobitism in Volume 7 forced me to postpone, a circumstance which I note with much regret.

The editor gratefully acknowledges the financial assistance of the Department of English and the School of Arts and Sciences of the University of Pennsylvania.

Paul J. Korshin

Philadelphia
August 1996

NOTES ON CONTRIBUTORS

JOHN L. ABBOTT is Professor of English at the University of Connecticut. He is author of *John Hawkesworth: Eighteenth-Century Man of Letters* (Wisconsin, 1982). With D. G. C. Allan he edited *"The Virtuoso Tribe of Arts & Sciences": Studies in the Eighteenth-Century Work of the London Society of Arts* (Georgia, 1992). He is currently editing volume 6 of *The Early Journals and Letters of Fanny Burney* for McGill University Press and Clarendon Press, Oxford.

JAMES G. BASKER, Professor of English at Barnard College, Columbia University, is the author of *Tobias Smollett, Critic and Journalist* (1988); his most recent publication, edited with Alvaro Ribeiro, S.J., is *Tradition in Transition: Women Writers, Marginal Texts, and the Evolving Canon* (Oxford Univ. Press, 1996). He is currently editing *Roderick Random* for the University of Georgia Smollett edition and finishing a book on *Samuel Johnson and the Common Reader.*

PETER M. BRIGGS is Professor of English at Bryn Mawr College. He is working on a longer study of the public uses of example in eighteenth-century British culture.

MICHAEL CALDWELL is a graduate student in English at the University of Chicago. He is completing a dissertation entitled "The Spotted Page: Danverian Satire in the Works of Gay, Pope, and Fielding."

JONATHAN CLARK is Joyce and Elizabeth Hall Distinguished Professor of British History at the University of Kansas; he was formerly a fellow of All Souls College, Oxford, and Peterhouse, Cambridge. *Samuel Johnson: Literature, Religion and English Cultural Politics from the Restoration to Romanticism* (Cambridge, 1994) is his seventh book.

JAMES CRUISE's most recent publication, on Sarah Scott, appeared in the Summer 1995 issue of *SEL*.

THOMAS M. CURLEY, a professor of English Literature at Bridgewater State College, has recently completed a biography, *Sir Robert Chambers: Law, Literature, and Empire in the Age of Johnson*, which contains major new information about Johnson and his age.

HOWARD ERSKINE-HILL is Professor of Literary History in the English Faculty of Cambridge University and Fellow of Pembroke College, Cambridge. Author of *The Social Milieu of Alexander Pope* (1975) and *The Augustan Idea in English Literature* (1983), he has two new books shortly to appear from the Clarendon Press, Oxford: *Poetry and the Realm of Politics: Shakespeare to Dryden* and *The Poetry of Opposition and Revolution: Dryden to Wordsworth*. The latter contains a substantial discussion of the earlier Samuel Johnson.

ERIC J. FRISCHHERTZ formerly taught English at the Jesuit High School of New Orleans and at Xavier University of New Orleans. He has recently decided to change his profession and is preparing to study medicine.

JEANNE GRIGGS teaches at Kenyon College and at Ohio State University's Newark campus. Her research interests include satire and parody, satirical prefaces and dedications, and rhetorical strategies, especially irony. Other current interests include her own poetry and Erasmus.

ISOBEL GRUNDY is Henry Marshall Tory Professor at the University of Alberta. She is editing Lady Mary Wortley Montagu (her *Romance Writings*, Oxford: Clarendon Press, 1996, and her *Selected Letters*, Penguin, forthcoming) and writing her life. Her next project is a major collaborative venture: a history of women's writings in the British Isles, appearing in printed and in electronic form.

GEORGE JUSTICE is Assistant Professor of English at Marquette University. He received his doctorate from the University of Pennsylvania and is working on a book tentatively entitled *The Production of*

Literature in Eighteenth-Century Britain, which examines relations between the material workings of the literary marketplace and the forms and meanings of literature from Addison and Steele to Frances Burney.

THOMAS KINSELLA is Assistant Professor of British Literature at the Richard Stockton College of New Jersey. He specializes in Boswell, bibliography, and bookbinding.

LAWRENCE LIPKING, Chester D. Tripp Professor of Humanities at Northwestern University, has written *The Ordering of the Arts in Eighteenth-Century England* (1970), *The Life of the Poet* (1981), and *Abandoned Women and Poetic Tradition* (1988), as well as a forthcoming book on Johnson's life as an author.

WILLIAM MCCARTHY, Professor of English at Iowa State University, is the author of *Hester Thrale Piozzi: Portrait of a Literary Woman* (1985) and co-editor of *The Poems of Anna Letitia Barbauld* (1994). Since 1987 he has been at work on a biography of Barbauld.

MAXIMILLIAN E. NOVAK is Professor of English at the University of California, Los Angeles.

ALLEN REDDICK is Professor of English at the University of Zürich and the author of *The Making of Johnson's Dictionary, 1746-1773.* He is now at work on a study of wit, poetics, and rhetorical violence in seventeenth- and eighteenth-century England.

JEROME M. REICH is a practicing pulmonologist at Bess Kaiser Medical Center, Portland, Oregon, who was "strongly impregnated with the Johnsonian aether" by Professor E. L. McAdam when he was an undergraduate student at New York University in the mid-fifties.

MARY MARGARET STEWART is Graeff Professor of English at Gettysburg College. She has published essays on Boswell, Fielding, and Collins, and is working on a biography of Collins. She is also studying father-daughter relationships, especially that of Charles Hanbury Williams with his two daughters.

HOWARD D. WEINBROT is Ricardo Quintana Professor of English and William F. Vilas Research Professor in the College of Letters and Science at the University of Wisconsin, Madison. His latest book is *Britannia's Issue: The Rise of British Literature from Dryden to Ossian* (Cambridge, 1993).

SAMUEL JOHNSON

AND

JACOBITISM

Editor's note. In the following exchange, the essays of Howard Erskine-Hill and J. C. D. Clark are in response to the contributions to the debate on Johnson and Jacobitism by Donald Greene, Thomas M. Curley, and Howard Weinbrot that appeared in *The Age of Johnson*, volume 7. Professors Erskine-Hill and Clark did not have the opportunity to consider the essays by Professors Greene, Weinbrot, and Curley that appear in volume 8. These three authors, however, in their essays in this volume, respond directly to the contributions of Professors Erskine-Hill and Clark that follow.

A KIND OF *LIKING* FOR JACOBITISM[1]

HOWARD ERSKINE-HILL

This briefly responds to Donald Greene's extravagant further contribution to the debate about "Johnson the Jacobite?" and Thomas M. Curley's supporting article (*The Age of Johnson*, 7). New light has been cast on Johnson, though I could wish we had achieved a more decent academic exchange. The degree to which Greene and Curley are disturbed is marked by their exaggeration and over-reaction. Something similar, I am afraid, must be said of the latter part of Professor Weinbrot's review of Professor Clark's *Samuel Johnson*. A strange creature, they feel, has intruded into their familiar pool. They hardly know its nature or size—some of them are all at sea as to varieties of Jacobitism, or what non-juring might mean—but they are resolved to drive it away at all costs. In Howard Weinbrot, one of the best factual scholars of eighteenth-century literary history, a champion has been enlisted who, at least in his fair-minded preamble, and in his citations of Nicholas Amhurst and Thomas Hearne, has done well but, on the evidence I have seen, is not without his own misunderstandings, and has been bested by Professor Clark on the question of Johnson and the Oaths.

My own overall position is not identical with that of Clark, though I have learned very much from him. To discover discrepancies between us is not (as Curley thinks) to show that both our arguments fall. Let me patiently explain what I have and have not been trying to do. I did not "attack" Greene, but only challenged some of his views. The modern rediscovery of Jacobitism is not my work but that of a generation of historians. The question of Johnson's relation to

Jacobitism is an old one; the allegation of crypto-Jacobite intent in certain of his works goes back to Sir John Hawkins. I do not claim that my articles on Johnson "tell the whole *truth*" (Curley's first quotation from the Vinerian Lectures, in *The Age of Johnson*, 7). I do claim to have begun to assemble a new case for Jacobite sympathy on Johnson's part which will, I hope, deserve consideration in future reviews of Johnson's total political record.[2] Most important of all, it will be clear to any impartial reader that I think Johnson changes and is not "marmoreally changeless and self-sufficient."[3] I think *The Vanity of Human Wishes* may mark the end of serious Jacobite inclination in Johnson; but on some future occasions he would still write and talk like a sometime-Jacobite sympathizer.

Professor Curley raises some new points. Having edited the Chambers/Johnson Vinerian Lectures, he naturally stresses their importance. I did not cite them, partly because they are much too late for my case,[4] and partly because this text is a collaboration. As late as 1986 Curley saw reason for caution in accepting all the attributions to Johnson proposed by E. L. McAdam, Jr. Indeed, some of the texts most insisted on by Curley, Greene, and Weinbrot are collaborative and therefore shaky evidence.[5] Curley does introduce new evidence on the pension episode (Wedderburn's statement), but misses the point of the current discussion: nobody is now arguing that Johnson was pensioned *because* he had Jacobite sympathy. What is significant here is that both the author of the pension letter and Wedderburn had to argue that the pension was awarded on literary merit alone, irrespective of Johnson's earlier political views. It is quite improbable that those defending Johnson's case for a pension would not have invoked his loyal disposition to the House of Hanover if clear evidence of such a disposition had existed. Curley also mentions Sir Robert Filmer and patriarchalism. Patriarchalism may indeed have remained a potent theory of the origin of government in an age more biblical and religious than our own, but it is wrong to think that an eighteenth-century Jacobite must by that token have been a Filmerian patriarchalist. Here, as elsewhere, Clark's and my opponents go astray through working with naively simple definitions. Reviewing Curley's case as a whole, however, I must say that it valuably reminds the modern reader what a serious matter it was to have had principled reservations about the *status quo* in church and state in eighteenth-century Britain. Not a few were beheaded, hanged, or transported for Jacobitism, not to mention being barred from the professions and suffering other disabilities. It is hardly surprising that Jacobite sympathizers did not readily go on record with plain statements of their real views. Curley too readily accepts what the late E. P. Thompson used to call "winners' history." The comprehensive historian will remember losers' history too. To assume that eighteenth-century British experience is coterminous with the orthodoxies of the Hanoverian/British state is quite wrong. To think that Johnson was mentally constrained by these orthodoxies is to think something quite improbable. It might also be interesting for Curley to reconsider the legal implications of what was done in 1688-89, and how far these could be accommodated within the principles and terms of the Vinerian Lectures. The *Course of Lectures*, far from commencing its account of the constitution from 1689, derives it from the long succession of English history. While it rehearses post-1689 legislation, it insists that "The Parliament used in the popular sense as comprising only the House of Lords and Commons is not strictly in the *legal* sense a Parliament. It may be termed a political body of which the king is the political soul, for unanimated by the regal authority they have neither voice nor motion."[6] What king concurred in settling the crown on William and Mary? Not, by definition, William and Mary themselves. Not James II. The decision of the Convention did not then constitute an act of parliamentary legality in the terms just expounded by the *Course of Lectures*. Rather it constituted what, according to Hawkins, Johnson had "almost asserted in terms": "that the succession to the crown had been illegally interrupted."[7]

In his new Introduction, Greene asked why we should trust Boswell on Johnson's opinions rather than Hawkins, Thrale, or Burney.[8] As the above quotation and other passages from Hawkins show, the current debate about Johnson and Jacobitism would probably have occurred if Boswell had never existed, for there would still be plenty of evidence on the side of Jacobite sympathy, not least Hawkins himself. Greene, however, does not hesitate to quote Boswell when Boswell supports Greene's argument. He also spends much time charging Boswell with a series of faults from "Sly editorial nudging" (new Introduction, p. xliii) to the reshaping or even invention of material. Boswell, then, continues to be of importance to each side of the debate, and if there is, to put it at its lowest, some deposit of truth in Boswell, then his revisions must be of interest. With the recent publication of the first volume of Marshall Waingrow's edition of the manuscript of *The Life of Johnson*, we begin to be able to study the intermediate material between Boswell's journal record and the first edition of the *Life*.[9] Where Jacobite material is concerned it is clear that Boswell's revisions soften and play it down, sometimes distancing both Johnson and Boswell himself from the events and views to which he refers.

Thus under 1746 the *Life* refers to "a civil war in Great-Britain, when a rash attempt was made to restore the House of Stuart to the throne" (Hill-Powell, I, 176). Waingrow's main text is the same in substance (p. 130), but the textual notes (p. 420 for l. 19) reveal that the words "when a rash attempt was made" previously ran "undertaken with a view." Boswell here departed in revision from a neutral earlier expression in order to decry the '45. If this is "sly nudging," it is nudging away from the most obvious Jacobite view.

Two revisions deal with the matter of Johnson's pension. Under 1762 the *Life* refers to this in the following (familiar) terms: "Johnson, having been represented to him [George III] as a very learned and good man, without any certain provision, his Majesty was pleased to grant him a pension of three hundred pounds a year" (Hill-Powell, I, 372). The MS is more revealing in several ways, but the relevant difference here is that Boswell had previously written: "Johnson though distinguished for his aversion from the first two princes of the House of Hanover was now considered to be a most proper object of Royal Munificence, and the King was pleased to grant him a Pension of £300 a year" (Waingrow, p. 258; see also p. 467). Here again Boswell moves in the direction of an acceptable Hanoverian viewpoint in his revision. Under 7 April 1773, Boswell in his original journal recorded a discussion with Topham Beauclerk much more explicit than anything in the *Life*. Boswell represents himself as saying: "Mr. Johnson's accepting a pension from a prince whom he had called an usurper was a circumstance which it was difficult to justify with perfect clearness." When Beauclerk then defended and explained Johnson's position, Boswell conceded that while Johnson had been "represented as a violent Jacobite," it was not clear that he would have wished for a Jacobite victory at Culloden (*Boswell for the Defence*, ed. W. K. Wimsatt and F. A. Pottle [London, 1960], pp. 173-74). Boswell's original remark was very strong, much stronger than any reference to Johnson's aversion to George I and George II. Not surprisingly, all this was omitted from the *Life*. If the corresponding moment in the *Life* be considered the discussion of the pension under 14 July 1763 ("I cannot now curse the House of Hanover": Hill-Powell, I, 429-30), then Boswell does here seem to have introduced new material in the very act of softening and extenuating Johnson's position.

I owe to the kindness of Tulin Faulkner, the Managing Editor of the Yale Edition of the Boswell Papers, the following further instance, from the period of the *Life* for which the MS is not yet published.[10] Boswell in the *Life* records Johnson's comment on the Oaths: "I own, that it is a very difficult question, when considered with respect to the House of Stuart. . . . I know not whether I could take them; but I do not blame those who do" (Hill-Powell, II, 220). The MS had exactly the same statement, but in indirect speech: "At the same time he owned it was a very difficult question . . .", etc. Here the change into direct speech makes Johnson's opinion come over slightly more assertive, but the substance is identical in the two versions. Tulin Faulkner also draws my attention to another case of omission in the composition of the *Life*. In Boswell's *Journal of my Jaunt* (1762), David Hume was recorded as having said of Johnson: "He is a man of enthusiasm and antiquated notions, a keen Jacobite yet hates the Scotch. He holds the Episcopal Hierarchy in supreme veneration" (*Boswell Papers*, ed. Isham, I, 128). Had Boswell perhaps merely forgotten this material? No, because he included from it in more decorous form the famous "Green room" episode, which was so close as to be on the same printed page in the Isham edition. Evidently Hume's report was deliberately left out of the *Life*. In his new Introduction, Greene rebukes me for not mentioning that the "brilliant piece of stage-direction" (p. xlvi) with which Boswell presents his last Jacobite exchange with Johnson (Hill-Powell, IV, 165) was not in the corresponding passage of the journal. He is right. This does not affect the point I was making ("Political Character," p. 116). But it is relevant to note here that the opinions Boswell credits Johnson with at this point in the *Life* are recorded as his at this point in the journal.[10]

All these instances are salient examples for the present debate. In so far as it is valuable to check Boswell against Boswell, they show a biographer who in drawing his original narratives either omits Jacobite material, or generally presents it in softened and extenuated form. They do not support the notion of a Boswell who frequently exaggerates the Jacobite remarks attributed to Johnson in the journals. If, furthermore, after all Boswell's diplomatic extenuations there remains a substantial body of Jacobite remarks in common between journals and *Life*, Boswell may by this token be held to be the more reliable. In the case of the remark about the Oaths, the close verbal correspondence ("I know not if I could take them . . ."/"He knew not if he could take them . . .") ought to be important to Weinbrot and Greene, since Johnson is not reported as just expressing generalized disapproval of the tendering of oaths, though he certainly is doing that; he is reported as clearly implying that he himself has not taken the oaths relating to the claims of the House of Stuart.

Of course we are not yet able to survey fully Boswell's processes of revision. Tulin Faulkner, a party to neither side in the current debate, scanned the Jacobite references in the MS of the *Life*, including some smaller instances not here mentioned, but in the long run we need to know how Boswell's revision treated comparably delicate matters. Meanwhile, on the Jacobite question, a provisionally positive view of Boswell the biographer seems in order. After all, so many other witnesses point in the same direction as he does: the anonymous author of the pension letter; Charles Churchill; David Hume (as attested by Boswell, to be sure); William Shaw (who speaks of Johnson as having, or being thought to have, "political principles inimical to the House of Hanover"[12]); Sir John Hawkins; Anna Seward; and—if I may be allowed to turn to our own time—a biographer whom Greene used to admire: J. L. Clifford, who wrote of "Johnson's known anti-Hanoverian position."[13]

Is there a substantial witness contemporary with Johnson on the other side? Perhaps, if Thomas Cooper can be considered one, but his specific report can best be appreciated when referred to the adjudication of a far more substantial witness, who saw both sides of the question but who yet felt that his full view could not be published, at least in his lifetime. Who is this valuable witness? It is, I am afraid, none other than Boswell, in the long passage he omitted from his published version of the *Tour to the Hebrides*, revealing as *that* was of his own and Johnson's interests by comparison with the self-restraint of Johnson's *Journey*. Boswell, who believed in succession by male primogeniture, could not agree to any *de jure* right to either the English or Scottish throne after James V of Scotland, if then. With the marriage of Mary, Queen of Scots, to Lord Darnley, Boswell sees only "the sons of Lennox on the throne." Because Boswell is thus the strictest possible believer in hereditary right, the Stuart posterity of Mary Queen of Scots has, in his view, paradoxically, no claim. He is thus free to recognize the House of Hanover on the British throne, a recognition he always makes clear. Johnson's position, on the other hand, is less theoretical, more pragmatic, more subtle, and no less principled. For Johnson it is more difficult than for Boswell. Johnson held that such right as there was to the throne of England, Ireland, and Scotland was gathered into the claim of the seventeenth-century Stuarts.[14] Johnson, Boswell explained, based the right of kings on

> long possession, which ought not to be disturbed on slight grounds. He said to me once that he did not know but it was become necessary to remove the King at the time of the Revolution; and after the present family have had so long a possession, it appears to him that their right becomes the same as the Stuarts had. His difficulty is as to the right still in some measure belonging to that unfortunate family. In short, he is dubious; and he would not involve the nation in a civil war to restore the Stuarts.[15]

This passage follows directly on from Boswell's statement that "Johnson is not properly *a Jacobite*. He does not hold the *jus divinum* of kings," on which Greene relies as part of his case that Johnson was no Jacobite. It is clear, however, that Boswell's discussion does not work well for either side of the present dispute, not, at least, so long as black-and-white political definitions are the bones of contention. Thus Johnson is not "properly" a Jacobite only according to the incredibly high-flying and theoretical criterion of Boswell himself, which made it so easy for Boswell to be a Hanoverian. By contrast, Johnson's criterion, long possession, must have meant that Johnson was pro-Jacobite in the earlier eighteenth century. Again, look at all of Boswell's carefully qualified language: "not properly," "it appears to him," "his difficulty," "dubious," "so dubious": what is the point of all these locutions if Boswell could truthfully have said, and published: "Mr. Johnson and I are loyal subjects of the kings of the House of

Hanover"? The same hesitating locution is used of Johnson's remark about the Battle of Culloden, in this passage and elsewhere: "If holding up his hand would have gained the victory for the Highland army in 1745, he does not know if he would have done it" (*Tour* journal, p. 163). This famous remark, which Topham Beauclerk attested to have heard Johnson make before he was awarded his pension, certainly shows that Johnson was not an all-or-nothing Jacobite—nobody has argued that he was. Can we, however, really credit that these were the words of a ready, sincere, loyal supporter of George II in the Hanoverian/British state, the legal structure of which Curley has so clearly displayed? ". . . Does not know if he would have done it": is that the language of convinced Hanoverian loyalty?

The problem Greene and Curley have in understanding these aspects of the eighteenth century—and perhaps Weinbrot shows it in another way—is that they are working with limited, black-and-white definitions which are insensitive to complexities and nuances of language. Johnson "did not know but it was become necessary to remove the King at the time of the Revolution": this reluctant, hesitating invocation of necessity is far from the language of one who believed that James II abdicated in 1688, or believed that in the ancient constitution the two houses of Parliament could depose the king. For Johnson as perhaps for Swift, 1688 took on in retrospect the features of a tragic dilemma. Many eighteenth-century Jacobites disagreed with the Roman Catholic measures of James II. A staunch Anglican such as Francis Atterbury was the leader of the English Jacobites in 1721. A Jacobite, one may suggest, was a person who sought to serve some major religious, political, or personal principle or interest through the restoration of the *de jure* royal line. Many active in the Jacobite cause were Jacobites in this broad sense. The constraints of a too cut-and-dried approach are seen elsewhere on the other side of this debate. Apparently neither Greene nor Weinbrot sees more in Johnson's poetic narrative of Charles XII of Sweden than yet another example of a conqueror who bites the dust. By neglecting (or not recognizing) its emotion, they reduce one of the most moving passages of eighteenth-century poetry to total banality.

Back to Boswell for the last time. He does not argue that Johnson ever fought or plotted for the Jacobite cause. Johnson was not a Jacobite in 1773 by Boswell's definition. Yet after all his exact explanation and significant disclaimers he comes to rest on a different point: "With all this, he and I have a kind of *liking* for Jacobitism, something that it is not easy to define" (p. 163). Perhaps the pro-Jacobite case should also rest there.

Impartial readers who succeed in fighting their way through all the contributions to this debate in the pages of *The Age of Johnson* should ask themselves four questions:

1. Which side has presented most new evidence about Johnson and his time?

2. Which side has more consistently addressed itself to the issues at stake?

3. Since no side is likely to be in total possession of the truth, which has shown itself the more willing to make concessions?

4. Which side has been the more ready to focus on a complex and perhaps strange-seeming eighteenth-century Johnson, rather than a Johnson acceptable to late twentieth-century views?

NOTES

1. *A Journal of a Tour to the Hebrides with Samuel Johnson, LL.D.*, ed. F. A. Pottle and C. H. Bennett (New Haven: Yale Univ. Press, 1961), p. 163.
2. One such review has already appeared, in John Cannon, *Samuel Johnson and the Politics of Hanoverian England* (Oxford: Clarendon Press, 1994).
3. Sir Robert Chambers, *A Course of Lectures on the English Law, 1767-1773*, ed. Thomas M. Curley, 2 vols. (Oxford: Clarendon Press, 1986), I, 77.
4. This reason applies also to Johnson's *Ramblers*. My argument would not lead one to expect crypto-Jacobite traces in Johnson's writings after 1749 though when, decades later, *The Lives of the Poets* afforded Johnson the opportunity to comment on the political history of the preceding century and a half, the language he used in discussing affairs of state after 1688 would take on a very studied restraint.
5. An example is the Preface to the *General Index to the Gentleman's Magazine* (1753). This was first, *but only partly*, attributed to Johnson by John Nichols in 1821, who then said: "the greater part of that Preface was written, and the whole of it corrected, by my illustrious predecessor, Dr. Samuel Johnson." The history of this brief collaborative item in the current debate is illuminating. Though quoted by E. L. McAdam in "Johnson, Walpole and Public Order," in Mary Lascelles *et al.*, ed., *Johnson, Boswell and their Circle* (Oxford: Oxford Univ. Press, 1965), pp. 93-94, and republished by Professor Arthur Sherbo (East Lansing: Michigan State Univ. Press, 1987) as a Keepsake for the Johnson Society of the Central Region, the contributors to the other side of the debate seem not to have been aware of it until I and Professor Clark featured it in our discussions ("Johnson the Jacobite?", *The Age of Johnson*, 7 [1995], pp. 16-17; J. C. D. Clark, *Samuel Johnson* [Cambridge: Cambridge Univ. Press, 1994], pp. 183-84). At this point the contributors on the other side of the debate fell upon it with cries of triumph, for it contains a hostile reference to the '45, but singularly failed to credit Clark for his integrity in quoting evidence apparently against his own case. More seriously, Weinbrot sees in Clark's discussion of this short and equivocal item "his crudest exercise in denial" ("The Historiography of Nostalgia," *The Age of Johnson*, 7 [1995], p. 188). While Sherbo, Clark, and I all quote Nichols's *partial* attribution, Weinbrot nowhere refers to this in his review of Clark's *Samuel Johnson*. To say the least, the unwary reader is likely to be misled by such practice.
6. *A Course of Lectures*, I, 138.
7. Sir John Hawkins, *The Life of Samuel Johnson, LL.D.* (London, 1787), pp. 70-72, 80.

8. Donald Greene, *The Politics of Samuel Johnson*, 2nd ed. (Athens, Ga.: Univ. of Georgia Press, 1991), p. xlii.
9. Marshall Waingrow, ed., *James Boswell's "Life of Johnson": An Edition of the Original Manuscript in Four Volumes*, vol. I: 1709-1765 (New Haven: Yale Univ. Press, 1994).
10. I wish to thank Professor Claude Rawson and Tulin Faulkner for their generosity in making available to me details from the Boswell Papers not yet published, and Tulin Faulkner's own expert survey of the Boswellian materials which bear upon the present debate. My conclusions, of course, are not necessarily theirs.
11. *Boswell: The Applause of the Jury, 1782-1785*, ed. I. S. Lustig and F. A. Pottle (New Haven and London: Yale Univ. Press, 1981), p. 74.
12. William Shaw, *Memoirs of the Life and Writings of the Late Dr. Johnson* (London, 1785), ed. Arthur Sherbo (Oxford: Oxford Univ. Press, 1974), p. 40.
13. J. L. Clifford, *Dictionary Johnson: Samuel Johnson's Middle Years* (New York: McGraw-Hill, 1979), pp. 263-65, 269.
14. *Boswell's Life of Johnson*, ed. G. B. Hill, rev. L. F. Powell, 6 vols. (Oxford: Clarendon Press, 1934-64), III, 157.
15. *A Journal of a Tour to the Hebrides*, ed. Pottle and Bennett, pp. 162-63.

THE CULTURAL IDENTITY OF SAMUEL JOHNSON

J. C. D. CLARK

The Cultural Politics of Johnsonian Scholarship

The politics of Samuel Johnson have now been established, beyond reasonable doubt, by the normal processes of literary and historical scholarship.[1] This has, inevitably, thrown a former orthodoxy into sharp relief; but a wider audience might still be astonished by the degree of rhetorical over-reaction on the part of the three representative respondents, Greene, Curley and Weinbrot. It might indeed be remarkable that a technical, scholarly thesis about a long-dead author should provoke this trio into such a response[2] in a "bullying, sometimes even a frenetic, tone." Given the nature of the evidence, it will continue to be possible to argue over some of the individual points recently raised; but the conduct of the debate by the three respondents suggests that no accumulation of evidence would move them. One repeatedly senses that something more general is at stake. In some areas of academe, it seems, Samuel Johnson has been inflated into a cultural icon, removed from his often fraught historical setting, raised above his polemical political and religious commitments and elevated into a seemingly timeless moralist. No such emancipation, of course, is really possible: Johnson has only been adopted by the respondents as a validating authority for a liberal-Evangelical or humanist mind-set which reached its apogee in the USA of the 1950s and which has been eroded by many forces in American society since that time.[3] In Britain,

Samuel Johnson, however widely admired, has no such significance as a cultural icon, and scholarly enquiry is less constrained.

The problem is compounded by the nature of the identity which recent scholarship has recovered for Johnson. For not only has Johnson recently been historicized; he has also been placed in categories (Tory, Jacobite, Nonjuror) that their history predisposes some Americans to regard as extreme, threatening and sinister, standing for "unacceptable authoritarian government" (Weinbrot, p. 164). It would make Johnson "a political reactionary devoted to absolutist Stuart monarchy" (Curley, p. 142). Such authors still seek instinctively, indeed violently, to defend their cultural icon from what can only seem like "charges" of un-American activities. In this sense, Jacobitism is the least important point at issue: the underlying issue being debated is the validity of a certain mind-set and its title to reinterpret other societies and cultures, eighteenth-century England among them.[4]

For reasons both of dominant methodology and of cultural politics, a characteristic of U.S. scholarship has been to tend to produce a usable version of past authors, especially those whose political writing can be given moral force. This tendency is at least as old as the Unitarian, ex-Methodist, minister Moncure Daniel Conway's attempt to create a "usable" Tom Paine, portrayed as a Founding Father who embodied the denominationally-unspecific moral altruism of late nineteenth-century American culture.[5] The central component of the process was the creation of a "usable" John Locke, largely in early twentieth-century America,[6] a project later adopted by that school of the history of political thought associated with Leo Strauss.[7] Part of the Straussian enterprise was the attempt to devise a "usable" Edmund Burke in the early 1950s.[8] Greene's invention of a "usable" Johnson in that decade was a lesser offshoot of a wider cultural phenomenon; but whereas the increasing sophistication of the methodologies applied to the history of political thought since the mid-1960s has allowed historians to move beyond these older interpretations, while remaining appreciative of much of value that they contained, one area of Johnsonian scholarship remained a backwater. The awakening, when it came, was all the more rude for being so long delayed.

The Classical Tradition

Not only are Johnson's dynastic politics anathema in some quarters; his chief literary idiom, too, is evidently culturally incorrect. To place him within a late-humanist tradition of Anglo-Latinity, according to Curley, is to argue that Johnson "has a diminished relevance to modernity in siding with lost causes like Jacobitism and classicism to the neglect of an emerging democratic vernacular culture holding sway in the West since the Romantic era . . . Surely it is wrong-headed to relegate Johnson to the ash-heap of a bygone classical tradition . . ." (p. 149). Greene, similarly, largely ignores the classical tradition and, when forced briefly to attend to it, seeks to disengage it from its specifically classical setting: in *The Vanity of Human Wishes* Johnson is depicted as "adapting a work of Juvenal's to the whole human situation at all times" (p. 69). Neither Greene nor Curley, in their other published writings, have given any extended consideration to Johnson's wide knowledge of classical literature, his linguistic ability in Latin and Greek, his Latin poetry, or his views on those seventeenth- and eighteenth-century authors who produced translations and imitations of the ancients. Curley's reference to "the ash-heap of a bygone classical tradition" tells us why. Weinbrot, who has given the tradition more attention, dismisses its importance as a myth sustained by "creaking clichés of eighteenth-century dominant classicism" (p. 169). Yet this will not make an ancient and distinguished literary idiom disappear.[9]

Its marginalization is, of course, not new. English developed as a subject within the universities as "a kind of poor man's Classics" to exclude Latin,[10] and attention to late humanist letters still more often comes from departments of classics.[11] Yet the dismissal of the classics has a special significance when it is argued that "the Moderns easily won their battle with the Ancients" (Weinbrot, p. 177), and that Johnson's politics are related to this alleged triumph. The historian must reply that the developing role of the vernacular (which is not in question) was neither as swift nor as easy as this rhetoric implies, and that it had a political dimension which has largely gone unnoticed.[12]

The politically sensitive nature of this culture disclosed itself by certain revealing characteristics. Late humanism sustained endless, seemingly trivial controversies over whether authors had committed small errors of vocabulary or grammar, over which side commanded the purer Latin or (less often) Greek. Such controversies were more momentous than they seemed, however, since they were struggles for the control of a whole tradition. The Jacobite William King's disputes with Hanoverians like John Burton and Benjamin Kennicott were precisely of this sort; Burton of Corpus, who had the temerity to challenge King's Latinity, was one of the Oxonians whom the firebrand Whig Nicholas Amhurst singled out to praise for reforms in the university's studies and discipline after 1719.[13] Such rivalries had the same significance, in the cultural politics of the 1750s, as the contests over the epistles of Phalaris in the 1690s between the Jacobite Charles Boyle, later 4th Earl of Orrery, with his donnish supporters Anthony Alsop, Francis Atterbury and George Smalridge, and the Hanoverian Dr. Richard Bentley. What matters to the historian is less which contenders really were the better classicists than the wider issues which these coded conflicts expressed.

I have argued against the picture of "a dominant and monolithic classical tradition" (Weinbrot, p. 178), and contended that Johnson was not in any simple sense an "ancient" at war with the "moderns": on the contrary, I recorded his expressions of sympathy with the latter.[14] I have, however, framed an argument about the characteristically Anglo-Latin high culture of his youth, and especially of his beloved Oxford; recorded that it was a prize fought over for cultural-political reasons by Whigs and Tories; and proposed a thesis about Johnson's equivocal and ambiguous role in the processes by which that tradition was damaged, partly dismantled, and modified in the middle decades of the century.[15] Johnson's *Dictionary*, *The Rambler*, *The Idler* and the edition of Shakespeare were important episodes in the "rise" of the vernacular, but Johnson's residual commitment to an Anglo-Latin canon is manifest as late as *The Lives of the Poets*.[16] The three respondents have largely ignored this thesis about the overall trajectory of a cultural tradition and its political engagement.

Weinbrot only replies (p. 202, n. 6) that "the High Anglican, classicist, and royalist mode in Britain is a post-1832 creation" rather than a cultural project most accurately located between Dryden and Johnson. This hypothesis cannot be reconciled with recent scholarship. Johnson's conception of Anglicanism, owed primarily to the Caroline divines,[17] was still represented by the "Old" High Churchmen of the early nineteenth century, but they found themselves denounced by the Oxford Movement that Weinbrot thinks created them.[18] The significance of the classics similarly continued to evolve into the early nineteenth century, with adherents in one area of the arts like Thomas Hope, Robert Smirke, James Stuart, and William Wilkins;[19] but this often denoted a cultural reorientation towards a Greek axis that had a political significance far removed from Johnson's patrician project.[20] A renewed cult of the monarchy developed around Victoria; but its historians have not confused it with the set of assumptions and values that informed the courts of the Stuarts or the Hanoverians.[21] Victorian classicism was a distinct cultural formation, understood as such by its students, and not confused with what went before.[22]

The Presentist Defense of Johnson

Greene and Curley have almost no new evidence to present in support of their own case. Their purposes are rather different, and Curley's essay offers a clue to them. He evidently believes that his task is to defend Johnson from a "subversive thesis," a "libellous charge" of "criminal disloyalty" which, if true, would diminish the greatness of a "great writer" of "legendary status." It would be to "stigmatize him with the very worst of English crimes," a "libel to his good name and fame." Curley sees himself rebutting a major threat: "The Jacobite thesis is thus more than a questionable re-reading of Johnson's politics and becomes a revisionist challenge to the enduring value of Johnson's vision of life and literature" (p. 149). Greene similarly depicts Johnson not as a man politically engaged, but as "one of a very small group of publicists . . . concerned to ameliorate the general condition of individual members of the human race" (p. 93).

There is in such pronouncements an acknowledged and strikingly ahistorical wish to treat Johnson as a moralist of timeless significance, "the premier moralist of English literature" (p. 157), standing outside and above his age, and a refusal to accept him as an historical figure like any other, rooted in his age's conflicts. In writing of Johnson, Greene revealingly employs the historic present tense: "he expresses . . . compliments . . . writes . . . is honoured." The universalizing claims of Greene's mind-set keep obtruding: he assures us that

Johnson's attitude towards English monarchs "seems to be much like the detached and independent attitude of an intelligent modern American towards his Presidents" (p. 71), as if a difference of two centuries and two cultures were negligible. This is not an isolated example: the essential timelessness of Greene's analysis is repeatedly illustrated by his lengthy attempted analogies with nineteenth- and twentieth-century America, as if the experience of that very different society could be easily transposed onto eighteenth-century England.

We might, alternatively, turn to recognized and accepted historical method to make careful and verifiable use of the diverse materials which are available. Much historical writing necessarily takes this form, partly because the rules of evidence which apply in a court of law during a trial for treason are (*pace* Curley) accepted as generally inappropriate or inapplicable in an historical enquiry where evidence is fragmentary or indirect. Moreover, the respondents have mistaken the point at issue. Curley attempts to acquit Johnson under the statute of 6 Anne c. 17 from a charge that he maintained "by writing or printing that the queen is not rightful queen, or that the pretended Prince of Wales has any right to the crown, or that any other person has any right to the crown." But neither Howard Erskine-Hill nor myself have argued that Johnson was guilty of treason under that Act: he shielded himself behind satire, innuendo and coded utterance from anything for which he could obviously have been prosecuted, although, in *Marmor Norfolciense*, he steered very close to the wind. The old *canard* that Johnson might have been personally involved in the rebellion of 1745 has been rejected too emphatically to need further examination. What is at issue is not treasonable action by Johnson but his cultural identity, his political and religious allegiance, and the ways in which his loyalty was divided as public life changed around him. We have no evidence that Johnson ever said "I am a Jacobite":[23] such a statement might meet Greene's and Curley's incantatory demands for "incontrovertible evidence" and "irrefutable proof" (though, given their nescient nominalism, it might not). Yet their own firm claims that Johnson was undoubtedly *not* a Jacobite and *not* a Nonjuror (that he had a "fundamental respect for Hanoverian church and state," p. 139) are highly vulnerable to their Occam's razor, the requirement of "decisive evidence." The "indubitable proof" for Johnson as a Hanoverian loyalist is non-existent, and none has been advanced in the present exchanges; the detailed circumstantial evidence for Johnson's opposite commitments continues to grow.

As Paul Korshin has observed, the sources for Johnson's life before the 1760s are so thin that "conjecture becomes a necessary tool if we are to make anything of the scant information at our disposal."[24] This is systematized as an historical methodology by the construction of intellectual contexts out of materials the relevance of which was hitherto unappreciated; contexts which then illuminate in a new way the content and purposes of an author's writing. That is also, ironically, what Greene himself did in his monograph of 1960 by constructing a novel Namierite context. The literary critic who disengages himself from the duty of historical reconstruction and falls back on an *ipse dixit* has only painted himself into a methodological corner.

The respondents repeatedly remind us that their training is not as historians. Yet there must now be some doubt as to the scale of Greene's and Curley's achievements as literary critics also, for the stranger absence from their work, here as elsewhere, is any synoptic account of the trajectory of Johnson's literary output *as literature*. Once Greene has established to his satisfaction that the political commitments and dynamics of Johnson's age were all illusions, imposed retrospectively on a Namierite reality, Johnson's literature is drained of political meaning. Indeed, in Greene's pages it seems peculiarly drained of almost all meaning, save for Johnson's diverse pronouncements on the complexities and difficulties of the human condition. Curley's briefly-stated thesis that Johnson in the *Lives of the Poets* "celebrates the growth of English civilization from primitive literary beginnings and takes pride in the impressive cultural strides made by writers of his own century" (p. 160) is his closest, but inadequate, approach.[25]

Here again, Johnson's Jacobitism is in some ways the smallest point at issue. One larger consequence of the case which has now been made is the thesis that English letters were, then as now, profoundly politicized.[26] I have suggested that cultural politics were at least as dependent on politics as they were autonomously cultural; the phases of English literary taste were not so much generated "from below" (Weinbrot's democratic vernacular) or "from within" (the autonomous sensibility), but to a larger degree worked out as part of a political conflict fought simultaneously and interconnectedly in many areas of national life. These implications make it illuminating to examine the development of Johnsonian studies in a wider historiographical context, both in the years between Johnson's death and Macaulay, and in the last few decades.

A key weakness of the respondents' approach is that its timeless pretensions have obscured the historiography of the chosen subject. Other historical interpretations were dismissed as error and therefore, by definition, uninteresting.[27] Greene observes that the real Johnson was obscured by subsequent layers of legend; but Greene's work has nowhere provided an historiographical account of how and why Johnson's reputation was reshaped after his death. By his definition, these were merely errors: they were therefore ignored. My *Samuel Johnson* attempts such an outline; by contrast, Namierite historiography has proved itself still unable to show why and when legends about Johnson's politics and religion were constructed, or to distinguish between perceptive contemporary comment and erroneous later legend (including, in Greene's pages, Johnson the Lockeian and Johnson the Evangelical).

There is, however, another point at issue involving the methodology of modern political historians. Much political historiography since the mid 1960s has emphasized how labels do not identify simple, unchanging commitments but encompass a spectrum of allegiances and tactical responses, evolving over time: it is an approach which literary scholars ought to find congenial and many now do. But this is not understood by the three respondents, in whose black-and-white world to contend that Johnson was not a "serious and active Jacobite," that he did not spend "much time lying awake at nights scheming ways" to bring about armed rebellion, openly "agitate for a repeal" of the Act of Settlement, publish a pamphlet in his own name calling in plain English for a restoration, or repudiate the patronage of George III, is to prove that he was never at any time or in any degree drawn to the rival dynasty. By reasoning like this, most such phenomena could be made to disappear.[28]

Johnson and the Oaths: Interpreting a Career

Greene (1995) adopts a querulously dismissive strategy throughout Johnson's career. He continues to question Sir John Hawkins's account of the political preferences of Johnson's father on the grounds that Hawkins cited no sources. Hawkins did not do so, but this does not establish that those sources (perhaps Samuel Johnson himself, and perhaps as early as 1739) were unreliable. Hawkins's lawyerly accuracy and attention to detail is manifest elsewhere in the *Life* and in his other writings, especially his *History of Music*.[29] Testimony from Michael Johnson's own hand is lacking; but the evidence now presented is sufficient to establish the point at issue, that the Lichfield of Johnson's youth contained a political culture fully attuned to the dynastic and ecclesiastical politics of the Tory high tide under Queen Anne. It was a culture very different from the Lockeian consensus which Greene depicted in 1960 as the norm for early-Georgian England and proposed as the appropriate context for Samuel Johnson's intellectual formation. This new point can now be avoided only by disregarding evidence, as when Greene (p. 63) claims that Johnson's childhood friend John Taylor was unaware of Johnson's Jacobite tendencies: there is clear evidence that Taylor was so aware.[30]

New work on Johnson seeks to locate him in that recently-recovered dynastic and ecclesiastical context. How far should we identify him with it? Was Johnson a Nonjuror? The evidence is substantial that Johnson consistently avoided situations in which he would be required to take the oaths of allegiance and abjuration; that they were never tendered to him; and that he explicitly disclosed in 1773 that he had never taken them.[31] As can be shown, it was not all oaths that Johnson refused, but these oaths in particular. Weinbrot (p. 191) embarks on an irrelevant argument by confusing the oaths at issue: it has never been suggested that Johnson refused to take the oath of supremacy. Such an oath would have posed no problems to any Stuart loyalists other than Roman Catholics.[32]

The nature of an oath, the force of its operation, and the senses in which men might be released from an oath by a prior obligation had been matters of careful enquiry as well as crude polemic since the Civil War; these questions were not ignored in any mood of pragmatism after the Revolution. The new oath of allegiance (1 W&M, s. 1, c. 8) directly challenged the consciences of the intelligentsia by requiring subscription before 1 August 1689 from "any person or persons now being master, governor, head, or fellow of any college or hall, in either of the two universities, or of any other college, or master of any hospital or school, or professor of divinity, law, physick, or other science in either of the two universities, or in the city of *London*" (clause VIII) on pain of deprivation.

The formidable oath of abjuration, imposed in 1701 by the statute of 13 W III, c. 6, was a matter for heart searching even among its advocates, in Scotland as well as England.[33] It was required of all civil and military officers, and "all ecclesiastical persons, all members of colleges and halls in either university, that are or shall be of the foundation (being of the age of eighteen years) and all persons teaching pupils in either university," together with other listed categories including schoolmasters, Dissenting ministers and lawyers, on pain of deprivation. After the accession of George I, new legislation, centrally the Act of 1 Geo. I, s. 2, c. 13, re-emphasized the degree to which the regime defended and defined itself with oaths. It was this Act which again imposed the oaths of allegiance and abjuration on "all ecclesiastical persons, heads or governors, of what denomination soever, and all other members of colleges and halls in any university, that are or shall be of the foundation, or that do or shall enjoy any exhibition, (being of the age of eighteen years) and all persons teaching or reading to pupils in any university," and others, on pain of deprivation. It also offered £100,000 for the Pretender, "alive or dead." This sweeping and draconian Act, perhaps an incitement to assassination, called forth a handbook to the legislation in force,[34] published in 1716 and reprinted in 1723 and 1744. Even into the 1730s, a strategic threat to the state, which materialized in the French invasion attempt of 1744, justified a defense of the oaths: "as we have the Word of Truth for it, *That a Kingdom divided against itself cannot stand*; I pray God open our Eyes, and make us duly sensible of the Danger of our Divisions, especially at this critical Juncture, when such powerful enemies are combined against us."[35]

The question of oaths related directly to denominational identities also. It was argued that "Popish" princes could not bind themselves by oath to safeguard the liberties of their subjects, being bound by the decision of the Council of Constance "That Faith is not to be kept with

Hereticks": this had led to "repeated Acts of Treachery . . . in all Ages . . . against the Protestants" and made it impossible ever to trust them. Jacobites who had sworn the oaths to the new regime (with mental reservations) would be equally untrustworthy to the Pretender: "how can he depend even on the Oaths of such Men, who tho they believ'd King *James* could not forfeit his Right to their Allegiance, and knew he claim'd it to the last, could yet swear Allegiance to his Enemy?"[36] The author of these remarks failed to point out that the greater number of Whigs who had sworn allegiance to James II and now did the same to William or Anne, and abjured the Pretender, were equally untrustworthy. In an age in which many were open to charges of hypocrisy, only the Nonjurors could be above reproach.

Was the oath of allegiance required of all Oxford undergraduates at matriculation? The respondents so argue, and build their model of the Hanoverian Johnson on the premise that he took such an oath at that time. On this technical point much else rests, and we must examine the evidence more carefully than they have done. In *Samuel Johnson*, I set out the evidence that Oxford's statutes and practice agreed in requiring the oath of allegiance only of foundationers on matriculation, not of commoners like Johnson. Weinbrot has cited in reply what appears to be evidence to the contrary from Nicholas Amhurst, an undergraduate at St. John's College from 1716 to 1719. On closer inspection, his evidence proves not to bear out Weinbrot's argument.

Amhurst (1697-1742) was a zealous, perhaps an obsessional, Whig. Even as an undergraduate, he had bid for the patronage of the great with a poetic compilation of the usual freethinking images into an eulogy of the Brunswick succession, Hoadly and Stanhope.[37] Even while an undergraduate, too, he satirized his university by impugning its loyalty to the new dynasty, fictitiously addressing the prior claimant:

> Your *Isis* Sons, a firm unshaken Train,
> Mourn your ill-fated Birth, and baffled Reign;
> Impatient of their Thraldom, they disown
> A foreign Lineage, and a plunder'd Throne;
> Nor Heav'n nor Earth, Laws Human or Divine,
> Shall disunite them from the STUART line.

This loyalty Oxonians expressed, he claimed, not by military valor but by sowing "delusive Fears" and "Suspicious Doubts" to "sap the Credit of the Government":

> Thus would our *Oxford*, with officious Love,
> Her Vows unbroken, and Allegiance prove.

It was a charge of personal dishonesty as well as political error:

> Nor Fame nor Conscience shake a *Tory's* Mind,
> Nor plighted Oaths his fixt Resolves unbind.[38]

On 29 June 1719, within sight of a Fellowship, Amhurst was expelled (as he claimed, for political reasons; as others recorded, for personal immorality).

Moving to London, he took his revenge in a series of 52 papers entitled *Terrae-Filius*, published from 11 January to 12 July 1721. Amhurst complained that the first experience of many freshmen, arriving at their colleges, was "perjury":

> If he comes elected from any publick school, as from *Westminster*, *Winchester*, or *Merchant-Taylor's*, to be admitted upon the foundation of any college, he swears to a great volume of [college] statutes, which he never read, and to observe a thousand customs, rights and privileges, which he knows nothing of, and with which, if he did, he could not perhaps honestly comply.[39]

I have argued that a foundationer would also, unlike a commoner, have to take the oath of allegiance with the other oaths at the university ceremony of matriculation. Amhurst himself had been educated at Merchant Taylors, and became a scholar of St. John's College, Oxford, on 16 June 1716.[40] His account of matriculation correctly recorded his own experience:

> Within fifteen days after his admission into any college, he is obliged to be *matriculated*, or admitted a member of the university; at which time he subscribes the *thirty-nine* articles of religion, though often without knowing what he is doing, being

> ordered to write his name in a book, without mentioning upon what account; for which he pays *ten shillings and six pence*.
>
> At the same time he takes the oaths of allegiance and supremacy, which he is praetaught to evade, or think null: some have thought themselves sufficiently absolved from them by kissing their *thumbs*, instead of the book; others, in the crowd, or by the favour of an *honest* beadle, have not had the book given to them at all.
>
> He also swears to another volume of [university] statutes, which he knows no more of than of his private college-statutes, and which contradict one another in many instances, and demand unjust compliances in many others; all which he swallows ignorantly, and in the dark, without any wicked design.

This was not the only absurdity:

> But I have not mention'd the most absurd thing in *matriculation* yet. The statute says, if the person to be *matriculated* is sixteen years of age, he must subscribe the *thirty nine* articles, and take the oaths of *allegiance* and *supremacy*, as also an oath of fidelity to the university: but, if the person is not *sixteen* years of age, and above *twelve*, then he is ONLY to subscribe the *thirty nine* articles.[41]

This was not what the statute said. Amhurst had evidently not consulted its text, and had probably relied on his own recollection of the event. He failed to notice that the great majority of his freshmen contemporaries, who were not foundationers like himself, did not take the oath of allegiance at matriculation; and because he was expelled from the university, Amhurst had no opportunity to discover his error at the ceremony of graduation.[42]

Amhurst did, however, half grasp one relevant fact: many people at Oxford had not taken the oath of allegiance. Amhurst merely interpreted this fact in the way most discreditable to them:

> It is hardly worth mentioning, amongst all these absurdities, that by this statute many persons avoid taking the oaths of allegiance and supremacy at all; for being, or pretending to be, under *sixteen* when they are *matriculated*, they are excused from

> it at that time; and I never heard that any body was ever call'd upon afterwards to take them, unless they take a degree; but how many are there who stay many years at OXFORD, without taking any degree?[43]

It is more likely that such persons were unwilling to perjure themselves on graduation than that they had done so by falsely claiming to be under age.

Later, Amhurst returned to the same theme, printing two anonymous letters in his issue of 15 March 1721. The first, from "J.R.," complained that Oxford's practice was undermining the sacred force of oaths by making matriculants swear to obey statutes they had not seen and subscribe the Thirty-nine Articles, over the meaning of which even learned clergy might disagree. The second purported to be a letter from "Philalethes," complaining of another abuse at matriculation:

> When I was *matriculated* I was about seventeen years of age, and consequently entitled to take all the oaths; accordingly I subscrib'd the thirty-nine articles of religion, *(though, by the bye, I did not know that I had done it till near six months afterwards)* and the then V[ice] Ch[ancello]r Dr. *B[aro]n*,[44] coming out of the convocation house, I took the oaths of *supremacy*, and of observing the *statutes, privileges*, &c. of the university. After which the doctor sign'd my *matriculation* paper, testifying that I had also taken the oath of *allegiance*, though not one word of it, or his majesty King GEORGE, was then mention'd.

Amhurst apologized for the two insertions:

> I confess an *anonymous* letter is not a sufficient voucher of the truth of any fact, and therefore I do not urge the *last* of these letters as such; let every reader believe or disbelieve it, as he thinks fit. All that I can assure him is, that it is a *genuine* letter, and came to my hands just as I publish it. But the *first*, which consists of *argument* upon the *known practice* of the university, is not liable to the same objection.[45]

On reconsideration, Amhurst was not willing to repeat what he had earlier claimed in *Terrae-Filius* no. 3 (confusedly linking the oaths of allegiance and supremacy) and again maintain that it was the "*known practice*" of the university to give matriculation certificates recording the taking of the oath of allegiance to candidates who had not in fact taken that oath. Amhurst would presumably have so maintained had he been able credibly to do so, for it would have been a major point scored against disaffected Oxford: his caution here is telling evidence. As a Whig, Baron had no reason knowingly to have handed out such inaccurate certificates to disaffected undergraduates.

The evidence currently available does not establish that foundationers received different certificates from commoners. All the surviving examples take the same form, recording the requirements common to both classes of undergraduate. The earliest so far located dates from 1722, a slip of paper printed with the formula:

> Oxoniae [blank] Anno Domini 1722
> QUO die comparuit coram me [blank] subscripsit Articulis Fidei, & Religionis; & juramentum suscepit de agnoscenda suprema Regiae Majestatis potestate; & de observandis Statutis, Privilegiis, & Consuetudinibus hujus Universitatis.

In one hand the first blank has been filled in with the date, "Feb: 20mo" and the second blank with the name of the candidate, "Eduardus Green e Coll: Aen: Nas: Arm: fil:." It is signed, in another hand, "Rob: Shippen Vice Can:."[46] This formula was unaltered on all the surviving examples. Joshua Ellis, a near-contemporary of Samuel Johnson, from Johnson's own Pembroke College, was given an identical certificate (the year only being changed) in December 1727, just as the statutes required; it similarly recorded only his assent to the Thirty-nine Articles, the statutes of the University, and the oath of supremacy.[47] Moreover, the Vice Chancellor concerned, John Mather, demonstrated his Hanoverian sympathies: he was one of only seven heads of houses, and a rather modest number of county gentry, who during the rebellion signed the Association in defense of King George on 14 October 1745 (together with staunch Whigs like Henry Brooke and John Conybeare).[48] Why would another Hanoverian Vice Chancellor also systematically have allowed disaffected undergraduates to matriculate by giving them deliberately falsified matriculation certificates? There is no evidence that he or any of his colleagues did so.

The possible confusion between the oath of supremacy, which all matriculants took, and the oath of allegiance, which only foundationers took, allowed some Whigs to make capital out of the claim that all undergraduates took both. Where Amhurst later shied away from that claim, however, Thomas Tanner made it in his edition of Anthony à Wood's *Athenae Oxoniensis* (1721). It was this which provoked Thomas Hearne's accurate correction: Tanner

> hath altered all things so, & made him talk in such a manner, as if M^r^. Wood had been a downright Villain, & had not known what even the most ignorant Scholar knows. How comes it, otherwise, to pass that more than once Gentlemen, when they are matriculated, are represented to take the Oaths of Allegiance & Supremacy? M^r^. Wood could not write so, since nobody knew better that the Oath of Supremacy only, & not the Oath of Allegiance, is then taken. Nor does the Statute require an Oath of Allegiance at that time. But this was added to bring a slur upon the University, and out of a Trimming Design. . . .[49]

Some Whigs did try to exploit this confusion and argue that, since all undergraduates had taken the oath of allegiance, an overt loyalty to the House of Hanover could be demanded of them. Richard Blacow so argued before the Vice Chancellor, Dr. John Purnell, Warden of New College, on 29 February 1747/8, urging the University to take disciplinary action against certain undergraduates who had, he claimed, uttered "treasonable Words" during a riot. Blacow was evidently unsuccessful, for Purnell was willing to punish the undergraduates only for their disorderly behavior, not for their having broken any oath of allegiance. Purnell deliberately avoided the point, according to Blacow's account:

> Upon asking the Vice-Chancellor, Whether such Treasonable Behaviour did not imply Perjury; he answered, *Mad young Fellows did not think about Perjury, and it would be hard to proceed against Boys in too severe a manner.* I answered, That, as the Wisdom of the University thought them of proper Capacities to take the Oaths, *four years ago*, they might be thought of proper Capacities *now* to be punish'd for the breach of those Oaths.

But Purnell refused to take legal depositions of evidence against the undergraduates; Blacow could only advance his cause by taking it to the highest authority, the Duke of Newcastle, a Secretary of State.[50] Evidently, the offending undergraduates were not proceeded against for a breach of the university statutes.

The same conflation of oaths was the tactic of a Whig partisan, Benjamin Kennicott, seeking to call in question the integrity of William King:

> Every person, at his *Matriculation*, if *sixteen* years of age, takes the Oaths of *Allegiance* and *Supremacy*; and every person, admitted either a *Clerk*, *Exhibitioner*, *Scholar*, *Fellow*, *Head* of any *College*, or, Sir, *Head* of any *Hall*, takes (or is requir'd by *Act of Parliament* to take) the Oath of *Abjuration*.

Kennicott then reprinted the text of the latter oath.[51] This was a fair point: King and others like him would presumably have taken the oath of abjuration with mental reservations. But Kennicott, like Amhurst, was probably also making a genuine mistake over the other oath, since Kennicott as an undergraduate had been a scholar of Wadham College,[52] and like Amhurst would indeed have taken the oath of allegiance at matriculation. The whole weight of Kennicott's argument thereafter rested on the oath of abjuration required of foundationers: it is likely that he introduced his reference to matriculation as an aside, for he did not subsequently return to it.

Although the oath of allegiance was not a requirement placed on all undergraduates at matriculation, testing oaths were certainly demanded of Oxonians at other stages of their careers, and another individual illustrates the difficulties created by these oaths for a scholar with a conscience. Thomas Hearne (1678-1735) took the oath of allegiance on graduation in 1699, and in 1700 had written a tract justifying his action for his patron, the Nonjuror Francis Cherry of Shottesbrooke. After Cherry's death in 1713 the manuscript passed to the Bodleian, from which archive one of Hearne's enemies, John Bilstone, later published it. Hearne had elsewhere argued from a utilitarian calculation that because of James II's conduct "the *Common GOOD of the NATION* requires either his Deposition, or, at least, that a Restraint should be put upon him some other way"; in the tract now published he produced arguments from medieval and ancient history, including a utilitarian interpretation of the binding force of an oath:

> The *Prime* End of an Oath is to be preferred to one which is *Inferior*. The *Prime* End of an Oath is *The Good of the Persons* concerned in it . . . *if the keeping of the Oath be really and truly inconsistent with the Welfare of a People, in subverting the Fundamental Laws which Support it; I do not see how such an Oath continues to Oblige*.[53]

Hearne's views were changing. On 19 January 1714/15 he was pressed by his friends to stand, and was elected, as Archetypographus and Esquire Bedell of Civil Law, but evaded the oath of allegiance tendered to him:

> All the Objections, such as being a Papist, a Non-Juror, &c. were started against the Writer. But these would not sway in the Case, so as to turn him [Hearne] by. The Abjuration Oath he is always resolved to refuse. Nor did he take the Oath of Allegiance, reading only the first Part of it, so as to suppose K[ing] J[ames] wch he did by speaking in a very low Voice, tho' this was not regarded, to his great Surprize.

The entry in the Register of Convocation therefore recorded that Hearne

> was accordingly admitted to both Offices having first taken the Oaths of Supremacy & Allegiance, without saying to whom ye Allegiance Oath was taken, in wch he [George Cooper, the Registrar] did right, it being certain y^t^ if he had said I swore to K. James (as I really did) he should be brought into a Scrape as well as myself.[54]

His views continued to evolve: he realized that even this evasion was unacceptable. According to Bilstone,

> Soon after this Promotion he acquired better Reasons (to *Himself* at least) for Non-Compliance with the Oaths that were necessary for *keeping* it, than he had before for the *Acceptance* of it; and accordingly resigns that advantageous Post, which the

> University had so lately conferred on him, as inconsistent with his *Conscience*, or at least, his Manner of *Thinking*.[55]

In November 1715, Hearne resigned the post of beadle to try to pre-empt a Whig maneuver to deprive him of his sub-librarianship. He would soon have resigned anyway, he recorded, rather than subscribe the oaths of allegiance and abjuration now required, by the statute of 1 Geo. I, s. 2, c. 13, to be taken by 23 January 1716:

> For indeed it was my full Resolution then, if the King [James] should not come in, not to act in either office, because I am fully resolved not to take the Oaths (i.e. as the Oath of Allegiance is clogg'd with the damnable Oath of Abjuration), come what will of it. Most people were surprized at this, especially considering that the Beadleship is worth ten times as much as the Office of Sublibrarian. But as I never was yet guided by Interest, so it fell out now, it being no other Principle than that of Honesty & Conscience & Regard for Learning that I went upon.[56]

By resigning, Hearne, aged thirty-seven, now gave the most convincing disavowal of the arguments for defeasible succession which, aged twenty-two, he had set out in *A Vindication*. Hearne's career illustrates both the evolution of a scholar's thinking on the oaths, the greater challenge imposed by the oath of abjuration in 1715, and also a poor man placing conscience before worldly advantage. Subsequently, he may also have refused appointment to other offices which would have required the oaths: the Camden chair of history in 1720 and 1727, the posts of keeper of the archives in 1726, and Bodley's librarian in 1719 and 1729. Weinbrot (pp. 208-209) misinterprets these passages by depicting an unchanging Hearne as duplicitous, or using pragmatic arguments to shrug off the significance of oaths. The reality was almost the reverse. Hearne explicitly condemned those who took the oaths with mental reservations:

> Notwithstanding the abominable Wickedness of the said Abjuration Oath, it is incredible w^t Numbers of all Kinds run in to swear, abundance pretending that as 'tis a forced Oath, they may do it, especially since the Imposers have no right to advance such an Oath, & they think, therefore, y^t all the Crime

> will fall upon them. But this reason will bring off any wickedness. . . .[57]

Refusal of the oaths was not a procedural scruple, but symptomatic of adherence to an extensive set of religious and political beliefs. There is evidence that this was a widespread culture, for it persisted among undergraduates as well as intellectuals. The poet Richard Graves, who matriculated at Johnson's Pembroke College on 7 November 1732, recorded being drawn into a circle of gentleman commoners there who "drank their favourite toasts on their knees,"[58] an easily-recognized Jacobite allûsion. On 23 February 1747/8 a minor disturbance in the streets of Oxford took on more serious overtones when undergraduates shouted Jacobite slogans at a junior Whig don, Richard Blacow. As a result of Blacow's information, three undergraduates were arrested and taken to London for trial. This persecution produced a piece of coded writing which revealed the issues of principle still embedded in men's awareness of the issues involved. One of Oxford's defenders, writing in 1749, offered no application of his pamphlet to the aftermath of the riot; but it was of the essence of a university, he argued, that it have "a power of determining causes, wherein any of its members were concerned, without appeal to the ordinary Courts of justice . . . that the youth might not be called away from their studies by litigious suits." Without specifying the present occasion, the anonymous author argued that "they who design to subvert the laws and liberties of any nation, commonly begin with the privileges and immunities of the Universities." Still without openly drawing the moral, he presented his modern readers with the Puritan attempt "*to subvert the Constitution of the Church and State of* England" in the 1640s, specifically the Ordinance of 1 May 1647 which threatened the deprivation of all in the university who had not taken the Solemn League and Covenant and the "Negative Oath," an oath not to assist the King in the civil war.[59] Oxford's reasons for refusing these demands in 1647 were now reprinted in full. On 1 June 1647, two days before the arrival of the commissioners, Convocation had produced a lengthy and defiant statement of principle against the Solemn League and Covenant; a disavowal of the "Negative Oath" (it could not be taken without "abjuring our *natural Allegiance*, and violating the *Oathes of Supremacy and Allegiance* by us formerly taken"), and a condemnation of the Directory of Public Worship.[60] These remarkable and courageous documents, for which many Oxford dons suffered swift retribution in the 1640s, were now reprinted with only the briefest of editorial introductions. The application was plain: the documents could be left to speak for themselves.

Whigs charged Tory dons with having perjured themselves in taking the oaths in order to hold their fellowships, but motives were open to challenge on the Whig side, too. Defending his role in the riot, Blacow later claimed that "I thought myself doubly bound to take notice of the Treason: because I had taken the Oath of Abjuration, and had been invested by the University with the authority of an officer in that particular Street," that is, the minor disciplinary office of Master of the Streets.[61] Yet Blacow's role attracted a variety of criticism. As well as acting as "*Delator* and *Informer*," one critic charged him with manipulating his testimony at the behest of the Duke of Newcastle to secure the acquittal of one of the three undergraduates charged, who happened inconveniently to be the son of "*John Luxmore* of *Oakhampton*, a Person who had the absolute Power of returning Two Members of Parliament for that Borough, which he had constantly been engaged to do by the Ministerial Party." Blacow was impelled by ambition, it was claimed, not by a sense of the inviolability of an oath.[62]

Even if Amhurst had been mistaken about the oaths in the early 1720s, had the university's practice changed (as the respondents claim) by a later date? A variety of evidence suggests that it had not. Other matriculation certificates of later dates survive, identical to Edward Green's of 1722 and Joshua Ellis's of 1727: one for William Drake of Brasenose on 2 October 1739, signed by John Mather as Vice Chancellor;[63] one for Thomas Holme of the same college on 5 June 1751, signed by Stephen Niblett, as the Vice Chancellor's deputy.[64] Richard Heber's, dated 14 December 1790, took the same form and used the same words.[65] So did Henry Ellis's, dated 27 June 1796.[66] So too did Henry Hoper's, dated 26 November 1806.[67] Had university practice evolved over time to include the oath of allegiance at matriculation for all candidates, the wording of the certificate would presumably have been amended to record this; no such change was made over many decades.

Clear evidence that practice was unchanged was given by Edward Bentham (1707-76), who matriculated at Corpus in 1724 and became a tutor of Oriel in 1732. On the basis of long knowledge both of the university's statutes and practice, he warned the imagined undergraduate of dubious loyalties in 1748: "Upon your Promotion to any *Degree*, you will be called upon to take the Oath of *Allegiance*; and if advanced to any Preferment, in or out of the University, any Office Military, Civil, or Ecclesiastical, you must necessarily take the Oath of *Abjuration*." He added an explanatory footnote: "All Heads, Fellows and Foundationers of Colleges or Halls, all Readers or Tutors in the University are required to take the Oath of *Abjuration*, under Penalty of forfeiting their Places, and Five hundred pounds, if exercising any Office."[68] Bentham's argument to encourage his wavering pupil would have been stronger had he been able to remind his undergraduate that he had already taken the oath of allegiance on matriculation; but he could not do so. Had Bentham, improbably, been in error on this point, he could have corrected himself, and advanced the stronger argument, in a subsequent pamphlet on similar themes; again, he did not do so.[69] He, and his intended audience, were familiar with what the university's rites of passage required.

Where university statutes allowed for exceptions, they made formal and explicit provision for them. Some doubt has surrounded the length of time required for an Oxford BA degree: this can now be clarified. The normal course for a first degree at Oxford was sixteen terms over four years; but of this time, "By an ordinary dispensation and liberal construction" only twelve terms were actually required to be spent in residence by those who were willing to pay the small fee required for obtaining a dispensation from Convocation.[70] Could Johnson have afforded to finance three years' residence on his slender resources? Estimates of the expenses of wealthy undergraduates were much higher than those of poor commoners: one estimate in 1760 suggested that "a frugal young man" could "live pretty well on the yearly sum of 20 gns."[71] Some evidence suggests that even less might have sufficed. Richard Newton, Principal of Hart Hall, in his careful draft regulations intended to secure for his society the grant of collegiate status, addressed the problem of the erratic presence of undergraduates and the need "to Secure their strict *Residence in Term Time*." His view was that "The only Objection against such a Residence is the Expense of it," and that non-residence would be cured by an adequate allowance. He proposed a scheme of scholarships of £10 per annum as "Sufficient to Answer this Expence," arguing firmly against scholarships of higher value.[72] Poverty at least meant self-discipline.

Although the evidence establishes that university practice did not vary without statutory authorization, it has been suggested that one college did change, but in the opposite direction of accommodating those with scruples about the oaths. At Christ Church, a student[73] who passed the necessary university examination for the BA or MA degree had also to obtain from the dean and chapter a "grace," signifying their consent and approval, before he could formally take his degree in the university's ceremony of graduation. These graces came to be treated within Christ Church itself as equivalent to the corresponding degrees. The college's most recent historian has argued that some students in the late seventeenth and early eighteenth centuries took advantage of this system, omitting to take their university degrees in order to avoid the oath of allegiance.[74] If so, no such option was available to Samuel Johnson at neighboring Pembroke. Yet even this degree of latitude is doubtful: students would have been obliged to take the oaths of allegiance and abjuration on admission to their Studentships, as did the Westminster student William Murray;[75] having done so, it is likely that few would have been helped by a procedural evasion of the oaths on graduation. Oxford's formal requirements were difficult to evade: practice conformed to the statutes.

Two later controversies also suggest that the university's practice had not evolved to require the oath of allegiance on matriculation. The first occurred in the late 1750s when Oxford was embroiled in legal controversy over whether it had, alone, the authority to alter its Laudian statutes of 1636, since these were a royal grant. In the point of substance at issue, Oxonians keeping their names on their colleges' books for many years merely in order to vote, it was argued that Statute III, section i "seems clearly to prove, that *Matriculation alone* doth not give a Right to the Privileges of the University" without admission to and residence in "some College or Hall."[76] The participants in the controversy did not argue that the statutory requirements at matriculation were not strictly complied with; on the contrary, the defenders of the statutes prided themselves on the success with which their framers had taken care in "securing the perpetual Observance of them."[77] The second controversy arose in the early 1770s over the question of subscription to the Thirty-nine Articles at matriculation.

None of the participants argued that the practice at that ceremony had evolved to include the imposition of the oath of allegiance also. Such a point would probably have been made by both the critics and the supporters of exacting oaths from matriculants had that been the current practice.[78]

The strongest evidence of all, which bears directly on Johnson's own conduct, comes from the archives of the University of Oxford. These records have always been held to establish that the procedure on matriculation had been laid down in the Laudian Statutes of 1636, and remained unaltered: if over 16, the candidate was required to subscribe the Thirty-nine Articles, swear to obey the university statutes, and take the Oath of Supremacy.[79] This is confirmed by the subscription register on matriculation for the years 1714-40.[80] A folio volume, it contains, pasted onto its first leaf, the relevant section from the printed university statutes, "Titulus II. De Matricula Universitatis. s. 3. *De Tempore & conditionibus Matriculationis*." This section set out the requirements to subscribe the Articles of Religion and the statutes of the university, and to take the oath of supremacy.[81] The oath of allegiance was not listed: it was imposed on foundationers only and by a subsequent act of Parliament (1 Geo I, s. 2, c. 13) rather than as a condition of admission in the Laudian Statutes. On the following page was pasted a printed table of fees for different ranks of undergraduate. Before the next page was tipped in the printed text of the Articles of Religion, evidently removed from a Prayer Book. The following pages are filled with the signatures of undergraduates, beginning on 8 October 1714. Under the entry for 14 December 1729 appears the signature "Samuel Johnson e Coll: Pembr: Gen: Fil:." Johnson chose to describe himself as the son of a gentleman, so paying a fee of 10s 6d, rather than of a plebeian, who paid only 6s 8d.

The subscription registers on graduation are missing for the years 1706-36, but the volume for 1736-63 survives.[82] It is a smaller volume, which could easily have been held in the hand. Like the matriculation register, it opens with two printed pages giving the texts of the requirements laid down in the Laudian statutes: "*The Oath of Allegiance*," "*The Oath of Supremacy*," and "*Three Articles in the 36th Canon*."[83] Unlike the texts in the matriculation register, however, these pages were specially printed to fit the volume, and printed on vellum. The reason for the choice of a more durable, though more expensive, material is suggested by the soiled state of these two pages, especially the wide black mark in the gutter at their foot: candidates evidently held the book open (often with their thumbs) at that page as they read aloud the texts of the oaths before writing their signatures on the blank pages which followed. After these vellum pages is inserted, once more, a printed copy of the Thirty-nine Articles. Candidates could not be unaware of the oaths that they were required to take.

That a reading aloud was called for on graduation is confirmed by a contemporary authority. In 1773 Charles Jenkinson, MP, anticipating a second attempt by Sir William Meredith, MP, to relax clerical subscription to the Thirty-nine Articles, took advice from Nathan Wetherell, Master of University College, on Oxford's current practice.[84] Wetherell's memorandum revealed, inter alia, that a candidate at matriculation *said* nothing to signify his assent to the Articles: "He writes his name only in a Book *to which the 39 Articles are prefixed.*" At graduation, however,

> the case is much altered. Each Candidate is obliged by Statute within three days of his petitioning for his Degree to read the Articles. He then subscribes in Congregation in the presence of all the Magistrates, after which one of the Proctors announces to the several members present that he has subscribed, & then the Candidate declares in the strongest manner his assent to the truth of the Doctrine in the words of the 36th Canon.[85]

Again, this evidence establishes conformity to the requirements of the statutes. Title IX ("Of the Congregations of the Regent Masters"), section 6 ("Of the Oaths of the Persons Presented"), chapter 3 ("The Common Form of Binding every Presentee by oath, and of exacting from all persons the Oath of Royal Supremacy and Allegiance"), included the passage:

> Next, the senior of the presentees shall read aloud, with a clear voice, so as to be heard by all the rest, the oath of royal supremacy, and shall kiss the book when tendered by the proctor, and after him all the rest who are presented on the same day; and lastly, while he distinctly leads with the oath of fealty and allegiance, all the others shall repeat it, word for

> word, and shall all, one after another, kiss the book in confirmation of the oath.[86]

This was not merely an academic matter; allegiance, as tested by oaths, bore directly on men's public conduct in a number of spheres. One such was military service, and it created for successive ministries the problem of organizing a reliable militia in a partly unreliable society. Charles Sackville, Earl of Middlesex, later 2nd Duke of Dorset, formerly a supporter of "*the best and most truly* PATRIOT PRINCE" Frederick, Prince of Wales, advocated a militia scheme in 1752 which anticipated George Townshend's abortive initiative of 1756 and William Pitt's scheme implemented in 1757. The plan would, wrote Sackville,

> (it is hoped,) meet with no Opposition, from any but Those, who would wish to see the present FAMILY indefensible, and without an Army.—*Party* is, or ought to be, out of the Question; and all Men, except the *Jacobites*, should unite in obtaining a Power for the Nation, that will make it rise again in Grandeur and Respect, to the Height it was at in the Reign of Queen *Elizabeth*.

His qualification—"ought to be"—was telling, for his plan acknowledged a difficulty: "All disaffected Persons, who refuse the legal OATHS of *Supremacy* and *Abjuration*, should be obliged to pay, but not permitted to SERVE."[87]

Johnson's reasons for avoiding service in the Trained Bands of the City of London also bear on the question of oaths. The Militia Act of 1662, 13 & 14 Car. II, c. 3, was the foundation document. It imposed the oath of allegiance on all officers and militiamen (clause XIX); it abolished England's ancient Trained Bands and replaced them with the reformed structures; but it left London's untouched (clause XXVII): the lieutenants for the City of London "may and shall continue to list and levy the trained bands and auxiliaries of the said city in manner as heretofore." They survived, governed by their own customs, until reorganized by the Act of 34 Geo. III, c. 81 in 1794.[88] The Militia Act of 1757, 30 Geo. II, c. 25, clause LXV, again specifically confirmed the City's ancient privileges, and it was in the City's Trained Bands that Johnson was, according to Boswell, picked to serve. It follows that

Johnson was not necessarily chosen in "mid-to-late 1759" (Weinbrot, p. 197), when Johnson (born 1709) might have been deterred from serving by his age. Age and ill-health are, of course, not implausible explanations, but more may also have been involved, and it is important that Johnson's selection be dated as accurately as possible. It could have been at any time after he moved into his first house, at 17 Gough Square (his tenancy of which lasted from some time after June 1746 until March 1759) and so, as a housekeeper, became liable to be chosen.[89] Johnson must have been aware from the late 1740s, while working on the *Dictionary*, that he might now be confronted with the moral dilemma which his selection for militia duty would have posed.

Where the militia established elsewhere by the legislation of the 1660s decayed,[90] London's Trained Bands continued on a regular basis into the 1740s and '50s, financed by the regular levy of the "Trophy Tax," picked from regularly-revised muster rolls, and exercised as regiments at the Artillery Ground on days of "General March and Muster" appointed by the Lieutenancy of the City of London: all this preceded the Militia Act of 1757.[91] Lists of those picked to serve do not survive, but it may be possible to infer a time frame for Johnson's selection. Boswell recorded that his colonel was Benjamin Rackstrow. This attribution of rank was clearly an error. The colonelcy of one of the City's six regiments was a senior rank, conferred as an honor only on men of high standing in public life: Rackstrow could never have aspired to it.[92] His career in the Orange Regiment can however be established: he was promoted ensign on 16 January 1745/6, Lieutenant on 2 July 1747, captain on 5 February 1752, major on 17 May 1759 and captain lieutenant on 17 May 1763.[93] Even if Boswell meant "Lieutenant Colonel," a later date for Johnson's service than the 1750s is ruled out by the fact that the lieutenant colonelcy of Rackstrow's Orange Regiment was held by Alexander Dalmahoy from 1 May 1763 to his demise in August 1778.[94]

Boswell probably meant that Rackstrow was Johnson's captain rather than his colonel: if so, the possible time frame is from 5 February 1752 to 17 May 1759. Within this period, the lieutenancy ordered new muster rolls to be drawn up on 23 April 1752 (there follows a probably-unimportant gap in the records between 9 August 1753 and 28 February 1754), 25 April 1754, 29 May 1755 and subsequently. On 15 November 1755 the lieutenancy received the government's instructions that the militia be put in readiness in response to the outbreak of hostilities with France. They accordingly "Ordered that Charging Tickets be forthwith printed in the usual Form and Delivered to the several Persons charged in the Muster Rolls to bear Arms to hold themselves in readiness fit for immediate service." Five days later, and on subsequent occasions, the lieutenancy ordered exercise days for the regiments.[95]

Charging tickets were the printed call-up papers, summoning named individuals to appear or send a substitute. Their text included the command:

> The Arms every Soldier is to appear with and with which you are charged are a Musket with a Bayonet to fix to the Muzzle thereof a Cartouch Box or Pouch a Belt of Buff Four Inches broad and a two Edged Sword.

This was evidently the equipment Boswell later discovered in Johnson's possession. An order of the lieutenancy in 1739 had laid down that all persons so chosen must obtain these arms within one month notwithstanding their having purchased exemptions from serving in person.[96] The charging ticket also indicated the acceptable ages for militiamen: "N.B. No Person in your Stead will be accepted by the proper Officer who shall be under the Age of Eighteen or above the Age of Fifty five."[97]

It is likely that Johnson's name first appeared on the muster roll not later than April 1752, that a charging ticket was delivered to him in Gough Square by Captain Rackstrow, and that Johnson was not disqualified by age from the summons to serve in person that Rackstrow had embraced during the rebellion of 1745. It is possible that this unwelcome demand came in 1752 or 1754, but the likely date is mid-November 1755. Johnson did complain of ill health a month later, from mid-December 1755,[98] but his health had not hampered him in his travels the previous summer, and as late as October he had written optimistically that "There is a kind of restoration to youth in the revival of old friendships."[99] Sir Joshua Reynolds's portrait of Johnson, painted in 1756, shows a robust, solidly-built man in his forties, not a valetudinarian. It is nevertheless possible that Johnson's health may have been one factor, just as his poverty was one reason for his leaving Oxford without a degree; what the date of November 1755 would establish is that it would not have been the only factor, for the date, if accurate, would show that Johnson's agonizings in the *Literary Magazine* for 15 May-15 June 1756 on the oath of allegiance required of militiamen[100] reflected his own recent experience. His distress, if so, can only have been compounded by his poverty, for on 16 March 1756 he was arrested for debt.[101] His decision to hire a substitute rather than serve in person is therefore still open to the explanation that he sought to avoid the oath of allegiance that service would have entailed. Nor was Johnson alone in these political reservations. Even at a later date, some of the riots of 1758 provoked across the country by the implementation of the 1757 Militia Act were interpreted as having Jacobite undertones;[102] many counties refused to participate in the 1757 scheme and some (including Tory Oxfordshire and Johnson's Staffordshire) held out until the American war in the 1770s.[103]

To sum up: on the evidence currently available, it is established beyond reasonable doubt that Samuel Johnson, while at Oxford, was faced with university practice which conformed to its statutes. Weinbrot's suggestion that the university's practice in this respect had changed is not substantiated. It follows that my hypothesis still stands: Johnson never took the oaths of allegiance or abjuration; throughout his career he avoided situations which would have required them; his remark to Boswell in April 1773 that "I know not whether I could take" the oaths correctly recorded that he had never done so.[104] Johnson, in the sense of the word defined in his own *Dictionary*, was a Nonjuror.

Johnson's Ideological Identity and the End of Party

Greene still declines a discussion of political theory or religion: he likens the question of Michael Johnson's political beliefs to the question of "the number of angels who can stand on the head of a pin" (p. 78), a phrase which discloses an astonishing estimate of the history of political thought. In the absence of such a discussion of Nonjuring religion *as religion*, the significance of Johnson's commitments will not be apparent.[105] In its absence, Greene still writes of Jacobitism as "a romantic nostalgia for the exiled Stuarts" (p. 100), a characterization that is probably the greatest single obstacle to academic research on the subject.[106] What is at issue is ideology, not sentiment. The evidence that Johnson rejected *jure divino* doctrine comes from the end of his life, and may relate only to the Hanoverian dynasty then on the throne; what can now be established is that the intellectual culture of Johnson's early years in Lichfield, Oxford, and London was saturated with *jure divino* doctrine to a degree unappreciated by Greene in 1960 or, indeed, 1995. Greene and Curley still adhere to the spurious timelessness of Namier's analysis: if they can find one piece of evidence from the end of Johnson's life or from the 1750s, when he reoriented his stance, that he did not endorse *jure divino* kingship, this is taken as secure evidence for Johnson's beliefs throughout his career.

We now know that Jacobitism combined many strands; it was not simply, or even chiefly, a replication of the doctrine of Sir Robert Filmer. Yet these complexities mean little to a method that seeks to deduce its results from a simple and unchanging definition. Curley, similarly, writes that "Allegiance to Jacobitism must have entailed . . .": the historian employs induction rather than deduction to discover the content of political doctrines and relate them to political institutions. Greene (1995) is still unwilling to address the extensive direct evidence which survives for the existence of an early-Hanoverian Tory party, first presented by Dr. Cruickshanks in the 1715-54 volumes of the *History of Parliament* (1970). To dismiss without examination a mass of evidence retrieved over three decades by a wide variety of researchers is unlikely to appear a plausible strategy to serious scholars, whatever its short-term effect as rhetoric. Greene (p. 87) cites Lord Blake's argument against a continuous Tory party having existed between Bolingbroke and Liverpool, but has here evidently misunderstood the historiography which he elsewhere boasts of knowing: such a discontinuity was precisely the first essential thesis of my work in the 1970s, and is basic now to my analysis of Johnson. To Greene (pp. 85-88, 114-15), the only two alternative scenarios of party structure are those of Macaulay and Disraeli on the one hand, and Namier on the other: this is a confession of ignorance. So is his attempt to produce "an entire vindication of Namier" by denigrating the history of ideas and arguing that the concrete social location of ideologies is "distasteful to British academics who preferred to dwell on the exalted plane mapped out by Macaulay" (p. 92). It could not be seriously maintained that this is what British historians actually do, and caricatures of this sort cannot strengthen Greene's argument.

Curley (p. 155) seizes the point that some leading members of the Whig opposition (Chesterfield among them, but others before him, including the great Duke of Marlborough himself) at times had contacts with the Stuart court as an insurance in case the times should alter. Curley interprets this to mean that no distinction can be drawn between opposition Whig and Tory or Jacobite rhetoric, indeed that any anti-Hanoverian rhetoric was securely Whig. This is a non-sequitur, unsupported by evidence, and disproven by recent literary scholarship.[107]

In an early-Hanoverian England so bitterly polarized between Whig and Tory, men had motives both for party loyalty and also, at times, for claiming to be of no party. Some asserted, for example, that a non-party "Country" opposition (in fact composed of mutually suspicious opposites, Tories and extreme Whigs) best embodied national ideals in their campaign against the corruption and misconduct of ministers. Others, increasingly from the 1750s and with dramatic practical analogues from the 1760s, claimed that the extinction of the Jacobite challenge and the increasingly pragmatic nature of Westminster party-politics in the 1750s provided the basis for party reconciliation (usually on Whig terms). Whigs argued thus in order to persuade Tories to renounce the Stuart allegiance that was residual in some of them; Tories argued likewise in order to secure admission to office and emolument. For that reason, contemporary assertions that party was dead cannot be simply read as reportage; such statements embody tactical political purposes.

Weinbrot cites three claims of the non-importance of party, by Nicholas Amhurst, Richard Newton, and Henry Brooke, but misinterprets the evidence for all three (p. 165). First, Nicholas Amhurst once more, on Oxford. In the Preface to the collected edition of *Terrae-Filius* in 1726, Amhurst set out in general terms "my charge of a *treasonable* spirit reigning in the university at that time," that is, the time of his residence as an undergraduate from 1716 to 1719. In 1726, he added:

> It must be confess'd, indeed, that this *seditious* spirit, and these *treasonable* practices have, of late years, so much *abated*, if not entirely *ceased* there, that it induced his majesty, out of his

> royal goodness, to distinguish his subjects at *Oxford* with several valuable *donations*, and marks of his affection

which the Oxonians acknowledged with two separate addresses of thanks. This "shews how far they have *alter'd* their sentiments, and how well they *deserve* his majesty's farther encouragement." To have said otherwise would have been an indictment of royal policy, which Amhurst naturally had to avoid; but he also thereby exploited the university's present professions of loyalty to contrast them ironically with their former disaffection, and to claim a hostage for the future. Their "being good subjects now does not prove that they were so *nine* or *ten* years ago,"[108] he argued, and left it to his readers to speculate on Oxford's possible conduct nine or ten years hence. Amhurst was far from being a simple reporter of a change of heart in his alma mater.

Second, Richard Newton (1676-1753). In 1735 he was Principal of Hart Hall, one of the university's ancient halls, and was campaigning for it to achieve collegiate status by the grant of a royal charter. In this he found bitter opposition from the notoriously Whig Fellows of Exeter College, and from the Whig Dean of Christ Church. In view of his practical goal, Newton wisely professed his neutrality: "if the Government *Itself* be not of any Party, that I am not can be no Mark of Disaffection in me to it." Weinbrot cites this without comment, but the Dean of Christ Church was less persuaded:

> Hath he ever declared his Respect to, and affection for, the *Present Government?* Not that I ever heard. And yet he now steps forth, and boasts, as though he, good Man, merited more from the Government than any of his Brethren. I am far from charging him with *Jacobitism*, or any Measure of *Disaffection* to the State: But this I will say, That, as far as I ever saw, or have been told, he hath contented himself to lie by quiet; and hath never employed his Eloquence, either to support HIS MAJESTY'S Title, or to recommend His Administration. He himself tells us, "Not being of any Party, I never pretend to be." I forbear to comment on these words, which I leave to be considered by the Reader.[109]

Plainly, Conybeare insinuated just such a doubt of Newton's loyalty.

In his reply of 1735, Newton included Chapter XIII, "Concerning Affection to the Government," in an attempt to clear himself from this insinuation in the minds of "the Great Men above." Newton went through the usual litany of professing support for the Revolution and the Hanoverian succession, pausing only to emphasize (what some would rightly interpret as an anti-ministerial slip) that Parliament should be "*freely* chosen." He boasted: "I have not only taken the *Oaths* to the Government *Myself*, but have also removed the *Scruples* which *Others* have had to do it."[110] Newton's politics were not so clear, however. On 9 February 1737, in the parliamentary by-election for the University caused by the death of George Clarke, Newton voted for William Bromley, Jr.[111] against the Whig Robert Trevor, a choice in which he concurred with Jacobites like William King of St. Mary Hall and Dr. Panting, Master of Johnson's Pembroke College. At Christ Church, Conybeare voted for Trevor.[112] Hertford eventually obtained its royal charter on 8 September 1740, with Newton its first principal, and with statutes based on his model of 1720. They provided for the text of the oath to be taken "upon Admission . . . By every student and scholar," but this was merely a promise not to do "*any thing Injurious to this College*," and had no wider content. Nor was it even an oath to obey the statutes: since a student "cannot have *Considered* them sufficiently . . . Young Men will often break them without *Adverting* that they do so. To them an *Oath* to observe the Statutes will be a Snare."[113] This was not Whig orthodoxy.

Nor did the Principal of Hertford sign the Association in 1745. Ten years later, Newton republished Chapter XIII of his tract of 1735 as a separate pamphlet, complaining in the Introduction: "*The reader will here perceive, that the spirit which at present prevails, long since had a beginning; but is now nursed up into full maturity by the indefatigable industry of* a society of informers."[114] The "informers" against whom Newton, like Samuel Johnson, were so indignant were the Whig dons like Richard Blacow who thought it their duty in the late 1740s and early 1750s to broadcast to the world what they claimed were the university's Jacobite loyalties.[115]

Third, Weinbrot cites a pamphlet of 1750 by Henry Brooke. Yet Weinbrot confuses the anti-Walpolian Henry Brooke (?1703-83), playwright and author of *Gustavus Vasa* (1739), with the Whig loyalist Henry Brooke (1697-1752), jurist and author of *A Letter to the Oxford*

Tories (1750), a conflation which makes *Gustavus Vasa* seem more securely Hanoverian than, in its proper context, it was. Weinbrot also fails to supply a context for the pamphlet of 1750. Brooke had held the Regius chair of civil law at Oxford since 1736.[116] His Whig zeal was not in doubt; his aim in 1750 was to promote the disintegration of the rival party.

Dating his anonymous pamphlet from the Middle Temple on 1 January 1750, Brooke professed, clearly for tactical reasons, to be ignorant of "the genuine and fixed Import" of the word Tory:

> I call you Tories, because such you call yourselves: But the real and essential Difference between a Church of England Academick *Tory* and a Church of England Academick *Whig* is, I confess, a Secret to my Understanding; both these having subscribed the same Articles of *Religious* Faith; Both having given to the same Government the same Solemn Security for their *Civil* Obedience: Surely then there can be no Absurdity in an Assertion that a Distinction made between Members of the same Communion, *sworn* Subjects to the same Prince, is a Distinction without a *Difference*, an Opposition in *Name* only, destitute of any Foundation in *Fact*.

But Brooke could only argue that Toryism was indistinguishable from Whiggism by disassociating Toryism from Jacobitism: "I see you eager on all Occasions to disavow the Creed of *Jacobitism*." As Brooke progressed in his argument, he disclosed his real purposes: not to advance evidence to prove the dissolution of the "Tory" identity, but to neutralize the Tories by vilifying the dynastic allegiances of a part of their number. Of Jacobites he wrote: "Men so absurd, so very impious place Themselves much below the Dignity of *Reason* and *Argument*." Brooke's prime target was that "*Chimera*, a *Juring Jacobite*":

> He, who can, for his own Temporal Advantage, enter into Obligations *sacred* and *solemn*, as the Oaths of *Allegiance* and *Abjuration* are, and This done, dispute, clamour, drink, or riot for the Person whom He has abjured, then wipe his Mouth and say "I have done no Wickedness," must have a System of Morality peculiar to Himself, must be deaf to the Voice of *Persuasion*, must be left to the Lashes of a Fiend of his own creating, an *Evil Conscience*.

Was the world not right to draw inferences from the conduct of the Tories? If, Brooke urged inconsistently,

> on a fair Scrutiny into the Conduct of the Leaders of your Party, for more than thirty Years last past, it shall appear that no one *Minister*, no one *Measure* of Government has obtained your Approbation, or escaped your Displeasure, can you, in such Case, expect that the World should have such a Partiality for your Sentiments, as to pronounce that the Rulers of *Great Britain* are always *wrong*, and the Rulers of *Oxford* always *right*? And will not Those, to whom you are not thoroughly known, be rather led to conclude, from such a Continuance, from such an Obstinacy of Opposition, that your Dislike to *every* Publick Measure proceeds from a determined Aversion to "*the Powers that Be?*"

Plainly, observers would infer this, and Brooke's phrase "to whom you are not thoroughly known" was irony. The pent-up resentment of an Oxford Whig over many decades now revealed itself: his opponents, he ranted, had manipulated fellowship and parliamentary elections, and heaped abuse on loyal Hanoverians. What Brooke intended as "a Defense of *your* Party against the heavy Charge of *Jacobitism*" ended as a diatribe against the Tories for giving too many grounds for such a charge.[117]

Benjamin Kennicott, Whig fellow of Whiggish Exeter College, pursued a similar tactic in his campaign against William King. He first acknowledged, reasonably enough, that "*The Old Interest*, in the County of *Oxford*, is a Political Interest, *long* supported there, partly by *Tories*, partly by *Jacobites*; but probably by a much greater number of the former than the latter, especially *since the late Rebellion*." His tactical task was to split the Tories from the Jacobites by a definition, and leave the latter politically isolated. Hence

> we must first settle the proper distinction between *a Tory* and *a Jacobite*. 1. A TORY then (I speak here of a *true, sober, thinking, systematic* TORY) is one, who wishes the true glory of *the Church of England*, in opposition to *Dissenter* on the one hand and *Papist* on the other; and who wishes the true glory of *British Liberty*, in opposition to *Licentiousness*, in being *free* to do every thing, and *Slavery* in being *permitted* to do nothing.

2. He has high notions of *Regal Authority*; but wisely distinguishes between *a King* and *a Tyrant*; and, tho' he believes it his duty, with *the most active Loyalty* to serve the *former*, he thinks himself not bound to *submit passively* to the *latter*, when he has brought the Religion and Liberty of his Country into extreme danger. 3. He is therefore an hearty friend to *the Revolution*; and consequently a sworn foe to the doctrine of *hereditary* Right, *absolute and indefeasible*. 4. He affirms, that *no claim of Right* properly belongs to the Descendants (even *admitting* Descendants) of *James the second:* but that the only rightful claim is *Lineal Descent*, limited by such *Conditions* as the wisdom of the Nation has fix'd, for the more effectual security of our sacred and civil Freedom. 5. He acknowledges this rightful claim in *His Majesty King* GEORGE *the Second*; to whom he readily swears Allegiance: and this the more readily, as he considers the present ROYAL FAMILY the most likely of all others, to perpetuate the Blessings of *Englishmen* and *Protestants*. 6. But tho' zealously loyal to *The* KING, he is perhaps dissatisfied with *the Administration:* he may think (and yet without proper foundation) that *the Ministry* frequently pursue such Measures, as tend to the detriment and disgrace of the Kingdom: and he may think himself the true *Friend* to *The* KING, for being an *Enemy* to the wrong Measures of *those*, who are, by their high Office, *The* KING's principal Servants. 7. But then—in order to preserve his Loyalty from suspicion—to perform his Oath of supporting His Majesty to the utmost of his power—and to give weight to his Censures of any Measure, that he apprehends to be wrong—he most zealously supports every Measure, that appears to be right: laying down this, as a fundamental Maxim, "No one, but *The* KING's *Enemy*, can censure, oppose and distress ALL the Measures of *The* KING's *Ministry*."

I must just remark here, by a necessary digression, that the preceding is the proper Character of *every honest* WHIG, as well as of *every honest* TORY; for I can *in these days*, perceive no real difference: excepting—that, as *the Tory* (in the Church) thinks more favourably of *the Papist*, than the *Whig* does; so *the Whig* (in the Church) thinks more favourably of *the Dissenter*, than *the Tory* does. But then, as every *Whig* is not a Dissenter, nor every *Tory* a Papist; so both *Whig and Tory* may be sincere

> *Members* of, and zealous *Friends* to, *The Church of England*. . . .[118]

Kennicott's choice of themes was shrewd; but as his rationale went on, it clearly became more prescriptive than descriptive, and ended by confronting without resolving a major source of lasting differences between Whigs and Tories, religion. Whigs and Tories were in the 1760s to find much more common ground, but not by the success of the ideological solvent Brooke and Kennicott tried to apply: it took the destruction of the parliamentary Tory party in the 1750s, the abandonment of hope of most of the Jacobite interest, and the accession of a monarch influenced by Bolingbrokean supra-party theories, to make the decisive change. Any who really believed in the growth of non-partisan behavior were quickly undeceived by the university's partisanship in the bitterly-contested Oxfordshire election of 1754. One Whig complained: "Out of twenty Colleges, nineteen, beside the Halls, have behav'd with what they call *great Decency*, and stuck close to what may properly enough be term'd the *Old Interest* of OXFORD. Such is the force of Academical Bigotry, and the *Prejudice of Education*!"[119]

To sum up: Samuel Johnson's early life, until the 1750s, was spent in a public culture still heavily polarized between the traditional parties of Whig and Tory. Claims that parties no longer existed were tactics which still addressed that polarized world. Johnson was a partisan, not a detached political skeptic.

The Years of Transition

A decade of particular difficulty is the 1750s, during which Johnson's political opinions may, to some degree, have adapted to changing circumstances. An important but unreliable document is the Preface to the consolidated index to the *Gentleman's Magazine*, published in 1753 and reportedly written by Johnson and another. Since we do not know the relative contributions of Johnson and his anonymous colleague, I judged that "the evidence is inconclusive,"[120] a view which, on reconsideration, I still hold, and based my argument on the more secure evidence of Johnson's *Dictionary* (1755). Since no new evidence has been produced on the authorship of the 1753 Preface, Greene's attempt to claim it as a "decisive" rejection by Johnson of Stuart claims (p. 74) is unsuccessful.

An equally insecure source for the assessment of Johnson is Sir Robert Chambers's course of Vinerian law lectures, delivered in the late 1760s. Very little evidence, direct or circumstantial, on the relative role of the two men in the process of the lectures' composition has been presented. Their modern editor formerly expressed the view that "In all likelihood Johnson lacked the detailed technical learning to play the leading role in drafting the Vinerian course," a judgment partly supported by Johnson's admission to Chambers in 1766 on the issue of the legality of an embargo on grain exports: "But I judge upon mere principles without much knowledge of laws or facts."[121] Chambers, suggested his editor, had a "heavy dependence" on the scholarship of his predecessor in the Vinerian chair, Sir William Blackstone, even where the style of a passage wrongly suggests Johnson.[122]

For the purposes of present controversy, the editor has reversed his earlier assessment of the relative contributions of the two men and now seeks to use Chambers's lectures as important evidence for Johnson's Hanoverian loyalties. Curley's assumption that some parts of the Vinerian lectures are reliably Johnsonian (pp. 143-46) is, however, subverted by his own demands for "facts," "hard extant data," "positive proof," and "incontrovertible evidence." No new evidence is presented by Curley about the carefully private processes of composition of those lectures or the strategic purposes of Chambers in writing as he did. Lacking such evidence, I made no use of Chambers's text in my *Samuel Johnson*;[123] but if an argument were to be framed, one might argue, against Curley's Introduction and by analogy with Blackstone's *Commentaries*, that Chambers's lectures reflect a further phase of the accommodation of early eighteenth-century Tory-Jacobite Oxford with the Hanoverian regime in and after the middle of the century. Curley's observation (p. 144) that Chambers's lectures, like Blackstone's *Commentaries*, are "unflinchingly loyal to the reigning Hanoverian monarchy" is strangely at variance with the content of both sets of documents.

Chambers's Vinerian course gave a decidedly un-Whiggish account of allegiance: since "by far the greatest part" of the common law "still retains the spirit of the feudal system," it followed that the king's "power over his subjects is conceived as more than political . . . For this reason allegiance is always by the law termed natural." It was a natural status, not a functional response to the command of the king; "allegiance being considered as a duty produced by causes with which the subject did not co-operate, and from which as he did not incur it by his own choice, he cannot by his own choice set himself free." It was exactly the theory of indefeasible allegiance which had caused such divided loyalties in and after 1688. Chambers even reminded his audience of the statute of 25 Edw. III, c. 2, by which "it is declared 'that the law is and ever has been that the *king's children* wherever born ought to inherit.'"[124]

Chambers acknowledged that the personal tie of "*Liege homage* or allegiance," evolved in feudal law and restated after 1603 by Sir Edward Coke, was somewhat modified by the new oath of allegiance introduced after 1688 by the Act of 1 W&M, c. 8: "As this oath was enacted by an assembly busy in re-establishing an unsettled government and therefore desirous to avoid all obstacles of scrupulosity; and by men who in political opinions differed very widely from each other, and agreed only in excluding the absent monarch, it is conceived in very general terms which the law is left to interpret." This fell far short of a glowing account of the Glorious Revolution and discarded the Whig theory of abdication in favor of a frank admission of deposition, though a reluctant deposition. It was immediately followed by a remark susceptible of an anti-Revolution interpretation: "While allegiance was considered as being the consequence of birth it is no wonder that our monarchs made great distinction between aliens and native subjects and were unwilling to grant property to those of whose obedience they did not think themselves secure." Alas, "The idea of natural allegiance has long been wearing away with other consequences of the feudal system. . . ."[125]

Chambers was ambiguous about Hanoverian kingship and the changes which the institution of monarchy was experiencing:

> coronation in ancient times procured to the king the homage of his nobles, with a public acknowledgement of his right, gave notoriety to his accession and conferred sacredness on his person. What are its effects at present it is not so easy to determine. It is perhaps now to the nation, the regular and legal notification of their sovereign. It is provided by the statute of the 11 Hen. 7, c. 1, that the king, reigning by

> whatever title, may so far legitimate the actions of those who carry his commission, that though he should be afterwards deposed, no man can be punished for having acted under him.[126]

This was consistent with the law of treason, which defended *de facto* as well as *de jure* monarchs:

> For though it be evident that *he* may be a king *de facto* who is no king *de jure*, that he may be in present possession of the royal authority by force of arms or otherwise who has no just title to the power which he exercises; yet since by the same right whatever it be by which he executes the laws for the protection of the people, he must be supposed to execute them likewise in vindication of himself, the peace of the world requires that those who accept the advantages of his protection should submit to the penalties of disobedience.[127]

That civil order required acceptance of *de facto* monarchs was close to Johnson's position after the early 1760s.

Despite this, the lectures offered little comfort to those who might have difficulties about their religious duty. An oath was "a direct appeal to the Supreme Being, by whom perjury will undoubtedly be avenged."[128] The one hopeful point was the claim that the problem was now receding. After explaining the penalties for treason under the Acts of 13 W III, c. 3, 6 Anne c. 17 and 17 Geo. I, c. 19, Chambers added that "the laws against popery being made in the animosity of contest and in dread of danger are now almost forgotten, and that the Protestant succession gathering daily strength all danger of committing treason by opposing it is now at an end."[129] To postpone such a gathering of strength to the late 1760s hardly suggested a strong confidence in the security of the first two Hanoverian monarchs; and Chambers's odd phrasing, suggesting that his audience might indeed have been in some danger of falling into treason, hardly credited them with an overriding zeal for the House of Hanover.

Chambers, like Blackstone and Johnson, illustrates aspects of the process of accommodation to the dynasty in the 1760s. Greene (pp. 75-76) still presents Johnson's growing links with the monarchy of George III as evidence against Johnson's Jacobitism *tout court*, an argument which could be plausible only if the structures of English politics in Johnson's lifetime remained constant: we now know they did not. Johnson indeed made critical remarks about many sovereigns, from Charles II to George III. But his critical remarks were couched in very different terms from monarch to monarch, and his opinions of individuals did not determine his opinions of the validity of their titles. Johnson was quite capable of making candidly critical remarks about Charles II and James II while remaining profoundly uneasy at the events of 1688-89. In no way has it been suggested that Johnson looked back on Stuart monarchy before 1688 as "the good old days" (Curley, p. 154): he was a realist, keenly aware of the limitations of governance at all times. Far from James II's being "the Jacobite darling" (Curley, p. 158), his Protestant sympathizers were often critical of his conduct and religion, and regarded him even more than his brother as having made the very worst of the opportunities with which they had been presented in 1660. None of these judgments involved Johnson's endorsing "the people's right to unseat James II" (Curley, p. 159). Similarly, Nonjurors had for certain purposes made common cause with juring Anglicans, even under the first two Georges but far more often after 1760;[130] it does not (*pace* Curley, p. 151) establish that Johnson was not a Nonjuror that he should have defended church and state against the threats which the establishment faced in the late 1760s and early 1770s.

This is familiar ground, on which the respondents have added nothing. Three small pieces of new evidence bearing on Johnson's reputation in his lifetime are however introduced into the discussion, and misinterpreted, by Curley. *The European Magazine* of May 1785 termed Johnson "a furious Jacobite." *The India Gazette* of 21 May 1787 similarly termed him "a *Tory* . . . but . . . a convert from the Jacobite faction." Curley brings no evidence to support his gloss that this was "a defamatory exaggeration." Third, Curley advances an exchange in the House of Commons in 1771. An opposition MP, Thomas Townshend,[131] angered by Johnson's pro-ministerial pamphlet on the Falkland Islands crisis, alleged that Johnson, "a Jacobite by principle, has been encouraged, fostered, pensioned, because he was a Jacobite." Alexander Wedderburn, the new Solicitor General, a Scot who had first entered Parliament in 1761 on Lord Bute's interest and had first proposed Johnson's pension to Bute, was forced to defend the author and, by implication, Wedderburn himself:

> From the course of my pursuits I have not seen the man four times in my life. This, however, I know,—that he was not pensioned because he was a Jacobite, nor on account of his political principles; that he was not pensioned from any such illiberal motive. The only motive for granting that pension was Dr Johnson's distinction in the literary world, and the prospect of approaching distress. The person [Wedderburn] who solicited it for him was totally unacquainted with any other circumstance. He [Wedderburn] knew that he [Johnson] was a man of letters by his great work, the Dictionary of the English language. The minister [Bute] to whom the application was made, and the man who made the application, never enquired into his political character. The pension was applied for solely upon the ground I have stated. Let it never be said in this House, nor in any other, that the royal bounty is to be so restrained. If he had been a papist, and at the same time a great mathematician, a great poet, and distinguished for literary merit in various ways, it would have been becoming to extend that bounty to him.[132]

Curley interprets this as Wedderburn's rejection of Johnson's Jacobitism. A parliamentarian or a political historian would appreciate that Wedderburn had markedly failed to deny that Johnson subscribed to such views; he had argued instead that Johnson's views had not been formally noticed, and that whatever they were, even if they were extreme and abhorrent, his pension was justified by his literary merits alone. This omission was an implicit admission. It was also, of course, at once seen to be implausible, as Thomas Townshend pointed out in his reply: "I desire not to be understood as wishing for literary rewards to be confined to party. How the learned gentleman could be acquainted with Dr Johnson's literary merits, and not be aware at the same time of his political conduct, is to me surprising."[133]

To sum up: the evidence of the 1750s suggests that Johnson's position was partly adapting to changing circumstances, not that he had been a lifelong Hanoverian. His reputation after 1760 correctly recalled his commitment against the dynasty under the first two Georges.

Professional Ethics and the Conduct of Academic Debate

Much new evidence on Johnson has recently been placed in the scholarly arena, evidence which previously dominant criteria had discarded as irrelevant. It deserves to be scrutinized in accordance with the normal canons of historical enquiry, and weighed against any new evidence which might point in a different direction. This the three respondents have scarcely done. They resort too readily to bluster and denigratory language in over-emotional defenses of their entrenched positions. Gross egotism needs no reply; but whatever might be their standards as literary students, the three respondents fall short as historians: they show various inabilities to weigh and use historical evidence, and an unprofessional preference for winning the argument, by whatever *ad hominem* rhetoric, over the recreation of complex historical situations. I must also record that at many points my work will be found to be advancing a case different from that which Greene and Weinbrot in particular claim that it makes. To follow such authors further into these thickets of their own devising would yield neither good sport nor public enlightenment.

None of the three respondents is known to me, and their *ad hominem* arguments were unwisely attempted on a basis of ignorance. When Greene depicts me as "brought up in the Macaulay-Trevelyan tradition" (p. 84), I feel justified in regarding other personal characterizations as equally inaccurate; and since he claims (p. 87) that I derived my conception of early-eighteenth-century dynastic ideologies from the novels of Sir Walter Scott, Benjamin Disraeli, Robert Louis Stevenson and John Buchan, it is only necessary to record that I had read none of them before writing *Samuel Johnson*. Personal partisanship is a charge hard to take seriously, and Weinbrot's attempt to make it may disclose only his unfamiliarity with historical practice. It is, for example, commonplace to refer to individuals in the past by their *de jure* and *de facto* titles where this aids identification; it would be merely pedantic to cease to write of "James II" after his flight to France in December 1688, and to refer to him thereafter only as "the Pretender." Some Tories refused to acknowledge the titles of George I and George II, referring to each only as "the Elector of Hanover," and some Whigs were equally resolute in referring to James Francis Edward Stuart only as "the Pretender"; to suggest that the historian endorses eighteenth-century claims by recording them is either a naive mistake or a disreputable stratagem.

Greene's essay indeed has its darker side. He makes reference to Senator McCarthy and the John Birch Society, evidently suggesting that scholarly attention to the phenomenon of Jacobitism derives from a strategy of improperly exaggerating or even inventing what the persecutors abhor. Yet elsewhere he suggests the opposite, that the reconstruction of Jacobitism derives from a covert endorsement of that reprehensible doctrine by modern scholars. Both cannot be true, but both can be false, and I have some authority for recording that false they both are.

That either charge should have been made, however, goes beyond attempted *ad hominem* discourtesy. If such strategies were widely adopted they would poison the wells of academe, and halt research into any subject condemned by the prevalent standards of our age. A constructively critical reply to a thesis seriously advanced and supported by substantial evidence would have been welcome, whichever way the verdict of professional opinion fell; an attempt to frustrate the ordinary progress of historical enquiry into the life and thought of Samuel Johnson would be worrying were it not, as it now appears, ineffectual and ultimately rather sad.

NOTES

1. Howard Erskine-Hill, "Johnson the Jacobite? A Response to the New Introduction to Donald Greene's *The Politics of Samuel Johnson*," *The Age of Johnson*, 7 (1995), 3-26, and Professor Erskine-Hill's earlier work there noted; J. C. D. Clark, "The Politics of Samuel Johnson," ibid., 27-56; idem, *Samuel Johnson: Literature, Religion and English Cultural Politics from the Restoration to Romanticism* (Cambridge: Cambridge Univ. Press, 1994).
2. Donald Greene, "Johnson: The Jacobite Legend Exhumed"; Thomas M. Curley, "Johnson No Jacobite; or, Treason Not Yet Unmasked"; Howard D. Weinbrot, "Johnson, Jacobitism, and the Historiography of Nostalgia," *The Age of Johnson*, 7 (1995), 57-211. Page references in parentheses are, henceforth, to this group of articles.
3. Allan Bloom, *The Closing of the American Mind* (New York: Simon and Schuster, 1987) was the best known defensive reaction from one part of this culture, though to challenges from a very different quarter. One of Bloom's central claims, that "Americans are Lockeans" (p. 167), suggests his affinity with the mind-set here examined.
4. I do not, of course, mean to diminish those many and fine American scholars whose outlook is not so influenced; but the presentist purposes of others are more evident from a distance.
5. Moncure Daniel Conway, *The Life of Thomas Paine*, 2 vols. (New York: G. P. Putnam's Sons, 1892) claimed (I, vii) that "Paine's 'deism' differed from Franklin's only in being more fervently religious"; *idem*, ed., *The Writings of Thomas Paine*, 4 vols. (New York: G. P. Putnam's Sons, 1894-96); *idem*, *Autobiography Memories and Experiences of Moncure Daniel Conway*, 2 vols. (Boston: Houghton Mifflin, 1904). For a more recent version see Jack Fruchtman, Jr., *Thomas Paine: Apostle of Freedom* (New York: Four Walls Eight Windows, 1994).
6. Carl L. Becker, *The Declaration of Independence: A Study in the History of Political Ideas* (New York: Alfred Knopf, 1922); Vernon Louis Parrington, *Main Currents in American Thought* (New York: Harcourt Brace, 1927, 1930); Preserved Smith, *A History of American Culture*, II (New York: Henry Holt, 1934); Merle Curti, "The Great Mr Locke: America's Philosopher, 1783-1861," *Huntington Library Bulletin*, 11 (1934), 107-51; Willmore Kendall, *John Locke and the Doctrine of Majority-Rule* (Urbana, Illinois: Univ. of Illinois Press, 1941); Max Savelle, *The Foundations of American Civilization: A History of Colonial America* (New York: Henry Holt, 1942); idem, *Seeds of Liberty: The Genesis of the American Mind* (New York: A Knopf, 1948); Michael Kraus, *The Atlantic Civilization: Eighteenth-Century Origins* (New York: Cornell Univ. Press, 1949); Clinton Rossiter, *Seedtime of the Republic: the*

Origin of the American Tradition of Political Liberty (New York: Harcourt Brace, 1953); Louis Hartz, *The Liberal Tradition in America* (New York: Harcourt Brace, 1955).

7. Leo Strauss, *Natural Right and History* (Chicago: Chicago Univ. Press, 1950).
8. Russell Kirk, *The Conservative Mind from Burke to Santayana* (Chicago: Henry Regnery, 1953); Peter J. Stanlis, *Edmund Burke and the Natural Law* (Ann Arbor, Michigan: Univ. of Michigan Press, 1958); Harvey C. Mansfield, Jr., *Statesmanship and Party Government: A Study of Burke and Bolingbroke* (Chicago: Univ. of Chicago Press, 1965).
9. For the concept in an earlier period, see A. G. Rigg, *A History of Anglo-Latin Literature 1066-1422* (Cambridge: Cambridge Univ. Press, 1992), esp. Introduction, pp. 1-8. English is not, of course, the only vernacular to have been influenced by the classics: cf. Michael Lapidge and Richard Sharpe, eds., *A Bibliography of Celtic-Latin Literature 400-1200* (Dublin: Royal Irish Academy, 1985).
10. D. J. Palmer, *The Rise of English Studies: An Account of the Study of English Language and Literature from its Origins to the Making of the Oxford English School* (London: Oxford Univ. Press, 1965), Introduction.
11. See, most recently, Barry Baldwin, *The Latin & Greek Poems of Samuel Johnson: Text, Translation and Commentary* (London: Duckworth, 1995).
12. I shall publish elsewhere an extended reply to Weinbrot on Johnson's classicism.
13. [Nicholas Amhurst], *Terrae-Filius: Or, the Secret History of the University of Oxford; in Several Essays. To which are added, Remarks upon a late Book, entitled, University Education, by R. Newton, D.D. Principal of Hart-Hall* (London, 1726), p. xix.
14. Clark, *Samuel Johnson*, p. 35.
15. It was not Weinbrot's thesis, in *Britannia's Issue: The Rise of British Literature from Dryden to Ossian* (Cambridge: Cambridge Univ. Press, 1993), or Robert DeMaria's, in *The Life of Samuel Johnson: A Critical Biography* (Oxford: Blackwell, 1993), but I did not think it appropriate to list every point at which I adopted a different interpretation from previous scholars. Nor did I then appreciate the degree to which my work called in question Weinbrot's.
16. Clark, *Samuel Johnson*, pp. 23-32, 74-77.
17. For a convenient summary, see Paul Elmer More and Frank Leslie Cross, eds., *Anglicanism: The Thought and Practice of the Church of England, illustrated from the Religious Literature of the Seventeenth Century* (London: SPCK, 1935); the best monograph is still George Every, *The High Church Party 1688-1718* (London: SPCK, 1956).

18. Peter Benedict Nockles, *The Oxford Movement in Context: Anglican High Churchmanship, 1760-1857* (Cambridge: Cambridge Univ. Press, 1994).
19. David Watkin, *Thomas Hope 1769-1831 and the Neo-Classical Idea* (London: John Murray, 1968); J. Mordaunt Crook, *The Greek Revival: Neo-Classical Attitudes in British Architecture 1760-1870* (London: John Murray, 1972); David Watkin, *The Triumph of the Classical: Cambridge Architecture 1804-1834* (Cambridge: Cambridge Univ. Press, 1977); idem, *Athenian Stuart: Pioneer of the Greek Revival* (London: Allen and Unwin, 1982). For a convenient introduction, see M. I. Finley, ed., *The Legacy of Greece: A New Appraisal* (Oxford: Oxford Univ. Press, 1984).
20. Clark, *Samuel Johnson*, pp. 251-52; William St Clair, *That Greece Might Still Be Free: The Philhellenes in the War of Independence* (London: Oxford Univ. Press, 1972); Palmer, *English Studies*, p. 3.
21. E.g., William Kuhn, *Democratic Royalism: The Transformation of the British Monarchy, 1861-1914* (London: Macmillan, 1996, forthcoming).
22. William St. Clair, *Lord Elgin and the Marbles* (London: Oxford Univ. Press, 1967); Richard Jenkyns, *The Victorians and Ancient Greece* (Oxford: Blackwell, 1980).
23. He did, of course, reportedly say almost that on a series of occasions: Clark, *Samuel Johnson*, pp. 139-40, 197, 200, 208-10, 222, 224.
24. In W. H. Bond, ed., *Eighteenth-Century Studies in Honor of Donald F. Hyde* (New York: Grolier Club, 1970), p. 42.
25. For a different interpretation see Clark, *Samuel Johnson*, pp. 23-32, 233-37.
26. For a major examination of this theme, see Howard Erskine-Hill, *Poetry and the Realm of Politics: Shakespeare to Dryden* (Oxford: Clarendon Press, 1995) and *Poetry of Opposition and Revolution: Dryden to Wordsworth* (Oxford: Clarendon Press, 1996, forthcoming).
27. This was, for example, the reaction of Namierites to Herbert Butterfield, *George III and the Historians* (London: Collins, 1957).
28. By an argument similar in form to Greene's and Curley's, it could for example be proved that there was no Marxian socialism in early twentieth-century Britain. Few people openly said they were Marxists; the Communist Party had hardly any representation in Parliament; there was not even one armed Marxist rebellion in England; most people were moderates and pragmatists; and so on. Such reasoning would miss the point that Marxism did have an influence in that period, despite never attaining its formal objectives.
29. Sir John Hawkins, *A General History of the Science and Practice of Music* (London, 1776) is a painstaking compilation in five large quarto volumes,

providing detailed and comprehensive references to original sources. For Hawkins's Whig allegiance, ibid., V, 315-17.

30. Clark, *Samuel Johnson*, pp. 111, 209.
31. Clark, *Samuel Johnson*, p. 123. See, too, Howard Erskine-Hill, "A Kind of *Liking* for Jacobitism," *The Age of Johnson*, 8 (1996), pp. 3-13.
32. Clark, *Samuel Johnson*, pp. 95-96. Curley cites L. S. Sutherland and L. G. Mitchell, eds., *The History of the University of Oxford*, V, *The Eighteenth Century* (Oxford: Clarendon Press, 1986) in support of his claim that an undifferentiated "eighteenth-century" Oxford was "only . . . pervaded" by Old Interest Toryism, not Jacobitism (p. 147). That long-delayed volume, in some ways admirable, is in this respect a late-Namierite memorial to a superseded phase in the historiography of the subject.
33. E.g., *The Case of the Abjuration Oath Endeavoured to be Cleared, To the Satisfaction of those who are Required to take it* (London, 1702); [Robert Wodrow], *The Oath of Abjuration Considered, both As to the Lawfulness and Expediency of it's being taken by the Ministers of the Church of Scotland, In a Letter to a Friend* (Edinburgh, 1712).
34. *A Summary of the Penal Laws relating to Nonjurors, Papists, Popish Recusants and Nonconformists. And of the late Statutes concerning the Succession, Riots, and Imprisonment of Suspected Persons. Collected and put into such a Method, that the Reader may at one View satisfie his Enquiry. The Offences and Penalties being ranged opposite to each other. To which are added, Several adjudged Cases, and Notes upon the most material Points: Wherein are contained all the Oaths, Submissions, Declarations, Confessions of Faith and Affirmations, Required by the Government, since the first Year of Queen Elizabeth, to this present Year 1716. With an Historical Introduction, giving an Account of the Behaviour and Practices of the Roman Catholicks and Dissenters, which gave rise to these Laws* (London, 1716).
35. *The Case of the Oaths of Allegiance and Abjuration Briefly Stated* (London, 1737), p. 17.
36. [John, Viscount Barrington], *A Dissuasive from Jacobitism: Shewing in general What the Nation is to expect from a Popish King; and in particular, from the Pretender* (2nd ed., London, 1713), pp. 5-6, 17, 26. *Pace* Weinbrot (p. 211, n. 44), Barrington did not simply argue "how worthless oaths are when taken by Stuarts and Jacobites."
37. [Nicholas Amhurst], *The Protestant Session, a Poem. Addressed to the Right Honourable Earl Stanhope. By a Member of the Constitution-Club at Oxford* (London, 1719).

38. [Nicholas Amhurst], *An Epistle from a Student at Oxford, to the Chevalier. Occasioned by His Removal over the Alps, And the Discovery of the Swedish Conspiracy* (London, 1717), pp. 4-5, 7.
39. [Amhurst], Terrae-Filius no. 3 (21 January 1721), in *Terrae-Filius* (London, 1726), pp. 11-12.
40. *Dictionary of National Biography*, sub "Amhurst."
41. [Amhurst], *Terrae-Filius*, pp. 13, 15.
42. Weinbrot (pp. 193-94) misinterprets Amhurst's testimony.
43. [Amhurst], *Terrae-Filius*, p. 15.
44. John Baron, a Whig, was Vice Chancellor in 1715-18.
45. [Amhurst], Terrae-Filius no. 17, 15 March 1721, in *Terrae-Filius*, pp. 86-93.
46. Brasenose College Archives, MPP 90 A1. I am grateful for the advice of the Archivist of Brasenose, Mrs. Elizabeth Boardman, for guidance on the college's collections.
47. Clark, *Samuel Johnson*, p. 94. This was printed in George Birkbeck Hill, ed., *Johnsonian Miscellanies*, 2 vols. (Oxford: Clarendon Press, 1892), II, 85, with the note that it was "pasted in at the end of a copy of *Parecbolae sive Excerpta e corpore Statutorum Universitatis Oxoniensis*. It was shown me by Mr Viner Ellis, a descendant of Johnson's contemporary at Pembroke to whom it had been given." The original volume does not survive in the Bodleian or at Pembroke College.
48. *An Authentick Copy of the Association entred into by Part of the Nobility, Gentlemen and Clergy of Oxford, at the Time of the late Unnatural Rebellion in the Year 1745, together with the Names of all the Persons who Subscribed thereto* (broadsheet, n.p., n.d.) College affiliations are not given in the list, but can be established from other sources. Of the heads of 21 colleges, only the following signed: John Conybeare (Christ Church); John Mather (Corpus); James Edgcumbe (Exeter); Thomas Pardo (Jesus); Euseby Isham (Lincoln); Joseph Smith (Queen's); George Huddesford (Trinity). Edward Butler, Whig President of Magdalen College, was probably prevented by terminal illness. In addition, the Principals of St. Mary Hall, Alban Hall, St. Mary Magdalen Hall, and New Inn Hall did not sign.
49. *Remarks and Collections of Thomas Hearne*, 11 vols. (Oxford: Oxford Historical Society, 1884-1918), 22 November 1721: VII, 300-301. This is misinterpreted by Weinbrot (pp. 192-93) as an argument against me; in fact it supports my case.
50. Richard Blacow, *A Letter to William King, LL.D. Principal of St Mary Hall in Oxford. Containing a particular Account of the Treasonable Riot at*

Oxford, in Feb. 1747 (2nd ed., London, 1755), pp. 14-24, at 17; cf. Weinbrot, p. 195.

51. [Benjamin Kennicott], *A Letter to Doctor King, Occasion'd by his late Apology; And, in particular, By such parts of it as are meant to defame Mr Kennicott, Fellow of Exeter College. By a Friend to Mr Kennicott, And lately A Member of the University of Oxford* (London, 1755), pp. 25-26.
52. Ibid., p. 16.
53. [Thomas Hearne], *A Vindication Of those who take the Oath of Allegiance, To His Present Majestie from the Perjurie, Injustice, and Disloyaltie, Charged upon them by Such as are against it . . . In a Letter to a Non-Juror* (London, 1731), pp. 1, 24. The editor, John Bilstone, is identified by mss. notes in the Bodleian's copy, Mus. Bibl. II. 101.
54. *Remarks and Collections of Thomas Hearne*, V, 17-18, 45-46, 139; Weinbrot, p. 209.
55. [John Bilstone], "To the Reader," in [Hearne], *A Vindication*, p. xv.
56. *Remarks and Collections of Thomas Hearne*, V, 137. On 24 January 1715/16, after the last day allowed by 1 Geo. I, s. 2, c. 13, he wrote to Bodley's Librarian declining to act as sub-librarian in an attempt to avoid the penalties to which he would have been liable had that post come within the scope of the Act (he was unsure that it did so): ibid., 164-65, 284.
57. *Remarks and Collections of Thomas Hearne*, 3 October 1723, VIII, 121-22.
58. [Richard Graves], *Recollection of Some Particulars In the Life of the late William Shenstone, Esq.* (London, 1788), p. 16.
59. *Reasons of the Present Judgement of the University of Oxford, concerning The Solemn League and Covenant. The Negative Oath. The Ordinances concerning Discipline and Worship. Approved by generall Consent in a full Convocation, 1. Jun. 1647. And Presented to Consideration* (London, 1749), pp. i-v. For this episode, see Edward Earl of Clarendon, *The History of the Rebellion and Civil Wars in England, Begun in the Year 1641*, 3 vols. (Oxford, 1702-1704), III, 56.
60. The text is reprinted in *Reasons of the Present Judgement*, pp. 6-51.
61. Richard Blacow, *A Letter to William King, LL.D. Principal of St Mary Hall in Oxford. Containing a particular Account of the Treasonable Riot at Oxford, in Feb. 1747* (London, 1755), p. 5. *Pace* Weinbrot (p. 210, n. 37), neither the first nor the second edition of this pamphlet reprints the oaths.
62. *An Answer to Mr B[laco]w's Apology, As it respects His King, his Country, his Conscience, and his God. By a Student of Oxford* (London, 1755), pp. 2, 14, 20. This work was "Printed for W. BIZETT in *St Clements Church-yard*," the same man who in 1753 had published a print (author's

collection) of the Jacobite Alexander Murray with a suitable legend from the opening lines of Horace, Book III, Ode 3 ("Justum, et tenacem propositi virum/Non civium ardor prava jubentium,/Non vultus instantis Tyranni,/Mente quatit solida"), and the same William Bizet who was examined in 1750 on a charge of publishing a seditious print: Public Record Office, SP 36/116/21 (I owe this reference to Mr. Richard Sharp). *Pace* Weinbrot, there is more to be discovered about the politics of the group of publishers and booksellers who lived in the London parish of St. Clement Danes.

63. Brasenose College Archives MPP 11 A1.
64. Bodleian MS Top Oxon b 42, f. 129.
65. Bodleian MS Eng misc c 406, f. 177.
66. British Library Additional MSS 42,506, f. 96. Henry Ellis had been elected as a scholar from Merchant Taylors on 11 June 1796: ibid., f. 95.
67. Brasenose College Archives MPP 25 A1.
68. Edward Bentham, *A Letter to a Young Gentleman of Oxford* (London, 1748), p. 26. Bentham printed the oaths of allegiance and abjuration, pp. 29-30.
69. Edward Bentham, *A Letter to a Fellow of a College. Being the Sequel of a Letter to a Young Gentleman of Oxford* (London, 1749), esp. pp. 56-58, 70.
70. *Considerations on the Residence usually required for Degrees in the University of Oxford* (Oxford, 1772), pp. 3, 16, 24. This was routine practice: for over 100 such applications for dispensation in the period 30 December 1721-28 February 1723/4, see Bodleian MS Top Oxon c 69.
71. Quoted by V. H. H. Green, "The University and Social Life," in Sutherland and Mitchell, eds., *The History of the University of Oxford*, V, 328.
72. [Richard Newton], *A Scheme of Discipline with Statutes Intended to be Established by a Royal Charter for the Education of Youth in Hart-Hall in the University of Oxford* ([?Oxford], 1720), Appendix, p. 1.
73. The term was not, as today, synonymous with "Fellow," but described both undergraduates and senior members who were supported on the foundation; it therefore covered the modern categories of "scholar" and "fellow" at other colleges. The governing body of Christ Church was, uniquely, not the students but the dean and canons.
74. E. G. W. Bill, *Education at Christ Church Oxford 1660-1800* (Oxford: Clarendon Press, 1988), pp. 141-45. In this learned and accurate study, Dr. Bill was briefly misled by the letter printed in Amhurst's *Terrae-Filius* (above, n. 45) into thinking that the oath of allegiance was taken at matriculation, and argued that, at that early stage, the oaths were

disregarded as mere formalities. Students were admitted to their Studentships some time after the university ceremony, in the case of Canoneer Students up to two years afterwards, and they were still commoners on matriculation. I am grateful to Dr. Bill for advice on this point.

75. Clark, *Samuel Johnson*, p. 117; Bill, *Christ Church*, p. 327. For an example of open Jacobitism by another Westminster student in 1750, however, see ibid., pp. 161-62. Dr. Bill suggests that it is unlikely that any student was deprived for failing to take the oaths at the accession of George I: ibid., p. 145, n. 3.
76. *A Representation of the Conduct of the Proctors, with respect to the Two Explanatory Statutes proposed by the Vice-Chancellor to Them and the Heads of Houses* [Oxford, 1759], p. 4; *An Answer to the Objections made in Convocation to the Representation of the Conduct of the Proctors* [Oxford, 1759]. Printed flysheets on this question are assembled in Bodleian Gough Oxf. 96.
77. *A Letter to a late Member of the U[niversit]y of O[xfor]d, with Respect to the Two Explanatory Statutes proposed to The C[onvocatio]n* [Oxford, 1759], p. 8.
78. E.g., *A Letter to the Reverend* **** ***** *MA Fellow of* ***** *College Oxford, On the Case of Subscription at Matriculation* (Oxford, 1772); *A Collection of Papers, designed to Explain and Vindicate the Present Mode of Subscription required by The University of Oxford, from all Young Persons at their Matriculation* (Oxford, 1772); *Reflections on the Impropriety and Inexpediency of Lay-Subscription to the XXXIX Articles, in the University of Oxford* (Oxford, [1772]).
79. G. R. M. Ward, ed. and trans., *Oxford University Statutes*, 2 vols. (London: William Pickering, 1845), I, xx; L. H. Dudley Buxton and Strickland Gibson, *Oxford University Ceremonies* (Oxford: Clarendon Press, 1935), p. 3.
80. Bodleian Library, Oxford University Archives, SP 12.
81. See Clark, *Samuel Johnson*, p. 94.
82. Bodleian Library, Oxford University Archives, SP 46.
83. See Clark, *Samuel Johnson*, p. 96.
84. Wetherell to Jenkinson, 20 February 1773: British Library Add MSS 38,207 f. 239.
85. Ibid., ff. 240-42.
86. Ward, *Oxford University Statutes*, I, 120; *Parecbolae* (1729), p. 128.
87. C[harles] S[ackville], *A Treatise Concerning the Militia, in Four Sections* (London, 1752), pp. 11, 62. In true patrician style, Sackville proceeded from a long dissertation (pp. 12-30) "*Of the* ROMAN MILITIA." A Bill

to give effect to his plan failed in the Commons the same year: J. R. Western, *The English Militia in the Eighteenth Century: The Story of a Political Issue 1660-1802* (London: Routledge, 1965), p. 121.

88. Ian F. W. Beckett, *The Amateur Military Tradition 1558-1945* (Manchester, 1991), p. 51; David Allen, "The role of the London trained bands in the Exclusion Crisis, 1678-1681," *English Historical Review*, 87 (1972), 287-303; C. Neville Packett, *Her Majesty's Commission of Lieutenancy for the City of London: A Brief History* (privately printed, London, 1987). Weinbrot (pp. 197-99) misinterprets this episode by believing that the Trained Bands of London were established by the Militia Act of 1757.
89. That this was the criterion for eligibility is suggested by the excuses accepted by the Court of Lieutenancy from defaulters, e.g., on 29 January 1746: "Ann Nash—Not a Housekeeper . . . John Catlin—Only a Lodger": Corporation of London Record Office, Lieutenancy Minute Book 1744-49, 441/A/7, p. 207. Even a woman, if a housekeeper in her own right, was liable to find a substitute.
90. Western, *English Militia*, pp. 52-74.
91. Lieutenancy Minute Books 1714-44 (441/A/6), 1744-49 (441/A/7), 1749-83 (441/A/8).
92. In 1756, the colonels were: Blue regiment, the Hon. Sir Joseph Hankey, Alderman, President of the Honourable Artillery Company; Orange, the Hon. William Baker, Alderman, Vice President; Red, the Hon. Sir William Calvert, Alderman, Vice Treasurer; Green, the Hon. Thomas Chitty, Alderman; Yellow, the Hon. Sir Samuel Fludyer, Alderman; White, the Hon. John Porter, Alderman: Packett, *Lieutenancy*, p. 68.
93. Lieutenancy Minute Books, 441/A/7, pp. 143, 234; 441/A/8, pp. 37, 202, n.p.
94. Ibid., 441/A/8, n.p., p. 244.
95. Ibid., 441/A/8, pp. 67, 115, 148, 170-71.
96. Corporation of London Record Office, Misc. MSS 11.20.
97. Lieutenancy Minute Book, 441/A/7, pp. 25-29.
98. Johnson to Lewis Paul, 29 December 1755: Bruce Redford, ed., *The Letters of Samuel Johnson* (5 vols, Princeton: Princeton Univ. Press, 1992-94), I, 116; "I have been ill for about a fortnight": Johnson to Lucy Porter, 30 December 1755: ibid., p. 118.
99. Johnson to Richard Congreve, 16 October 1755: *Letters*, I, 114.
100. Clark, *Samuel Johnson*, pp. 120-21.
101. Johnson to Samuel Richardson, 16 March 1756: *Letters*, I, 132.
102. Charles M. Clode, *The Military Forces of the Crown; their Administration and Government*, 2 vols. (London: John Murray, 1869), I, 38-40; Tony Hayter, *The Army and the Crowd in Mid-Georgian England* (London:

Macmillan, 1978), pp. 38, 81, 103; Paul Kléber Monod, *Jacobitism and the English People 1688-1788* (Cambridge: Cambridge Univ. Press, 1989), pp. 197-98; Western, *Militia*, pp 147-48.

103. Beckett, *Amateur Military Tradition*, p. 65; Western, *Militia*, p. 158.
104. Clark, *Samuel Johnson*, p. 123.
105. Donald J. Greene, in "The Via Media in an Age of Revolution: Anglicanism in the 18th Century," in Peter Hughes and David Williams, eds., *The Varied Pattern: Studies in the 18th Century* (Toronto: A. M. Hakkert, 1971), pp. 297-320, showed a commendable desire to rescue the subject of eighteenth-century Anglicanism from scholarly nescience, but little knowledge of what it was he sought to rescue.
106. Curley similarly writes of "the credo of a Jacobite waxing nostalgic for the good old days of pre-Revolution Stuart monarchy," etc.
107. Notably Christine Gerrard, *The Patriot Opposition to Walpole: Politics, Poetry, and National Myth, 1725-1742* (Oxford: Clarendon Press, 1994). For Dr. Gerrard's views of Johnson, see pp. 230-47.
108. [Amhurst], *Terrae-Filius*, pp. iv, vi. In re-publishing his periodical, Amhurst was bidding for an official reward for his Hanoverian loyalty. He followed these arguments with the Scriptural parable of the laborers in the vineyard, to claim that those (i.e., himself) who had "*borne the* heat *and* burthen *of the day*" should be not less well rewarded than those hired "*at the* eleventh *hour*," p. viii.
109. John Conybeare, *Calumny Refuted: or, an Answer to the Personal Slanders Published by Dr Richard Newton, in his Letter to Dr Holmes, Vice-Chancellor of the University of Oxford, &c.* (London, 1735), p. 78.
110. Richard Newton, *The Grounds of the Complaint of the Principal of Hart-Hall, Concerning the Obstruction given to the Incorporation of his Society, by Exeter-College and their Visitor . . . In Answer to the Misrepresentations of Dr C[onybear]e . . .* (London, 1735), pp. 59-60.
111. William Bromley, Jr. (?1701-37), son of the Jacobite William Bromley (?1663-1732), MP for Oxford University 1701-32. Bromley Jr. had apparently paid his respects to James Francis Edward Stuart while on the grand tour in 1721: Romney Sedgwick, ed., *The History of Parliament: The House of Commons 1715-1754*, 2 vols. (London: HMSO, 1970), I, 493-95.
112. *An Exact Account of the Poll, As it stood between The Honourable Mr Trevor, and Wm. Bromley, Esq.* (London, [1737]), pp. 9, 37, 39.
113. Richard Newton, *Rules and Statutes for the Government of Hertford College, in the University of Oxford* (London, 1747), pp. 16, 98.

114. [Richard Newton], *The Principles of the University of Oxford, As far as relates to Affection to the Government, Stated* (London, 1755), Advertisement.
115. Clark, *Samuel Johnson*, pp. 182-83. Newton was the subject of a lengthy attack by the undoubtedly Whig Nicholas Amhurst, in the collected edition of his periodical: [Amhurst], *Terrae-Filius . . . To which are added, Remarks upon a late Book, entitled, University Education, by R. Newton, D.D. Principal of Hart-Hall*, pp. 289-335.
116. On Brooke's death in 1752, the ex-Jacobite William Murray reportedly proposed to the Secretary of State, the Duke of Newcastle, that William Blackstone should succeed: Newcastle allegedly disbelieved his political reliability, and gave it to Dr Jenner instead: John Holliday, *The Life of William late Earl of Mansfield* (London, 1797), pp. 88-99; Lucy Sutherland, "William Blackstone and the Legal Chairs at Oxford" in Sutherland, *Politics and Finance in the Eighteenth Century* (London: Hambledon Press, 1984), pp. 551-62.
117. [Henry Brooke], *A Letter to the Oxford Tories. By an Englishman* (London, 1750), pp. 3-7, 11.
118. [Benjamin Kennicott], *A Letter to Doctor King, Occasion'd by his late Apology; And, in particular, By such parts of it as are meant to defame Mr Kennicott, Fellow of Exeter College. By a Friend to Mr Kennicott, And lately A Member of the University of Oxford* (London, 1755), pp. 118-21. Weinbrot (p. 202, n. 2) quotes one sentence of this passage (inaccurately) and evidently fails to see its tactical intent. This is the historian's justification, here as elsewhere, for presenting his evidence *in extenso* rather than inaccurate and tendentious fragments of it.
119. [Benjamin Buckler], *A Proper Explanation of the Oxford Almanack For this present Year M.DCC.LV* (London, [1755]), pp. 27-28.
120. Clark, *Samuel Johnson*, p. 184. The respondents proceed as if I had argued the opposite.
121. Thomas M. Curley, "Johnson's Secret Collaboration," in John J. Burke, Jr. and Donald Kay, eds., *The Unknown Samuel Johnson* (Madison: Univ. of Wisconsin Press, 1983), pp. 91-112, at 95-96. This editorial sentence is repeated in Thomas M. Curley, ed., *A Course of Lectures on the English Law Delivered at the University of Oxford 1767-1773 by Sir Robert Chambers*, 2 vols. (Oxford: Clarendon Press, 1986), I, 10; "The likelihood is that Chambers almost always had sole charge of amassing and digesting the usually technical information synthesized in the series": ibid., p. 36.
122. Chambers, *Lectures*, I, 7-8, 22. For Curley's erroneous interpretation of Blackstone as a Whig, ibid., pp. 30-31, 68.

123. Curley implies that this was due to ignorance of his work. I had however already cited Curley's edition of Chambers's lectures in another context: J. C. D. Clark, *The Language of Liberty 1660-1832: Political Discourse and Social Dynamics in the Anglo-American World* (Cambridge: Cambridge Univ. Press, 1993), pp. 89, 110.
124. Chambers, *Lectures*, I, 268, 270-71.
125. Chambers, *Lectures*, I, 270.
126. Chambers, *Lectures*, I, 155.
127. Chambers, *Lectures*, I, 350-51.
128. Chambers, *Lectures*, I, 155.
129. Chambers, *Lectures*, I, 374-75.
130. J. C. D. Clark, *English Society 1688-1832: Ideology, Social Structure and Political Practice during the Ancien Regime* (Cambridge: Cambridge Univ. Press, 1985), pp. 277-99.
131. This was Thomas Townshend (1733-1800), MP for Whitchurch, who had resigned from Grafton's administration in June 1768 after an imagined slight and thereafter indulged his resentment.
132. Misquoted by Curley (p. 146) from William C. Townsend [*sic*], *The Lives of Twelve Eminent Judges of the Last and of the Present Century*, 2 vols. (London, 1846), I, 177-78. The original source, which also gives relevant text omitted by Curley, is J. Wright, ed., *Sir Henry Cavendish's Debates of the House of Commons, during the Thirteenth Parliament of Great Britain*, 2 vols. (London, 1841-43), II, 456-57. Curley transcribes "We knew that he was a man of letters . . . ," implying a general awareness of Johnson's political detachment; Cavendish's text reads "He knew . . .": Wedderburn was making a personal statement.
133. Cavendish, *Debates*, II, 457.

JONATHAN CLARK AND THE ABOMINABLE CULTURAL MIND-SET

DONALD GREENE

Well, it's all over now. There's really little point wasting any more print on what must be a useless discussion about Samuel Johnson's politics. Jonathan Clark has spoken *ex cathedra* and determined that:

1. "The politics of Samuel Johnson have now been established, beyond reasonable doubt"—by Clark himself. This is one of the most amazing feats of scholarship on record. Dozens of writers are recorded in the Clifford-Greene-Vance bibliography of Johnsonian studies as having earlier tried to grapple with the problem. Clark's name is not among them. Yet in one article and one book in 1994 he has finally solved it, once for all. (Though surely a little credit ought to be given to his predecessor, Jack Wilkes, who in 1762 first brilliantly intuited that the government pensioner Johnson was a crypto-Jacobite.)

2. Alternative points of view foolishly put forward by Professors Curley and Weinbrot and myself—and no doubt by many others who have thought that their published studies of Johnson entitled them to express opinions on the subject—are ruled out, since whatever we may write is invalidated by the fact that we have all been infected with a vicious "cultural mind-set"—"a liberal-Evangelical or humanist mind-set which reached its apogee in the USA of the 1950s."

I must have suspected something of the kind when I wrote, in my earlier reply to Clark's and Howard Erskine-Hill's attacks on my work, that we seemed to be dealing here not with a *bona fide* attempt to contribute something of scholarly value to our comprehension of Johnson's complex political thinking, but with the promotion of some as yet undisclosed ulterior objective of the authors. Clark now confirms that suspicion: he writes, "It will continue to be possible to argue over some of the individual"—scholarly—"points recently raised," but "one repeatedly senses that something more general is at stake." One does indeed, and Clark's latest contribution triumphantly demonstrates it.

The beginning, at least, of the demonstration, the disclosure, came, if it had not come earlier, in the London *Times* of 27 June 1995, p. 3, where Clark publicly announced his support of John Redwood's candidature for the leadership of the British Conservative party, with the eventual possibility of the Prime Ministership. Redwood has been described as heading a small ultra-right-wing faction of the party, intent on ousting its present leader, Prime Minister John Major. A week after Clark's endorsement, Redwood was shot down in flames by an overwhelming majority of his fellow Conservative MPs. But not to worry, Dr. Clark: in Wilkes's first excursion into national politics, he came (as Johnson gleefully reports in *The False Alarm*) at the bottom of the poll in the election for MP for the City of London. But a short time afterwards, with the help of the cry "Wilkes and Liberty!", he attained a stunning victory in the election for MP for the county of Middlesex. Eventually, with Clark as campaign manager raising the cry of "Redwood and Religion!" (the correct variety of religion, of course), we may yet see a party in power dedicated to eliminating whatever erroneous mind-sets may still remain to corrupt eighteenth-century scholarship. Possibly a "cultural revolution" like Mao Zedong's is needed to correct such undesirable mind-sets, with wrong-thinking professionals and intellectuals sentenced to hard labor on backwoods farms and constant brain-washing seminars until they realize and confess the error of their ways.

Meanwhile, the best Curley, Weinbrot, and I can do is to try humbly to understand just where our thinking has gone wrong. There may be some difficulty here, since the cultural backgrounds of "this trio," as Clark contemptuously calls us (wouldn't "the Gang of Three" be even better?), are so diverse that it is hard to see how we could all have picked up the infection of "liberal-Evangelicalism" or "humanism." Our backgrounds are those of three different historic "mainstream" religious bodies, none of which would accept the designations "Evangelical" and "humanist" in the pejorative sense Clark uses them. I myself happen to be an Anglican, as my family has been for generations, and have published articles and reviews dealing with Anglican doctrine and church history in reputable theological and other learned journals; what Dr. Clark's qualifications are for criticizing my theology, I do not know. Incidentally, like my family, I have always been a supporter of the Canadian Conservative party, opposed, for a century and a half, to its traditional rival, the Canadian Liberal party.

I don't happen to know Curley's and Weinbrot's ages, but I would guess that neither is old enough to have picked up much of the "liberal-Evangelical mind-set" that "reached its apogee in the USA of the 1950s." My own case is still more confusing, since I am not a United States citizen, but a loyal subject of Her Majesty the Queen. At the time I finished my dissertation, I had lived in the U.S. a total of nineteen months, and was kept too busy with it to read many of the large collection of deleterious works of the 1940s and 1950s that Clark lists in his notes. In fact, I had earlier spent a longer time in Britain, though without absorbing much if any of the Socialist ideology then fashionable under a Labour government. I must apologize for all this autobiography, but it seems the only way to defend oneself against Clark's extensive *ad hominem* charges. Although he rebukes me severely for using *ad hominem* argument (and thus breaching "professional ethics" and the proper "conduct of academic debate"), it was of course legitimate for Clark, given his magisterial status, to do so, as when he wrote, "For Greene, the [scholarly] clock stopped in 1960."

As in Kafka's *Der Prozess*, it is difficult for the accused here to understand exactly what the crime is with which they are being charged. "Liberal-Evangelical" in current American parlance is a contradiction in terms, an oxymoron. In its primary sense, of course, "evangelical" refers to the teaching of the four Gospels ("god spell"="good tidings"=New Testament Greek *eu angellion*) as the indispensable source for Christian morality; Johnson, who so uses the term in his *Sermon* 13, was equally an "evangelical" with his good friend and admirer, John Wesley. In recent years, however, in the United States "Evangelical" has come to be applied, with their own approbation, to a fairly large and powerful group of political activists, sometimes called or calling themselves the "Religious Right" or the "Christian Coalition." They seem to be becoming well organized, and even have at the moment (mid-1995) a prospective candidate, Patrick J. Buchanan, in next year's election for the presidency of the country.

The extremely puzzling thing is that what Clark is denouncing in Curley, Weinbrot, and me as "Evangelicalism" is precisely what he seems to be championing in his own pronouncements. In the "profile" of John Redwood in *The Times*, cited above, in a section bearing the headline "Influential Thinker with a Feeling for Christian Values," Clark praises a book by him that discusses, among other things, religion in "The Age of Enlightenment." It is, Clark avers, "a pioneering work, which put the ideological debates of the 18th century back into their theological context and rejected the mind-set of the mid-1970s, still dominated by J. H. Plumb and his pupils"—who as "Plumbites" thus join the "Namierites" in Clark's well-populated historiological demonology.[1] "The book was so different that it was largely ignored." (Alas, in spite of Clark's valiant P.R., it still is. Let us hope that the same fate will not meet Clark's own book, which has so definitively established Johnson's political position.)

Clark has been quoted as censuring "the real American values (divorce, abortion, homosexuality, consumerism)."[2] One might of course query whether statistics show that these activities occur relatively oftener in the United States than in Britain. But "Evangelicals" in the US deplore them (with the possible exception of "consumerism") equally with Clark, as prejudicial to "traditional family values." (Johnson's view of these "values," as expressed in *Rambler* 148 and in his account of family life during his own childhood, seems rather different.) Many Americans who call themselves "liberals," vague and protean as that term is, tend to oppose "Evangelical" demands, such as those which one supposes Clark would support, for eliminating, or at least drastically modifying, legislation which seems to give women the right to undergo abortion at their own choice and which decrees that employers may not penalize employees or applicants for employment because of a homosexual or bisexual rather than an exclusively heterosexual orientation. So it is difficult to understand how Clark can condemn Curley, Weinbrot, and me for being "liberal" and "Evangelical" at the same time.

There are similar problems connected with his charge that we are "humanists," a term which he seems to use as synonymous with "liberal-Evangelical." Johnson's *Dictionary*, 1755, defines "humanist" as "A philologer; a grammarian"; in 1773 he adds "a term used in the schools of Scotland," where indeed university professors of Latin are still officially "professors of humanity." The original humanists of the Renaissance (the "revival of learning," as Johnson admiringly called it) were those who recovered much of the Latin and Greek literature that had been lost or forgotten during the Middle Ages, when its content of pagan polytheism often made it an object of suspicion to churchmen. Undoubtedly Johnson himself cannot be denied the term in that sense: one of his earliest projects of writing was a life of the famous early humanist Politian, together with a history of Renaissance Latin poetry from Petrarch to Politian, and the extent of his reading in both classical and Renaissance literature, as indicated by the sadly abbreviated sale catalogue of his library, almost equals that in Christian theology. (Not, surely, that it can be held to equal the vast erudition of professional classical scholars of his time, such as Bentley, Markland, or Elmsley, and he would surely not have claimed to be ranked with them as a Latinist or Grecian; like others, I am baffled by what Clark means by calling him an "Anglo-Latinist.") But Clark no doubt has his own pejorative definition of "humanist," which he will perhaps reveal to the world in his own good time.[3]

But perhaps Clark and Erskine-Hill overestimate the probative force of *ad hominem* argument. Even if their agents—they must have had investigative agents to discover so much about our intimate mind-sets—were to find that "this trio" are Tibetan monks in disguise, conspiring to convert the world to Buddhism, the revelation would hardly affect certain formidable evidence that seems to show that Johnson was a loyal subject of the government under which he lived, if sometimes critical of its actions, as every loyal subject in a constitutional democracy has the right to be, and had no hankering to see it replaced by a French-controlled Stuart absolutist monarchy.[4]

Such pieces of evidence have to be dealt with firmly, and Clark and Erskine-Hill deserve much credit for the skilled detective techniques they have brought to these problems. One stems from the statement in the opening article of Johnson's new *Literary Magazine*, 1756, the first of what was intended be a series giving a synopsis of the history of England from the reign of Elizabeth onward, with a view to explaining how the country had become embroiled in the Seven Years' War, which had just begun. In the course of it, Johnson asserts that "the necessity of self-preservation had impelled the subjects of James [II] to drive him from the throne." It was not merely that James's ambition was to convert the English to his own Roman Catholicism, but that he was attempting to do so by "suspending" or "dispensing with" the laws of the land passed by the people's representatives in Parliament—in short, transforming what had been a constitutional monarchy into an absolute one, like that of Louis XIV in France.

Like a blood-stained glove, the presence of this statement has to be explained away, and Erskine-Hill does so in time-honored fashion: it was planted on Johnson by the authorities; he was constrained to write it by *force majeure*, by fear of being charged with sedition if he didn't. What a fascinating picture it is—a pair of jack-booted government thugs standing menacingly behind Johnson as he sits at his desk in Gough Square, snarling, "Look here, Sam, you write what we dictate, or it's off with you to the Tower and the headman's block"! (Since Erskine-Hill declines to consider the fact that when, a year later, Johnson was in fact delated to the Attorney-General on a charge of seditious libel for something he had printed in the same magazine, he thumbed his nose at it and went on to print more of the same, I suppose we must conclude that that incident is irrelevant.)[5]

Clark's method is more sophisticated. When in 1753 Johnson was commissioned to write as the introduction to the index for the first twenty-three volumes of the *Gentleman's Magazine* a short account of the principal events that had occurred during the magazine's existence, and, among other things, condemned the French-aided Jacobite uprising of 1745, thanking Providence for its successful suppression, Clark feels that he could not have expressed such sentiments as these, and, although the essay was accepted as Johnson's by such eminent Johnsonian scholars as L. F. Powell and E. L. McAdam,[6] believes these passages to be interpolations, no doubt by one of the government spies that he and Erskine-Hill find so ubiquitous at the time. He whips out his forensic microscope, carefully examines the droplets of blood, and finds—aha!—an anomalous DNA molecule: a sentence, or part of a sentence, that he pronounces to be ungrammatical, and therefore unJohnsonian. Just what the error in grammar is, he does not disclose, and no one else has so far been able to find it. But it would be shocking to suppose that am ex-Fellow of All Souls College, Oxford, is unable, as fourth-form schoolboys used, at least, to be able, to analyze the grammatical structure of a straightforward English sentence correctly.[7]

Erskine-Hill's latest contribution is less searching and outspoken than Clark's; indeed, he seems to be rather drawing away from his association with him—"My own position is not identical with that of Clark." Although he still emphasizes the importance of Boswell as our prime authority for Johnson's political thinking ("We turn first and finally to Boswell for the goods on Johnson," as a critic rebuking an earlier misguided essay of mine once wrote), he deserves praise for now studying earlier versions, in his *Journals* and in the manuscript, of what he printed in the *Life*, 1791. Erskine-Hill finds that in the *Life* he very often tones down passages that in earlier versions seem to emphasize the strength of Johnson's Jacobite tendencies. Erskine-Hill thinks that these changes may indicate that those tendencies in fact were weakening in the later part of Johnson's life. But it might also be considered that in the late 1780s and early 1790s Boswell had ambitions to enter the national political arena—unsuccessful ones, unfortunately: he came at the bottom of the poll in the by-election in 1789 for MP for his home constituency of Ayrshire, and his attempt to gratify the Prime Minister, Pitt, by singing a song of his own composition in praise of him at the Lord Mayor's Feast in 1790, failed to do so. It could be that when he deleted from the printed *Life* the assertion he had made to Topham Beauclerk that Johnson had called the reigning sovereign, George III, a usurper, he feared that his readers and potential supporters in a Parliamentary election might feel some skepticism about his credibility, considering how much Johnson had published in praise of the King. I'm afraid that in trials for promulgating treasonous statements, the courts tend to take a "black-or-white" view; a plea by "Lord Haw-Haw" that he was only expressing "a kind of liking for Hitlerism" would not have been accepted.

I fear that I have yet no defense to Erskine-Hill's charge that I "see no more in Johnson's poetic narrative of Charles XII of Sweden than yet another example of a conqueror who bites the dust," and feel no emotion for poor Charles, thus "reducing one of the most moving passages of eighteenth-century poetry to total banality." That Johnson, who wrote the passage, shared Erskine-Hill's emotion seems extremely dubious when we consider that, in *Adventurer* 99, four years after he published the poem, his only wish for Charles is that he and his opponent Peter of Russia might be "huddled together in obscurity and detestation": for ruthless aggressors, whether defeated or victorious, whose ambitions result in the needless deaths of thousands (nowadays, thanks to improvements in technology, millions), Johnson feels no emotion but detestation and contempt. One recalls Hannah Arendt's remark about "the banality of evil": she was referring to Adolf Eichmann, the henchman of Adolf Hitler, another once successful aggressor whose hopes ended in a banal suicide in a squalid subterranean bunker. Johnson indeed closes his account of Charles with a description of the banality of his end and his place in history—"a barren strand,/A petty fortress, and a dubious hand" (the rumor was that he had been killed by one of his own men, disgusted at last with the futility of Charles's homicidal adventuring):

> He left the name at which the world grew pale
> To point a moral or adorn a tale.[8]

Anticlimax, banality indeed. After all, the title of the poem in which Charles appears is "the *vanity*" of egocentered human wishes for wealth, political power, military glory, and so on, not their *pathos*. The only alternative to these deceptive human values, as Johnson states emphatically at the close of the poem, is the transcendent ones of faith, hope, and love, which alone can produce true human happiness. Well, perhaps, given enough instruction by Erskine-Hill, I (and Johnson, if he were alive) might learn to increase our emotional sensitivity enough to be able to be "moved" by the "tragic" end of Charles of Sweden. And no doubt by the equally banal conclusion of Charles Edward Stuart's life, perhaps even by Adolf Hitler's. One understands that there exist small groups here and there calling themselves "neo-Nazis" or something similar who are able to do so.

Of course I appreciate Erskine-Hill's exhortations in his attacks on me—sorry, "only challenges to some of my views"—that in defending myself from them I should use more "civility" and "decency," and refrain from indulging in ungentlemanly *ad hominem* argument. I know they are well meant, and I shall try to profit from them, though it may be difficult, after so many decades of publishing books and articles in scholarly and other journals, for so old a leopard to change his spots. I'm afraid, though, that after pondering what some unsympathetic readers might describe as Pecksniffian sermonizing by Erskine-Hill and Clark on "professional ethics" and proper "conduct of scholarly debate," I've not been edified by an example in Erskine-Hill's latest contribution of what some might call a "cheap shot," a bit of below-the-belt infighting; it is certainly at least *ad hominem*, and I fear I rather resent it. I mean his remark about "a biographer whom Greene used to admire: J. L. Clifford, who wrote of 'Johnson's known anti-Hanoverian position.'" When my dear teacher Jim Clifford was writing his *Dictionary Johnson*, which Erskine-Hill is quoting—I compiled the index to it after Jim's sudden death—he often consulted his former pupils and other scholarly friends for advice, sending them drafts of sections about which they might have special knowledge.

Jim used to call on me to look over portions dealing with Johnson's political views, and I offered no objection to "anti-Hanoverian," which of course means "anti" the electorate of Hanover. Johnson had made his position clear in *Marmor Norfolciense*, when he wrote about the white horse of Hanover sucking the (economic) blood of the English lion. So did dozens of other political polemicists of the time, such as the elder Pitt, who raved away about how much the support of Britain's Continental allies was costing her, until, of course, having taken charge of the conduct of the Seven Years' War, he subsidized them lavishly, and was later to boast that, thanks to this policy of his, the war had been won on the plains of Germany. It is symptomatic, I'm afraid, of a "mind-set" on Erskine-Hill's part that he should mistake the phrase as meaning "anti" a monarch of Hanoverian descent. (For that matter, Johnson not infrequently dropped "anti-Scottish" remarks about poverty-stricken Scots coming in "droves" down the high road to England, their "noblest prospect," no doubt with a view to their economic advancement at the expense of the English. Should we infer from this that he objected to monarchs of Scottish descent?) At any rate, for Erskine-Hill's information, I not only "used to" admire Jim Clifford; I still admire him, and, since Erskine-Hill seems dubious, *in re* Charles of Sweden, of my capacity for emotion, more than admire him.

But I must compliment Erskine-Hill (hoping I won't get into trouble again for complimenting him—see note 6) for referring to Boswell's musings in the MS of his *Tour to the Hebrides* (the existence of which

I was aware of, though I had not pondered them as closely as Erskine-Hill has). Clark (*AJ* 7, p. 40) wrote of "Jacobites in the sense of acknowledging some title [to the throne] in the exiled house. . . . In this sense Johnson was a Jacobite." As Erskine-Hill points out, Boswell was certainly not a Jacobite in that sense: he firmly denies that James VI and I and his descendants had any title whatsoever to either the Scottish or the English throne, since they were not Stuarts at all—James's father was Lord Darnley, and therefore he and his progeny were "Lennoxes." If Boswell was not a Jacobite, can we seriously argue that Johnson was? Erskine-Hill now writes, "[Boswell] does not argue that Johnson ever fought or plotted for the Jacobite cause"—though Clark did, or at least very strongly implied it, when, in 1985, he wrote, "Johnson's whereabouts during 1745 and 1746 are unknown" and went on to mention the mysterious musket, sword, and belt that Boswell saw in Johnson's closet. Clark now retracts what he calls "The old canard that Johnson might have been personally involved in the rebellion of 1745," though without apologizing for having promulgated it himself.

"Johnson was not a Jacobite in 1773 by Boswell's definition," Erskine-Hill continues—nor has it been shown that he was at any other time by anyone else's definition, except Wilkes's and Charles Churchill's. He concludes by quoting Boswell's assertion that "he and I have a kind of liking for Jacobitism," though I doubt that he could even establish this as regards Johnson except by continuing to insist that, in *The Vanity of Human Wishes*, Johnson feels sorry for poor Charles of Sweden (with the corollary that he wishes that Charles had been more successful in his military conquests) and that somehow this sympathy carries over to Charles Stuart. He goes on, "Perhaps the pro-Jacobite case should rest there"—one at least of the prosecuting counsel seems to be announcing his withdrawal from the case. It is a pity it was ever exhumed from its peaceful grave—had it been allowed to rest *there*, how much scholarly time, energy, and paper would have been saved!

But Johnson, as I suggested at the outset of the present discussion, is not the real concern of Erskine-Hill and Clark—indeed, the very sketchy knowledge they display of Johnson's writings leads to that conclusion. From time to time the word "religion" comes up in their discussion. Is there the slightest indication that either has ever looked at his sermons—Volume XIV in the Yale Edition of his works—or his religious prayers and meditations—Volume I?

Johnson for them is only a vehicle, a pawn, that they wish to use in a much greater enterprise—essentially a political one, and an ominous one, that of ridding scholarship of certain widespread pernicious "cultural" tendencies that they have detected in it and believe should be eradicated—a crusade, in short. Erskine-Hill is cannier, more discreet, about this than Clark, but we see it in his final question, where he announces that the object of his "side"—how ready he and Clark are to assert that when three experienced Johnson scholars independently disagree with what they have written, they are a "side," perhaps even a conspiracy!—is to discredit "a Johnson acceptable to late twentieth-century views." *What* views? And why are they unacceptable? Which "late twentieth-century views" are, to Erskine-Hill and Clark, politically "correct" and which are politically "incorrect"? Clearly much more than Johnson is involved here.

Clark is, happily, more explicit and ingenuous. He makes it quite clear in his latest contribution that his chief concern is with the deleterious presence of certain unacceptable "cultural mind-sets" in the modern world, especially in the USA, and by implication, with the need to extirpate them—Johnson is of interest only as a handy tool that might be used in that great undertaking. He wants to get rid of the Johnson who is no more than a "moral icon," especially in North America—though why less so in Britain is hard to understand: I am a past president of the senior British Johnson Society, that headquartered in Lichfield, a long time member of the Johnson Society of London and a contributor to its journal *The New Rambler*, and an honorary member of the venerable Johnson Club, and have noted over the years no less attention being paid in North American Johnson societies to which I belong than in the British ones to Johnson as poet, biographer, literary critic, lexicographer, letter writer, Parliamentary reporter, sermon writer, and much else, as well being a "moralist"—and many, including myself, would argue that political thinking is inseparable from moral thinking.

Successful politicians over the centuries have devised many effective methods of suppressing cultural mind-sets that they think inimical to the one official mind-set they rely on to sustain them in power, few more effective than those of Hitler, Stalin, Mao Zedong, and the Ayatollah Khomeini in the twentieth. Orwell generalized it all nicely in *Nineteen Eighty-Four*, where the various methods of eradicating "crimethink" or "oldthink" and replacing them with "goodthink" are described in detail, methods that succeeded at last in causing Winston Smith to love Big Brother. But then I suppose Orwell was a "liberal humanist," although, after *Homage to Catalonia*, the Communists called him a Fascist, and after *Animal Farm* and *Nineteen Eighty-Four* even worse names. (By the way, where have all the Communists, like the flowers of the 1960s, suddenly gone?)

So what has Clark in mind for the future? In his unheralded attack on me in the *Times Literary Supplement*, he reported, correctly, that I had found Johnson to be "a moderate, a realist, intelligent and well informed, responding to the political complexities of his age, neither a bigot nor a reactionary," and complained that by doing so, I had "homogenized" him. The metaphor puzzled me at first, but finally I got it: in unhomogenized milk the cream rises to the top and is carefully preserved and tended; the skim milk that remains is, on many farms at least, discarded or mixed with the pigs' mash. In Orwell's dystopia, the "proles" are kept uneducated and powerless, or if they show any signs of rebellion, are vaporized. It is only the small bureaucracy at the top, the *nomenklatura*, who have the privileges. The parallels in Hitler's Germany, Stalin's Russia, and Mao's China are obvious. I asked Clark if he could be more explicit about he meant. He obliged a few days after I had sent my response to the *TLS* by announcing his support for John Redwood as possible future Prime Minister. Shall we see him some day as Redwood's Secretary of State for the Extirpation of Undesirable Cultural Mind-sets, in charge of re-education therapy for the unenlightened, with those of us who fail to be born again condemned to wear, say, purple hexagons to warn the innocent against being contaminated by us?

Absurd? I devoutly hope so. But some of us remember how absurd it once seemed that funny little powerless Adolf Hitler could ever be taken seriously—we laughed and laughed at Charlie Chaplin's cinematic putdown of him. Let's treat him like a British gentleman and surely he'll respond in kind, argued Neville Chamberlain and many others. Only Winston Churchill recognized the implications of Hitler's "cultural mind-set," as stated explicitly in *Mein Kampf*, and oh, how

Churchill was denounced for stirring up needless controversy, for breaching the niceties of diplomatic decorum, for "overreacting" (though that term, so useful for abusing those who disagree with one, like "state-of-the-art" for self-promotion, hadn't yet been invented)!

One begins to grow weary of Erskine-Hill's and Clark's constant whining that people who think the opinions they express are wrong-headed, unsupported by fact, and potentially dangerous aren't nicer to them. I remind them again that they initiated this controversy, suddenly mounting an attack on my book a quarter-century after its publication, declaring that I know no scholarship after 1960—a flagrant piece of defamation of my professional character—satirizing my relationship with my teacher James Clifford, stigmatizing all those who disagree with them as infected with a pernicious "cultural mind-set" that disqualifies them from expressing any opinion on Johnson differing from their own. And when such an infidel points out that some statement of theirs is factually mistaken, their defense is to cry angrily or plaintively, "You're no gentleman!"[9]

Is this scholarship—the eternal struggle to try to arrive as close as humanly possible to the truth about things as they are or were? Of course it isn't; it's a crude and cheap kind of propaganda for some vaguely defined but ominous "new model" of culture in which the propagators will be the enforcers of the "correct mind-set." Of some popular political slogans current in his lifetime, Johnson commented that they were "thundered in our ears, sometimes by interested faction and sometimes by honest stupidity."[10] One thinks of Clark's public support of the Redwood faction in current British politics and Erskine-Hill's ignorance of the readily accessible facts of the publication of McAdam's important essay. I suspect that these outbursts about Johnson by Clark and Erskine-Hill will be as quickly forgotten by serious Johnson scholars as Clark laments that John Redwood's "pioneering" book about the "theological" substructure of eighteenth-century British politics has been.[11]

NOTES

1. Clark didn't take up my challenge to explain how he would go about correcting Sir Geoffrey Elton for his perverse view that the great English historians are Maitland and Namier. I wonder whether he noticed in the *TLS* for 1 September 1995, p. 28, that the eminent American historian Bernard Bailyn, reminiscing about his professional career, singles out for special praise four historians, two American, Perry Miller and Charles Andrews, and two British, Ronald Syme, and Lewis Namier. I sometimes wonder too whether Clark, for all his bluster, hasn't as sketchy a knowledge of general historiography as he has of Johnsonian scholarship, and whether he has ever looked at Sir Ronald Syme, O.M.'s magnificent *The Roman Revolution* (Oxford, 1939). If he has, he will have discovered that its methodology is straight Namier—current and later ideological cant about the preservation of historic republican Roman "liberty" under Augustus is contemptuously dismissed, and instead we are presented with an enormously complicated account of family, adoptive, and economic relationships that resulted finally in the Mafioso-like dictatorship of the Julio-Claudian clan. Syme indeed, in his introduction, apologizes to the reader about how complex and indeed boring all this information is, and that, no doubt, is why Namier is not popular with the casual student of history, who much prefers to read of "freedom broadening down," thanks to Whiggish "libertarianism" and in spite of Tory obstructionism. So Clark advises us to drop Namier, Plumb, and no doubt Syme down the glory hole to oblivion (Winston Smith's daily job in *Nineteen Eighty-Four*) or, as Marx put it, in "the dustbin of history"—Marx to be sure had a spectacular run, over seventy years, 1917-1989. There is nothing new about this, of course: "He who controls the present controls the past; he who controls the past controls the future." Johnson noted that "In one of the parliaments called by Cromwell it was seriously proposed that all the records in the Tower should be burnt, that all memory of things past should be effaced, and that the whole system of life should commence anew" (see Greene, *Politics*, p. 29). Similarly, Clark argues, historiography, by a natural law, moves inexorably forward to better things, discarding Sir Lewis, Sir Jack, and Sir Ronald (and no doubt Thucydides, Tacitus, Guicciardini, and Froissart), to culminate in—J. C. D. Clark and his "New Model."
2. See Ian Buruma, "Action Anglaise," *New York Review of Books*, 20 October 1994. One wonders what Clark made of the scene in David Lodge's hilarious novel *Changing Places* where one of its heroes finds himself the only male on board a jumbo jet flying from San Francisco to Heathrow, all the other passengers being American women intent on making use of the free abortion facilities of the British National Health Service. There are no statistics about it of course, but one gets the

impression, from reading recent biographies and memoirs, that in expensive British private boys' schools and among male undergraduates at the older British universities, homosexuality is almost routine in a way that it doesn't seem to be at similar American institutions. Redwood (*Times*, 27 June) is quoted as in favor of "privatization," which would seem to imply "consumerism," and one suspects that this appeals to American "Evangelicals."

3. Some light is cast on the term by a remark by Patrick J. Buchanan in an address to a convention of the Christian Coalition (*Los Angeles Times*, 13 September 1995, p. E2): "We don't need some secular humanist in sandals and beads at the Department of Education telling us how to educate our children." Clark might find useful in his future oratory a description of their opponents by another delegate to the convention: "You're secular extremists from the culture of death" (p. A4).

4. See, e.g., Johnson's advice to young students as to what to study, in his preface to *The Preceptor*, 1748:

> The principles of Laws and Government come next to be considered, by which men are taught to whom obedience is due, for what it is paid, and in what degree it may be justly required. This knowledge by peculiar necessity constitutes a part of the education of an Englishman, who professes to obey his prince according to the law, and who is himself a secondary legislator, as he gives his consent by his representative to all the laws by which he is bound, and has a right to petition the Great Council of the nation whenever he thinks they are deliberating upon an act detrimental to the interest of the community. This is therefore a subject to which the thoughts of a young man ought to be directed.

That "he may obtain such knowledge as may qualify him to act and judge as one of a free people," he is advised to study such books as "Locke on Government," "Harrington's Oceana," and "Hooker's Ecclesiastical Polity." (Johnson, *Prefaces and Dedications*, ed. Allen T. Hazen [New Haven, 1937], pp. 187-88.)

5. See Ruth K. McClure, "Samuel Johnson's Criticism of the Foundling Hospital and Its Consequences," *Review of English Studies*, 27 (1976), 17-26.

6. As someone has said, no good deed goes unpunished. In my earlier reply (*AJ* 7, p. 73) to Clark's and Erskine-Hill's attacks, in an effort to live up to Erskine-Hill's high standard of civility in scholarly debate, I

complimented him on citing this too little known work, writing, "Erskine-Hill deserves commendation for calling attention" to it. What was my reward for this well-intended little pat on the head? This: "Though quoted by E. L. McAdam in *Johnson, Boswell and Their Circle* (Oxford, 1965) . . . and republished by Professor Arthur Sherbo [in 1987] . . . the contributors to the other side of the debate seem not to have been aware of it until I and Dr. Clark featured it in our discussions. . . . At this point the contributors on the other side of the debate fell upon it with cries of triumph, for it contains a hostile reference to the '45. . . ."

I could hardly have failed to be aware of the existence of McAdam's article since it precedes by only a few pages my own contribution to the *Festschrift* for L. F. Powell that Erskine-Hill cites. Nor could Howard Weinbrot, as Secretary of the Johnson Society of the Central Region, have been unaware of Sherbo's reprint, which was contributed to that Society as a "keepsake." It is apparently Erskine-Hill who is unaware that (as McAdam of course points out) it was Powell who first called it to the attention of modern Johnsonian scholars. Powell's article is listed in the Clifford-Greene *Samuel Johnson: A Survey and Bibliography of Critical Studies* (Minneapolis, 1970), an expansion of Clifford's pioneering bibliography of Johnsonian scholarship, 1951, as item no. 25:44, and given one of our rare asterisks to indicate its importance; McAdam's is item no. 20:93. If Erskine-Hill were to condescend to make use of this work and its continuation to 1985 by Greene and Vance, which most serious Johnson scholars find indispensable guides, he might be helped to make less of a fool of himself than he does here. I refer of course to the Powell and McAdam articles in my note 16 to the passage in *AJ* 7, but evidently Erskine-Hill does not condescend to consult source notes either.

7. Another statement of Johnson's that is carefully ignored is this, from his review of the memoirs of that venomous old Whig Sarah, Duchess of Marlborough (Greene, ed., *Samuel Johnson* [The Oxford Authors], 1984, p. 115). Sarah lays about her at almost every member of the royal family with whom she has had contact, including Queen Anne, and especially at William III, whose manners were so deplorable—once, when dining with Sarah, Anne, and others, and a dish of the first green peas of the season was served, William took every one himself and left none for the others! Shocking!

Johnson records Sarah's reports of William's unlovability with some seriousness, maintains that it is important for the leader of a country to cultivate the art of good public relations, and sums up: "Charles the Second, by his affability and politeness, made himself the idol of the nation that he betrayed and sold"—to the French. "William the Third was, for his insolence and brutality"—"churlishness," Johnson's *Dictionary* defines it—"hated by that people which he protected and enriched." Is this evidence that Johnson wished the restoration of a superficially attractive regime that would betray and sell his country, and deplored the "revolution" that replaced it with one which, though superficially unattractive, protected and enriched it? Or is it a warning, as applicable to "democratic" nations today as it was to the British when it was published in the *Gentleman's Magazine*—in 1742, three years before the 1745 insurrection—not to be disastrously taken in by glamorous P.R. by those who would rule them?

8. It is well to consult the original of Johnson's "imitation" here, Juvenal's *Satire* X. Juvenal's Charles XII is the Carthaginian Hannibal, whose invasion of Italy (from the north) struck terror into its inhabitants. But in the end he was defeated and died, a disgraced exile, by a banal suicide. The lines in Juvenal corresponding to Johnson's couplet here are "i demens, et saevas curre per Alpes/ut pueros placeas et declamatio fias"—"Go, madman, and rush through the rugged Alps, in order that you may please schoolboys by becoming a topic for their declamations." (Juvenal's use here of zeugma, so often confused by modern students with syllepsis, might be noted.) It is hard to detect any signs of sympathy in Juvenal's account of this ambitious aggressor.

9. Or, when an awkward question is raised, such as "How did the 'Non-juror' Johnson get around taking the required oaths when he became a schoolmaster?", or "If Johnson was so frightened of a charge of sedition in 1756 that he felt he needed, contrary to his true beliefs, to print a repudiation of James II, how do we come to find him, in 1757, treating an actual charge of sedition with the utmost contempt?", simply ignoring it. Or, when a recent scholarly work, such as Edward Gregg's biography of Queen Anne, is cited as disagreeing with their views, brushing it aside as "mistaken," with not even a gesture at trying to explain why it is mistaken (see Clark, *AJ* 7, pp. 42 and 55, n. 71). In spite of Gregg's reiteration of the well-known fact that Anne was anything but a "Jacobite"—she and her husband had been among the first to desert her father James II for her sister and cousin, Mary and William, and would never admit that "James III" was actually the child of her detested stepmother Mary of Modena—the myth still persists that Johnson's having been "touched for the evil" by her when an infant somehow corroborates the theory of his own Jacobitism. I think that, as the legend of Johnson's having joined Charles Stuart's army in the Highlands in 1745 was given a great boost by John Buchan's novel *Midwinter*, so the

legend of Anne's crypto-Jacobitism was greatly aided by Thackeray's *Henry Esmond*. His novelistic skill makes its climactic scene unforgettable—James Edward has been smuggled over to England, and, through the intrigues of Bolingbroke and Lady Masham, Anne is suddenly confronted with her "baby brother," the rightful possessor of her throne. Presumably overwhelmed with guilt at her usurpation, Anne goes into a hysterical fit and dies a few days later. Bolingbroke and his allies had hoped at that point to be able to place James on the throne, only to find that he is off in the country, in pursuit of Beatrix Esmond. So George of Hanover is proclaimed instead. Delightful nonsense, that has charmed many generations, and conceivably influenced some historians, such as Clark. A book might be written on the impact of "historical" fiction on historiography; it could start with Virgil's *Aeneid*.

10. Johnson, *Political Writings*, ed. Greene (Works, Yale ed., Vol. X), p. 418.
11. When *Samuel Johnson and the Politics of Hanoverian England* (Oxford: Clarendon Press, 1994) by the distinguished veteran political historian John Cannon appeared, I resolved to put it away unread until I had finished writing the present article. I have now read it through quickly and doubt that it will provide much comfort for Clark's and Erskine-Hill's "side" (Erskine-Hill's term) in the present controversy. The final paragraph of Cannon's introduction begins, "Far from being, as Macaulay suggested, an extreme and outlandish figure, Johnson was to be found, on most questions, putting forward a moderate case, and one which he shared with many of his fellow countrymen." Clark, in his attack on me in the *Times Literary Supplement* of 14 October 1994, denounced me for similarly finding that "Johnson was a moderate, a realist, intelligent and well-informed, responding to the political complexities of his age." By doing so, I had "homogenized" him and demonstrated that I had "embraced . . . that historical school most likely fundamentally to misinterpret Johnson." So it looks as though Professor Cannon will be joining me, Professor Weinbrot, Professor Curley, Sir Lewis Namier, and Sir Jack Plumb in Dr. Clark's rogue's gallery, which is showing signs of becoming seriously overpopulated.

JOHNSON AND JACOBITISM REDUX: EVIDENCE, INTERPRETATION, AND INTELLECTUAL HISTORY

HOWARD D. WEINBROT

Readers now may wish the exchange regarding Johnson and Jacobitism sent to the devil. If so, they are seconded by at least one contributor who gladly shall move to fresh woods and pastures new. J. C. D. Clark's "The Cultural Identity of Samuel Johnson" includes Injured Virtue's hyperbolic catalogue of personal and national insults to critics who, at the least, must consort with Beelzebub. Howard Erskine-Hill's "A Kind of *Liking* for Jacobitism" portrays him as brave Polydamas helping the fallen Hector. Professor Erskine-Hill nevertheless sees the preponderance of evidence and has moved slightly away from his colleague's excesses. Neither side, though, is likely to move much further, and Dr. Clark threatens his candid messengers with yet another fusillade.[1]

The electricity here generated is surprising. Few readers of Johnson care about his putative Jacobitism. His extraordinary achievements remain whether he was a Stuart, Williamite, Hanoverian, Whig, or Tory; whether in opposition, administration, or some combination thereof. He did not have *the* cultural identity any more than he had *a* prose style. Those insisting on Johnson's Jacobitism seem to be satisfying their own political and emotional needs rather than assessing evidence, drawing appropriate inferences, or plausibly and overtly speculating. We thus still need to examine the Jacobite apologists' latest effort. The news is less good than one hopes. I start with a very few of Dr. Clark's, and where appropriate Professor Erskine-Hill's, continuing errors.

1. Putting Down Stakes in Error's Cave

Dr. Clark reasserts old errors and invents new ones. Henry Brooke's *Gustavus Vasa* (1739), he still argues, was not "securely Hanoverian" (*AJ* 8, p. 48). One then wonders why this demonstrably anti-ministerial work praises "those *Gothic* and *glorious* Nations, from whom our Form of Government is derived." The last words in Brooke's Epilogue are "*hail* *Nassau!—*The Deliverer of our Country."[2] Dr. Clark also still fancies that an Index is an appropriate place to make polemical points and polemical dynastic distinctions—like Richard Blacow as a "Whig Toadeater" (*SJ*, p. 257), George I and George II as only "styled" kings of Great Britain, George III as king "in the Hanoverian succession," and all three primarily as invasive German dukes (*SJ*, p.260). A spot check of two dozen books by British historians reveals only one other with this practice: J. C. D. Clark's *English Society 1688-1832* (Cambridge, 1985).

We continue to hear that Johnson refused to serve in the largely surrogate-manned and illiterate Trained Bands because he refused the oaths of Allegiance and Abjuration. An amusing foray into art history demonstrates that Johnson was fit enough to have served. Dr. Clark sees Dictionary Schwarzenegger in an amply girthed gray-wigged forty-seven year old sedentary lexicographer: "Sir Joshua Reynolds's portrait of Johnson, painted in 1756, shows a robust solidly-built man in his forties, not a valetudinarian" (*AJ* 8, p. 42). This medical examination of a man in a painting overlooks one of its prominent traits. H. W. Liebert remarks the obvious: "In Johnson's blind left eye Reynolds has reproduced the bland appearance of sightlessness." Other portraits by Reynolds, Frances Reynolds (?), and James Northcote all show Johnson with a book inches from his good eye. He could barely see or hear actors on the stage much less determine a target or whether that target was shooting back. Arthur Murphy met Johnson during the summer of 1754, and later observed that "Mr. Johnson's health had been always extremely bad since I first knew him."[3] That bad health included his compulsive rocking and touching. What fellow-soldier would want such a man next to him? We do not need an argument on non-juring to explain why Johnson was not in the Dogberry Brigade.

Dr. Clark is more deterministic than ever. *If* young Johnson in Lichfield experienced the "Tory high tide"; *if*, to mix Dr. Clark's metaphor, that tide was monolithic; *if* it was as Dr. Clark describes it; and *if* he was "saturated" in *jure divino* doctrine (*AJ* 8, p. 44), it scarcely follows that he therefore believed in divine right—which he never supported. Johnson always regarded government as a human construct that evolved to meet each nation's needs. Individual humans in communities and families also evolve to meet their own needs. Lancelot Addison, Dean of the Cathedral Church in Lichfield was a Tory; his son Joseph was a Whig. Lichfield produced Tory Jacobite Bishop of Bristol George Smalridge; it also produced Whig Hanoverian Thomas Newton, Johnson's close contemporary (1704-82), preacher of loyalist sermons during the '45, chaplain to George II in 1757, and Bishop of Bristol in 1764. Small Staffordshire issued four loyal addresses and provided one "Military Association in support of the Hanoverian dynasty" before the Jacobites' retreat north.[4]

Johnson's presumed Toryism was not a function of environmental osmosis. Nor was his Anglicanism "owed primarily to the Caroline divines" (p. 4) rather than to his mother, the Bible, the Book of Common Prayer and, probably, *The Whole Duty of Man* and *Pilgrim's Progress*. Dr. Clark praises Sir John Hawkins' "lawyerly accuracy and attention to detail" (*AJ* 8, p. 23). Accurate Hawkins observes that Johnson's religion stemmed from "the perusal of St. Augustine and other of the fathers, and the writings of Kempis and the ascetics, which prompted him to the employment of composing meditations and devotional exercises."[5]

Alas, there is no end of fertile correction in point-to-point refutation of Dr. Clark's errors. Readers who have come thus far, though, may wish for larger issues and new contributions rather than debate-like ping-pong. I thus propose yet once more to examine the main assertions regarding Johnson and the oaths. I then will grant several of the Clark-Erskine-Hill hypotheses and see how this concession affects belief regarding Johnson as a Jacobite. I will conclude with an overview of French activity before the '45, and relate this to his "General Preface" to the *Gentleman's Magazine* Index in 1753. Such a review makes plain that Johnson never was a Jacobite and certainly could not so have been after the early 1740s.

2. Johnson, Oxford, and the Oaths

For J. C. D. Clark, and for Howard Erskine-Hill, Johnson the conscientous Jacobite refused to take the Oath of Allegiance to the Hanoverian dynasty and the Oath of Abjuration rejecting the Stuart exiles and accepting the Hanoverians. There are three presumed bases for this argument: 1) Johnson did not leave Oxford because of poverty, since the modest fees were affordable; 2) those then and now who argue that matriculants took the oaths of Allegiance and Supremacy are wrong and Johnson never could or would have taken the Oath of Abjuration; 3) the literal wording of the statutes determined actual conduct in eighteenth-century Oxford.

The traditional argument on Johnson's poverty, Dr. Clark claims, is wrong. Even in more expensive 1760 one could live frugally on twenty guineas a year, and even £10 per year would do: "Poverty at least meant self-discipline" (*AJ* 8, p. 37). No doubt. But poverty is a state of mind as well as purse and is especially galling when suffered among the offended affluent. Even when Johnson enjoyed a handsome pension of £300 per year he complained in "Post Lexicon Anglicanum" (1772) that poverty so limited him. As two earlier eighteenth-century poems make plain, life for the poor was far worse at socially repulsive eighteenth-century Oxford.

The unsigned *The Servitour* (1709) uniformly mocks a gross upstart country bumpkin whose speech, appearance, and intelligence are inappropriate for Oxford. This wretch's family is "not worth a Farthing"; he himself is without "a Rag his Arse-on," should not live among gentlemen, and steals food from the kitchen or goes hungry before returning to his meager four room-mate chamber. *The Commoner* (1710 and 1717), probably by the same author, is equally contemptuous of a class of students closer to Johnson's.[6] The poet intends to expose the many Commoners who "can scarce live handsomely [as] *Servitours* . . . and yet wou'd willingly impose on the World." These impoverished "scandalous" men "foolishly indulge themselves in a stupid Fondness of being counted Genteel" (sig. A2^{r}).

In the poem itself an ambitious farmer-father refuses to enroll his boy as a servitor at the cost of a mere "three or four Shillings a week" (p. 7). No No, says the father whose dunned and ridiculed offspring soon will hide in college and be ostracized by his betters, "For Vorty Pounds a Year I have been told/A Mon might live" and still keep something for contingencies. Yes, his interlocutor says, "He may live on't, if he'll but careful be,/But somewhat shabby (as you'll quickly see)" (p. 8).

Things had changed little in 1729, when Johnson refused to return to Christ Church because "his shoes were worn out, and his feet appeared through them." Johnson's "indignant" rejection of the anonymous gift of shoes often is regarded as pride in the face of good intentions.[7] It may have been rage in the face of continuing insults: "These are the men to be exposed," *The Commoner* says, seeking to make such a misfit "learn to know himself and act according" (sig. A2^r). Johnson's poverty and Oxford's intolerance remain essential reasons for his departure long before graduation.

Dr. Clark also remains unpersuaded that Oxford undergraduates regularly took both the oaths of Supremacy and Allegiance upon matriculation. Only scholarship students were so required, he insists. Contemporary experiential evidence continues to refute this assertion. In 1721 Thomas Hearne complains that the Oath of Allegiance was added after 1714 "to bring a slur upon the University, and out of a Trimming Design." In 1721 as well Nicholas Amhurst twice repeats his experiences in 1716—that all students take the oaths of Allegiance and Supremacy upon matriculation. He describes the casual ceremony in which he hardly knew what was being mumbled. Richard Blacow in 1747 said of seven drunken Jacobite undergraduates shouting treasonous slogans: "as the Wisdom of the University thought them of proper Capacities to take the Oaths *four years ago*, they must be thought of proper Capacities *now* to be punish'd for the breach of those Oaths." Benjamin Kennicott in 1755 says that "Every person, at his *Matriculation*, if *sixteen* years of age, takes the Oaths of *Allegiance* and *Supremacy*."[8]

However definitive, terms like "oaths" and "Every person" do not suit Dr. Clark's needs. Those who participated in the process must have been ignorant of their own experiences. Since Nicholas Amhurst was, we hear, a vengeful, zealous, obsessional Whig firebrand (*AJ* 8, pp. 18, 25), Dr. Clark dismisses him as too dim to be accurate (*AJ* 8, p. 26). The Whig Benjamin Kennicott "like Amhurst, was probably also making a genuine mistake" (*AJ* 8, p. 31). The "Whig toadeater" Blacow needs to be further sullied. His "motives were open to challenge," and he "was impelled by ambition" (*AJ* 8, p. 35). The historian of Christ Church E. G. W. Bill argued that Amhurst's experience reflected reality but was "misled" by him (*AJ* 8, p. 65, n. 74). Other recent historians of Oxford who do not agree with Dr. Clark have committed a cardinal sin—they wrote "a late-Namierite memorial to a superseded phase in the historiography of the subject" (*AJ* 8, p. 62, n. 32). Perhaps we can put aside so defensive and tendentious a mode of proceeding. I offer some reasons why the oaths of Allegiance and Supremacy seem to have been taken by matriculants above the age of sixteen.

The statute itself was 1 George I. s. 2, c. 13 (see Appendix), and in one relevant place specified oaths for the following university residents: "all Ecclesiastical Persons, Heads, or Governours, and all other Members of Colleges and Halls in either University, being of the Foundation, or who enjoy any Exhibition (being of the Age of 18)." The statute ends by specifying the oaths to be taken in England: "the Oaths of Allegiance and Supremacy, and the Abjuration Oath."[9]

Custom generally wins the battle between custom and law, otherwise automobiles would never exceed fifty-five miles per hour, perjurers would be as common as jockeys on Houyhnhnms, and a certain weed never would be smoked or certainly never inhaled. So far as I can tell, in practice the Oath of Abjuration was not normally given to undergraduates of Johnson's kind either upon matriculation or graduation. Oxford graduation subscription books, for example, include the oaths of allegiance and supremacy, the relevant articles in the 36th canon, and the 39 Articles. The matriculants' subscription book, which Samuel Johnson signed on 16 December 1728, contains the Oath of Supremacy and the 39 Articles.[10] As contemporaries said, however, the Oath of Allegiance was really or apparently added for matriculants as well.

There were at least five reasons for this embellishment of the statute. One is Hearne's sound belief that Hanoverian legislation pressured the university. A second is the administrative difficulty in the statute that singles out only foundationers and exhibitioners for both oaths. The vice-chancellor generally administered the oaths to several matriculants at once. Rather than separate students into scholarship or non-scholarship, servitors or gentleman, it was easier to have, or seem to have, one indifferent ceremony. Amhurst, for example, was informed that he took the Oath of Allegiance without having done so.[11] A third reason is that the perfunctory process avoided career-threatening squabbles. The ultimately non-juring Thomas Hearne once took the oaths of Allegiance and Supremacy in order to accept two Oxford positions. He did so, he admits, "without saying to whom ye Allegiance Oath was taken, in wch he [his examiner the Registrar George Cooper] did right, it being certain y^{t} if he had said I swore to K. James (as I really did) he should be brought into a Scrape as well as myself."[12]

The fourth reason is both political and logical—namely, since the part was regarded as the whole, parts may as well be added to the whole. The Latin preamble to the Oxford subscription book that Johnson signed translates as all matriculants "shall take their corporal oath to acknowledge the superiority of the king."[13] Like the 37th of the 39 articles, the Oath of Supremacy rejecting any "Foreign Prince, Person, Prelate, State or Potentate" is an implicit Oath of Allegiance and an epitome of the Oath of Abjuration.

As early as 1689 the brilliant Huguenot emigree Peter Allix argued that "The Oaths of Supremacy and Allegiance, were at first prescribed only in opposition to this Doctrine of the *Romish* Church," whose pope "has arrogated to himself a Right of deposing Kings."[14] Nicholas Carter draws an appropriate conclusion in 1716 in his sermon on *The Obligation of an Oath; and particularly of the Oaths of Allegiance, Supremacy, and Abjuration* (London). Carter knows that the Oath of Supremacy connotes allegiance to the Protestant Hanoverian monarch. To declare against a foreign power requires withstanding that power: "how can they be said to oppose the Exercise of the *Pope*'s pretended Power here [in the Oath of Supremacy] who endeavour to bring in one to be their Governor, who ('tis very likely) would oblige all his Subjects, under Pain of Death or Banishment, to acknowledge it, and live in Subjection to it?" Anyone, he says after quoting the Oath of Supremacy, who would swear to that but "would assist a *Papist* Pretender" lies regarding "the plain meaning and Design of the Oath so taken" (pp. 18-19). A commentator in 1772 makes a similar remark —namely, that "in truth the oath [of supremacy] has always been understood by the legislature to imply *the supremacy of the King in all matters ecclesiastical and spiritual*; a claim erected upon the downfall of the Pope's usurpation."[15] There is no reason not to make explicit what the legislature and others "always" thought implicit.

The fifth reason also is based upon experiential evidence. Unless one fussed about not taking or about violating the oaths—as Hearne and Blacow's drunken Jacobites did to their peril—Oxford preferred to ignore or even subvert their intention. Nicholas Amhurst seconds Thomas Hearne's remark, but from a loyal Whig point of view. He himself gladly took the oaths which, he says, matriculants were "prætaught to evade, or think null: some have thought themselves sufficiently absolved from them by kissing their thumbs, instead of the book; others, in the croud, or by the favour of an honest beadle, have not the book given them at all."[16] *The Grub-street Journal* for 8 July 1736 (No. 341), more generally suggests how irrelevant, irreverent, and "notorious" oath-taking had become: "the repeated Swearing required by law in all branches of the Constitution, has taken away the terror and awe, that should naturally accompany an Oath." The *Journal* laments "the indifference" with which oaths impossible to perform are taken in court, for state occasions, and in the universities. The oath-takers "scarce ever read, or heard them read, till the time of swearing"; but "there is nothing to be done . . . without Swearing; there can be no admission into any College or Hall without an Oath; no matriculation without an Oath, and subscription." He asks whether this is not a dangerously cynical practice "and whether learning cannot be acquired without Oaths that are never intended to be kept."[17]

Samuel Johnson agreed and resented the moral and intellectual basis of such obligatory and thoughtless oath taking. Parts of Johnson's "Remarks on the Militia Bill" (1756) sound like the Oxford oath-taking ceremony Nicholas Amhurst described. Oaths intended to trap "men of disloyal principles" do not work. "They know, what every one knows, that mean men called before those whom they have always regarded with veneration, will be more afraid of man than God, and will take the oath taken and offered by their betters, without understanding, without examining, perhaps without hearing it."[18] For practical purposes, however, one may have to accept such folly.

Johnson thus makes two other germane remarks regarding necessary accommodation to coercive power. On 27 March 1775 he says that a non-juror may have been "less criminal in taking the oaths imposed by the ruling power, than refusing them." After all, a man must live, and he probably will "be reduced to very wicked shifts to maintain himself" (*Life*, II, 321). In the "Life of Fenton" he praises Fenton's adherence to conscience and willingness to suffer by not taking the oaths; but he refers to such action as a "perverseness of integrity" that did his professional life unnecessary material damage.[19]

Several instances of experiential, theological, political, and social commentary together with Johnson's own remarks, suggest that like other post-Hanoverian matriculants he indeed took the oaths of Supremacy and Allegiance. In later life he had numerous opportunities, as with Boswell and with Fenton above, for example, safely to say that he himself had never taken the oaths. Since that did not happen both Dr. Clark and Professor Erskine-Hill creatively read Johnson's remark on 15 April 1773.

In *Samuel Johnson* Dr. Clark says that "Johnson accidently disclosed his own refusal to subscribe in April 1773" (*SJ*, p.123). He now repeats that Johnson "explicitly disclosed in 1773 that he had never taken" the oaths; he says again that Boswell "correctly recorded that he had never" taken the oaths (*AJ* 8, pp. 26, 43). Professor Erskine-Hill is more tentative: "he is reported as clearly implying that he himself has not taken the oaths relating to the House of Stuart" (*AJ* 8, p. 8). Johnson, though, neither explicitly nor implicitly said what the Jacobite scholars claim he said.

Johnson and Boswell speak about the Stuarts. Johnson observes that it seems as if the Hanoverians now have "as good a right as the former family" and that it might be regarded as culpable to disturb them. Still, Johnson adds with his characteristic ability to see both sides of a question, the matter is more difficult "when considered with respect to the house of Stuart. To oblige people to take oaths as to the disputed right, is wrong. I know not whether I could take them: but I do not blame those who do" (*Life*, II, 220). Johnson's remarks limit the question "with respect to the house of Stuart"—that is, from their point of view Hanoverian longevity and stability do not replace hereditary right. The present dynasty wrongly forces those who dispute the government to take the oaths but, Boswell's Johnson says, "I know not whether I could take them" and excuses "those who do." He speaks in the present not the past tense; he says that he might not be able to take such oaths if required to do so now. He does not say or hint that he refused such oaths in the past. Samuel Johnson clearly states what he means: by 1773 he need not have feared repercussions if he said that he had not taken oaths some forty-four years earlier.

Perhaps one reason that he did not so say is that he took the oaths that 1 George I s. 2, c. 13 required. It specifies who takes which university oaths and when. For Dr. Clark, the law is the law. What the statute says is what the people do. Let us see what happens when we grant that proposition.

The statute requires the oaths of Allegiance and Supremacy and the Oath of Abjuration for "all Persons teaching or reading to Pupils in the Universities or elsewhere; all School-masters, Ushers, Preachers, and Teachers of Separate Congregations."[20] Samuel Johnson was an Usher at Market Bosworth from March to July of 1732 and applied for another ushership at Ashborne on 27 July 1732. It seems reasonable to suppose either that, if such a procedure was possible, he informed his master that he had taken the oaths at Oxford, that he took them at Market Bosworth, or that he did both. He also had his own school at Edial from about July of 1735 to December of 1736. Dr. Clark dismisses this portion of the statute he elsewhere finds so compelling: there was no automatic ceremony that required subscription; the tyrant Sir Wolstan Dixie was not scrupulous about such matters; oaths probably were regularly tendered only during emergencies (*SJ*, p. 118). Apparently for J. C. D. Clark all parts of a statute are equal but some are more equal than others.[21]

The statute, evolving practice of it, and contemporary evidence regarding the timing of oaths refute Dr. Clark's notion that Johnson was a non-juring Jacobite who refused to take the oaths of allegiance and abjuration and thus could not graduate from Oxford. Dr. Clark's own reading of texts in support of Johnson's Jacobitism generally is wrong. Dr. Clark's insistence upon the intimate relationship between statute and conduct is inadequate on the face of it, is selectively employed, and is abandoned when it becomes inconvenient.

I now enlarge upon what I offered a moment ago—granting some of Dr. Clark's and Professor Erskine-Hill's assumptions regarding Samuel Johnson and his contexts. These are 1) that Johnson did not take the Oath of Allegiance at Oxford; 2) that he speaks in codes requiring elucidation by reclaimed political context; 3) that the Charles of Sweden passage in *The Vanity of Human Wishes* (1749) is about Charles Edward; and 4) that the *Dictionary*'s (1755) definitions and illustrative quotations express Johnson's true feelings. What changes as a result of these generous concessions?

3. Granting the Jacobite Hypotheses

Let us assume that Dr. Clark soon announces what either he has not sought or cannot find—early eighteenth-century experiences in which Foundationers and non-Foundationers regularly were separated and each was required to take a different set of oaths in different ceremonies. Scholarship Jones takes the oaths of Allegiance and Supremacy and thereafter Gentleman Smith takes only the Oath of Supremacy. In such a case, congratulations and mea culpa. Yet the larger argument remains the same: there still is no evidence that Johnson left Oxford because he would have refused the Oath of Allegiance upon graduation; there still is no evidence that he avoided service in the Trained Bands because he refused its Hanoverian oaths; there still is no evidence that he ever said or implied that he had not taken them; there still is no evidence that Samuel Johnson was a Jacobite, however much young Johnson enjoyed opposition, and however much his sympathetic imagination and respect for subordination recognized some value in some aspects of Stuart politics.

Let us also assume that Johnson spoke in codes, as Dr. Clark says he does in *London* (1738). That poem, we hear, adapts Juvenal's third satire and laments "by implication, the passing of a former regime" (*SJ*, p. 41). It is "profoundly political" as the Scottish Jacobite Thomas Ruddiman "would naturally have recognized" (*SJ*, pp. 145, 146). A broken code, however, may show a political Johnson uncongenial to the Clarkean-Jacobite mode.

All readers of *London* know Johnson's hostility to the French in lines 91-157. It is an extraordinary outburst without apparently extensive historical warrant; but it is redolent of political implication. The French, especially French servants, prey upon and corrupt the state, turn free British warriors to slavish French beaus, practice mock-submission that enslaves their British masters, replace British air with their own and in general replace the British with the French political system. "Their air, their dress, their politicks [they] import" (line 110). No wonder Johnson's narrator Thales "cannot bear a French metropolis" (line 98) and must leave London for the British countryside and its British politics.[22]

Johnson's concern with the dangers of French servants and values soon becomes a trope of anti-Stuart rhetoric. *Common Sense* for 10 March 1744 is reprinted in that month's *Gentleman's Magazine*. Its title indicates its contents: "*On* French *Papists being preferred to* English *Servants*" complains that Britons should worry about more than the overtly pending French invasion from Dunkirk. The best families are taking those adder-like alien Papists into their bosoms. Such stinging behavior "cannot but tend to strengthen that Nation and weaken our own, and propagate the Popish Religion" (XIV, 147). Shortly thereafter the same harsh wisdom comes from a fabricated French source, the unsigned *Cardinal Tencin's Plan, Presented to the French King, For Settling the Pretender's Family Upon the British Throne, and Compleating the Long-concerted Scheme of Universal Monarchy in the House of Bourbon* (London, 1745).[23] Britons knew that Tencin was an architect of the Dunkirk invasion, but like Thales and *Common Sense* before him he regards servants as a domestic fifth column that subverts the British state. He advises Louis XV how to divide and conquer the heterogeneous British nation and how to encourage native servants to be dismissed for hating their superiors. Seemingly servile French servants then occupy the best homes, speak to their masters of French needs and values, and "introduce the manly and obliging Behaviour of * * * * * *Charles*" while still passing "for true Protestants, and Lovers of King GEORGE." These emissaries will extract British secrets, make the British servants seditious, and "by this means, be made very proper Tools for our experienced Workmen" (pp. 20-21). In this context, Johnson's attack upon French servants in *London* attacks the French attempt to install their Stuart puppet on the British throne. Johnson's poem is anti-ministerial and occasionally anti-certain Hanoverian policies; once the French servant-code is broken, it emerges as demonstrably anti-Jacobite as well.

Such emergence also is clear in *The Vanity of Human Wishes* (1749), which Dr. Clark thinks includes Anglo-Latin-Tory-Jacobite suggestions, and which like *London* was "reprinted in Oxford in 1759," where "their significance would be correctly appreciated" (*SJ*, p. 103). Professor

Erskine-Hill is even more insistent. He has long argued that the Charles of Sweden passage is about Charles Edward, and he scolds Dr. Clark's critics for thinking otherwise:[24] "Apparently neither Greene nor Weinbrot sees more in Johnson's poetic narrative of Charles XII of Sweden than yet another example of a conqueror who bites the dust. By neglecting (or not recognizing) its emotion they reduce one of the most moving passages of eighteenth-century poetry to total banality" (*AJ* 8, p. 10). I blush.

As Moderation Itself I am pleased to support Professor Erskine-Hill's hypothesis: though there is virtually no similarity between the two Charles' careers, let us assume that Johnson indeed had him and the '45 in mind. After all, in the manuscript of what now is line 34 in *The Vanity* Johnson originally wrote that rebellious times find "the bonny traytor in the Tow'r" (*Poems*, *YE* VI, 93n). Johnson probably wrote his *Dictionary* (1755) description of "bonny" at about the time that he was writing the *Vanity*: "It is a word now almost confined to the Scottish dialect."

"Scottish traitor" for bonny Prince Charlie does not bode well for a sympathetic Jacobite response to the poem. Nor indeed does Charles of Sweden's reputation among key opposition men of letters before and after the *Vanity* was written. In "Epistle IV" (1734) of *An Essay on Man* Pope characterizes great heroes as pretty much the same "From Macedonia's madman to the Swede." Their entire "strange purpose" is "to find/Or make, an enemy of all mankind!" Thereafter, the twentieth of George Lord Lyttelton's *Dialogues of the Dead* (1760) is an exchange between Alexander and Charles. Charles complains that the insulting Pope has come to Elysium and should be thrown "down to the bottom of Tartarus, in spite of Pluto and all his Guards." The arrogant Swedish bully wants to do in the underworld what he briefly did in the overworld; but he finds that Alexander thinks as little of this bad strategist and bad general as Pope did: "I must say, that such Follies gave Pope too much cause to call you *a Madman*" (p. 210). Johnson agreed with the judgment of Charles as an overrated, murderous crackpot. His *Adventurer* No. 99 (1753) again couples Alexander and Charles and urges that they, Caesar, Catiline, Xerxes, and Peter be "huddled together in obscurity or detestation."[25]

This remark emphasizes Johnson's hatred for war in general and civil war in particular. He would have applauded the moving "*Exhorta-*

tion of the Church of Scotland" in *The Gentleman's Magazine* for December of 1745:

> Of all wars, a civil war is the most dreadful; in which countrymen shed the blood of their countrymen, children rise up against their fathers, and those of the same house are divided against each other; laws are silent, justice banished, and violence is unrestrained. Such is the dreadful war the rebels have kindled in this nation. (XV, 634).

A "rebel," Johnson's *Dictionary* says of the noun, is "One who opposes lawful authority by violence." His first illustrative quotation is from *Macbeth* and isolates a Scottish name in a tragedy about a Scottish regicide:

> The merciless Macdonel
> Worthy to be a *rebel*; for to that
> The multiplying villanies of nature
> Do swarm upon him.

Johnson's fourth definition of the adjective "civil" is "Not foreign; intestine." His illustrative quotation of this word probably defined late in the 1740s again reveals much: "from a *civil* war God of his mercy defend us, as that which is most desperate of all others. *Bacon to Villers*."

Grant, then, that Charles of Sweden is designed to recall Charles Edward. The contexts and Johnson's own words urge us to regard him as a destructive sanguinary rebel. He raised his own name on others' blood and achievements, and as a loveless, heartless man he deserves to be further punished after death as a schoolboy's exercise or a preacher's sermon. The Charles of Sweden passage indeed is not banal and indeed is moving—as Johnson expresses his anger at the Jacobite disrupter. His deserved ignominy helped to stabilize a united kingdom unnecessarily placed at risk.

As the examples cited above suggest, I also grant Dr. Clark's notion that one can base arguments "on the more secure evidence of Johnson's *Dictionary*" (*AJ* 8, p. 51). Here are two examples from words of immediate relevance for Dr. Clark's view of Johnson's "cultural politics." For Johnson, the noun "government" is "1. Form of a community with respect to the disposition of the supreme authority." His illustrative quotation is from Temple: "There seems to be but two general kinds of *government* in the world: the one exercised according to the arbitrary commands and will of some single person; and the other according to certain orders or laws introduced by agreement or custom, and not to be changed without the consent of many." One need not look far to see which form of government Johnson preferred. "Arbitrary" is "Despotick; absolute; bound by no law; following the will without restraint." Johnson's definition of "King" is equally instructive: "A contraction of the Teutonick word *cuning*, or *cyning*, the name of sovereign dignity. In the primitive tongue it signifies stout or valiant, the kings of most nations being in the beginning, chosen by the people on account of their valour and strength. *Verstegan*."

Johnson has conflated key opening sections of chapter 10 of Richard Verstegen's *Restitution of Decayed Intelligence* (1605, 1628, 1634, 1653, 1673). The passage Johnson quotes captures two of Verstegen's main, Tacitean, points: the English people had Germanic roots; the Germanic roots denoted kings chosen by the people, not anointed by God, and not subject to inheritance. Johnson makes even clearer that British government is secular and evolutionary. He quotes lines 209-14 from Pope's "Epistle III" (1733) of *An Essay on Man*. For Pope as for Verstegen, the origin of monarchy is social not divine:

> Thus states were form'd; the name of *king* unknown,
> 'Till common int'rest plac'd the sway in one;
> 'Twas virtue only, or in arts or arms,
> Diffusing blessings, or averting harms,
> The same which in a sire the sons obey'd,
> A prince the father of a people made.

The outcome is about the same whether one rejects or accepts many of the Jacobite scholars' assumptions: at best they offer a series of hypotheses that want significant reconsideration; for the most part those hypotheses want a decent burial.

4. Jacobitism, France, and Johnson's "General Preface"

J. C. D. Clark argues that Johnson's politics began to moderate only after the accession of George III in 1760. Even that is an implicit negative comment on the insufficiency of the German dukes wrongly ruling Britain: "The evidence that Johnson rejected *jure divino* doctrine comes from the end of his life, and may relate only to the Hanoverian dynasty then on the throne" (*AJ* 8, p. 44). As the "secure evidence" of the *Dictionary* demonstrates, Johnson never supported *jure divino* doctrine. His anti-Roman review (1756) of Thomas Blackwell's *Memoirs of the Court of Augustus* (1753-63) also insists that no "one form of government is preferable to another." So long as people "continue to obey their lawful governours, and attempt not to make innovations . . . they may differ for ever, without any just reproach from one another."[26] Much of Johnson's own reproach of government came in his pugnacious opposition voice. Christine Gerrard thus well argues that the *Compleat Vindication of the Licensers of the Stage* (1739) was written from "a radical patriot Whig position." She adds that "Patriot Whigs were rarely Jacobites" and "had a remarkably consistent level of support for the House of Hanover" and the Protestant succession "irreconcilable with Jacobitism."[27]

Howard Erskine-Hill is more temperate than Dr. Clark regarding a change in Johnson's political attitudes: "I think *The Vanity of Human Wishes* may mark the end of serious Jacobite inclination in Johnson" (*AJ* 8, p. 4). However wrong the assumption of Johnson's serious Jacobitism, the year 1749 is in the correct decade for change. We need to consider the traumatic events not only of the '45, but of its context in the War of the Austrian Succession's dress-rehearsals, in which the exploited Stuarts played supporting roles. This discussion necessarily includes Johnson's comments upon the '45 and his own politics and, in turn, the Jacobite scholars' own comments on Johnson. I am speaking of Johnson's General Preface to *The Gentleman's Magazine* Index for 1753.

Messrs Erskine-Hill and Clark insist that the attribution "first, *but only partially*" was made by John Nichols in 1821. A collaborative text, they claim, can only be "inconclusive" (*SJ*, p. 184), or "equivocal" (E-H, *AJ* 8, p. 12, n. 5). Professor Erskine-Hill relies on Arthur Sherbo's reprinting of the Preface in his Johnson Society of the Central

Region keepsake for 1987. Their critics, he says, did not know about this text, but he and Dr. Clark do and deserve praise for featuring it in their discussion: "While Sherbo, Clark and I all quote Nichols' *partial* attribution, Weinbrot nowhere refers to this in his review of Clark's *Samuel Johnson*" (*AJ* 8, p. 12, n. 5). Of course I do refer to and dismiss Dr. Clark's notion that the attribution is partial, and with very good reason (*AJ* 7, p. 188).

As Secretary-Treasurer of the Johnson Society for twenty-five years, I long have had Arthur Sherbo's keepsake for our excellent meeting at Michigan State University on 16-17 May 1987. I thus know that Professor Sherbo does not partially attribute the Preface to Johnson; he quotes Nichols' initial judgment regarding Johnson's extensive role in the Preface and then adds Nichols' final opinion which Professor Erskine-Hill withholds from his readers: "the General Index was published in 1753, with a Preface by his [Cave's] incomparable Friend" Johnson.[28] The Jacobite scholars misrepresent Sherbo and Nichols. They also misrepresent, or are innocent of, the date, source, and history of the original attribution.

The attribution was not made in 1821, but at least as early as 1807 and perhaps based on reliable earlier information. Edmond Malone's edition of Boswell's *Life of Johnson* then includes this footnote to Boswell's chronological catalogue of Johnson's prose: "To this List . . . Mr. Alexander Chalmers, with considerable probability, suggests to me that we may add the following:—1753 Preface." Chalmers himself repeats the attribution in his 1822 edition of Boswell's *Life*.[29] By then the attribution may long have been circulating within the small group of printers and writers concerned with and known to Johnson, Boswell, and their associates—including John Nichols who was a printer of the *Lives of the Poets* and became Johnson's good, admired, friend.

Johnson sent for Nichols on 7 December 1784 in order to return "some of the early volumes of the [Gentleman's] Magazine, with a professed intention to point out the pieces which he had written in that collection . . . particularly those which contained his share in the Parliamentary Debates."[30] Nichols apparently did not keep a record of those pages; we do not know how far forward Johnson's efforts went; and the books themselves must have been destroyed in a calamitous fire; but memory and knowledge remained. Either John Nichols or his son John Bowyer Nichols wrote in *The Gentleman's Magazine* files:

"Johnson, Samuel: Anon. S. Preface to this vol. xxiii (1753)."[31] We may at least plausibly conjecture that the attribution comes from Johnson himself in 1784. However that may be, the Preface was verbally attributed to Johnson before Malone published his edition of Boswell in 1807. Nichols provides the sole modest caveat regarding Johnson's authorship of "the greater part of that Preface" and also states that Johnson "corrected" all of it (I, iii). He is not likely to have published either views hostile to his own or to have made the grammatical error that Dr. Clark fancies he sees and fancies a disqualifier for Johnsonian authorship (*SJ*, p. 184).

Johnsonians on either side of the Atlantic long have known this Preface and have supported Nichols' final judgment: L. F. Powell does so in 1943 and cites Nichols' authority; Powell repeats the attribution in his edition of Boswell's *Life of Johnson*; Arthur Sherbo affirms it in 1962, and cites Alexander Chalmers' authority; E. L. McAdam, Jr. accepts and elaborates upon Powell in 1965; J. D. Fleeman, Johnson's distinguished late bibliographer, accepted the attribution as does O M Brack, Jr., editor of Johnson's forthcoming shorter prose.[32] There thus is some two hundred years of almost unanimous late contemporary recognition, perhaps even Johnsonian testimony, and later expert opinion that the General Preface to the 1753 *Gentleman's Magazine* is Samuel Johnson's.

The Jacobite scholars nevertheless need both to discredit the attribution and to repeat the inaccurate mantra of "*partial*," for Johnson's Preface bluntly rejects their views. The Preface itself is part of larger developments. It relates to Johnson's role in the Parliamentary Debates, his disillusion with the Patriot Opposition, and his recognition of Walpole's achievement and the stability of the Hanoverian dynasty. It also reflects his awareness of what by 1743 being a Jacobite meant—supporting a French invasion of Britain with foreign troops on behalf of two foreign monarchs and a foreign prince. These were Louis XV, his client the nominal Jacques III, and his son Charles whom the French called Monseigneur le Prince de Galles. That young man also was known as the Young Pretender, the Franco-Italian-Polish Charles Edward Louis Philip Casimir Stuart who claimed Britain as the birthright he planned to seize with a French army.

There were many reasons for Johnson's early and enduring antipathy to France. One relates to the years between 1743-46 when France threatened invasion and supported the Jacobite civil war. Johnson's view of the political world permanently changed as Jacobitism no longer was a matter for intellectual or political debate, but a matter with potentially terrible consequences. Britain was faced with a major expeditionary force: Catholic France planned violently to replace a Protestant monarch with Catholic Charles Edward on behalf of his father Catholic James domiciled in Rome since 1719. Whatever the Jacobite fantasies of liberating a grateful nation, Johnson knew that an attempt to return Britain to the pre-1688 world would mean dreadful bloodshed for the nation and perhaps for his own ancestral city and family: when Cumberland returned from Flanders he established headquarters in Lichfield, in which some 12,700 troops were quartered.[33] Moreover, a change in dynasty under such circumstances must mean long subordination of the British to the French crown and interests. Grumbling about apparent subservience to small and distant Hanover would pale in comparison.

The press fed relevant information to its readers. From late 1743, virtually each month's *Gentleman's Magazine* reports its own or others' confirmed fears that a Stuart-inspired French invasion was imminent.[34] There were frequent reports of Anglo-French naval engagements. Several religious sects, guilds, cities, and branches of government affirmed their loyalty in the face of "undoubted Intelligence of the Pretender's eldest Son being arrived in *France*, and of the Preparations which are making there to invade this Kingdom, in concert with disaffected Persons here" (14 [1744], p. 104). Such an invasion will first "lay waste our dear Country, and having made *England* a Scene of Blood and Devastation, then . . . reduce it to the wretched Condition of a *French* province" (XIV, 164; and see XIV, 519, 627).

The fears were based upon large French troop concentrations at Dunkirk under either the formidable Count Lowendahl or the yet more formidable Maurice Comte de Saxe, soon created a Marshall of France. These prominent foreign generals in the French service would defeat the Duke of Cumberland and his allies at the Battle of Fontenoy on 11 May 1745. In the meanwhile, the number of French troops was variously estimated as 15,000 (XIV, 79, 111), between 20,000 and 30,000 (XIV, 105), or nearly 20,000 (XIV, 583) plus supporting ships, muskets, and other materiel. Fear is a great multiplier. Fielding's *Tom Jones* (1749) is set during the '45. There an unreflective English Jacobite sings the correct number—"ten thousand honest *Frenchmen* are landed in *Suffolk*. Old *England* for ever!" Shortly thereafter Sophia Western's maid Honour magnifies the rumor, adding sexual necrophagy to French conduct: "several hundred thousands of them are landed, and we shall be all murdered and ravished."[35]

Fortunately for Honour's honor, the invasion itself was thwarted; but part of its planning-history has been recorded by the French miltary historian and archivist J. Colin. His *Louis XV et les Jacobites* (1901) transcribes enlightening letters between several of the lead actors in this sorry event encouraged by cardinals de Fleury and de Tencin, together with British and European Jacobites.[36] French archives demonstrate some of the responses and fantasies of, among others, the key English Jacobite, the nominal, Jacobite created, Francis Lord Sempill, the dreams of Jacques III and of his brutally self-centered son Charles Edward, the negotiations of the Marquis d'Argenson and, the hero of the near tragedy, Saxe himself. He was the best European general between Marlborough and Frederick the Great; by the age of thirteen he had fought in the appalling battle of Malplacquet (1709); and he understood that an invasion to overthrow a well-supported monarchy on its own soil meant an expensive butcher's bill. Britons did not know the specifics of the letters epitomized below; if only by assumption they would have known much of the letters' general contents.

Both the Old and Young Pretenders asserted that Britain would embrace them without a battle and that France will be rewarded for its trouble. On 22 May 1743 Jacques writes to Louis XV seeking his aid and assuring him that the Stuarts would be "une famille qui serait autant attachée à Votre Majesté que celle de Hanovre lui est contraire" (p. 14). On 15 March 1744 the angry Charles Edward complains to Sempill that the French should not have aborted the invasion whose consequences would have been "si avantageuses à la France." Once the Elector of Hanover had fled, Louis XV would be able "régler la paix d'l'Europe suivant ses justes désirs, puisqu'il n'est point douteux que le Roi mon père, rétabli sur le trône de ses ancêtres, n'agisse en cela et en toutes choses essentielles de concert avec un ami si généreux" (p. 178).

Sempill was an appropriate recipient for this visionary Francophile epistle. As a chief British architect of the invasion, he assured the French court that there were no more than 16,000 troops guarding all of the United Kingdom (p. 19)—a figure he later reduced by almost 10,000 (pp. 28-30); that an invasion of England would be easy; that the French can acquire horses and rations once they arrive on so accommodating a shore; and that they can land in Maldon, Essex, triumphantly drive south to disaffected London, and establish a Jacobite government. Once Jacques III masters London he masters England (pp. 21-22). So long as the invasion plans are kept secret, success is inevitable: the House of Hanover will quickly flee, the ministers will promptly submit, and best of all is "qu'il n'y aura pas un seul coup frappé, ni peut-être un seul homme detué dans l'affaire." Indeed, "En un mot, il n'y a rien qui puisse empêcher le succès de cette entreprise" (p. 27).

This destructively naive fantasy was ruined by several forces, including Louis XV's strategic realpolitik. The Jacobite scheme was foolish and secondary to French continental interests; but it was a diversion from which France would profit. On 10 December 1743 Louis writes to his ally, uncle, and Louis XIV's grandson, King Felipe V of Spain:

> le débarquement fait, la révolution ne soit pas aussi subite et aussi générale qu'on le promet; mais, le moins qui puisse en résulter, est une guerre civile qui rappellera les troupes anglaises de Pays-Bas. Les cours de Vienne et de Turin ne recevront plus de subsides de l'Angleterre, et, en agissant vigoureusement de toutes parts dans cette conjoncture critique, ces deux cours, abandonnés à leure propres forces, subiront la loi qu'on voudra leur imposer, pour peu qu'on ne veuille pas en exiger les conditions trop dures. (pp. 35-36)

Meanwhile, the British government was awaking from its slumber, had regular reports concerning French movements in Dunkirk, and had agents in Rome and elsewhere monitoring Jacobite activities. The French had their own agents, and Saxe received reports of British shore batteries and naval manoeuvres under Admiral Norris (pp. 109-10) and of British knowledge of the French Brest squadron sent to escort the troop transports at Dunkirk (pp. 112-13). Saxe was aware that there was no evidence of a British rebellion, that a secret invasion was impossible, that free Britons liked to slang one another without being seditious, that there were at least twice the 6,000-7,000 regular troops now claimed for Britain, that, Sempill's silence to the contrary (p. 118), there were between thirty and forty ships of the line arrayed against them in the English Channel, and that a large body of British troops was forming near London. The experienced and realistic Saxe knew what the Jacobites chose to ignore: if the British army and navy opposed their government they would have overturned it without a foreign invasion—against which they now were galvanized to fight.

Saxe politely informs his superiors that he will follow orders and, he trusts, will succeed; but pending disaster hovers behind the formulaic politesse. He makes and remakes several points, always aware of Jacobite myth and military fact. He will fail if the British do not welcome him, if his troops are badly blooded or diminished, and if they therefore should not leave their ships. Norris can blockade Dunkirk; if Saxe eludes him, Norris can destroy the French squadron and his transports at sea; if they traverse the Channel and reach the Thames, the British can block it; if they do not block it, the French nevertheless have no pilots and will be beached and defenseless; if they do not get to the Thames but are driven northward, presumably to Scotland, how can he feed 10,000 men? Saxe anticipated a catastrophe which he hoped to prevent and about which the British and European Jacobites were uncomprehending if not indifferent. Saxe preferred not to have his men martyred for le Prétendant. The maritime gods agreed.[37]

The storms that so punished Saxe's men deeply troubled him. Many without transports were soaked and ill as they stood on the docks; others on small canal boats (bilanders) also suffered serious health problems. The storm of 6-7 March 1744 killed numerous men and horses, broke masts, tore anchors and cables, beached ships, and ravaged supplies: "Les armes, les tentes, le bagage, je crois tout perdu. . . . Beaucoup de ces soldats et matelots vont tomber malades, parce que sortant de l'eau, ils ont été exposé longtemps au vent froid et très fort" (p. 145). The more Saxe saw of his inadequate naval cover and of Jacobite ignorance, the more concern for his men grew. Admiral de Barrailh's sailors, for example, "sont plus propres à conduire une charrette qu'à naviguer"; "Le prince de Galles ni les sienes n'en savent pas plus que nous" (p. 148). He frankly tells d'Argenson on 8 March "ce n'est pas en aveugle que je m'embarque dans cette affaire" (p. 149).

Saxe's franchise did not go unrewarded. The invasion was canceled and the troops sent to Flanders. On 9 March d'Argenson both so informs Charles Edward and describes the true situation in Britain: "toutes les nouvelles de Londres s'accordent à dire que la ville, le milord maire et le Parlement ont fait de nouvelles assurances de fidélité au roi Georges et je n'ai vent de nulle part que quelqu'un ait levé le masque en Angleterre" (p. 153). Saxe and d'Argenson avoided the French bloodbath and humiliation that the Jacobite inspired invasion would have entailed.

Charles Edward learned nothing. His letter to Sempill of 15 March 1744 but briefly alludes to the men who already had died and his own "affliction bien sensible" (p. 178). All evidence to the contrary, he still believes that the remaining transports and French squadron can elude or defeat Norris's flotilla (pp. 178-79); he still fantasizes that he will lead a great French army to victory (p. 181); and he resents not being paid more attention: "je suis ici tout à fait inconnu et . . . il n'y a pas le moindre soupçon à mon égard" (p. 179). He then constructs Plan B—a French-supplied invasion of Scotland, which will be easy and economical: "tous ces articles ensemble ne monteront pas à la dépense qu'il faudra faire pour la prise ou la défense d'une seule bonne forteresse en Flandres" (p. 180). As usual, Charles Edward ignores other mens' blood as part of the dépense of "une expédition en Ecosse" (p. 180). So much for the Jacobite notion that "il n'y a rien qui puisse empêcher le succès de cette entreprise." So much for J. C. D. Clark's notion that "the realities were that a French invasion had every prospect of success in 1744."[38] And so much for his other notion that "Charles Edward Stuart, acting on his own initiative" left for Scotland and started the '45 (*SJ*, p. 173)—which perhaps ought to be called the failed '44.

Let us now return to Samuel Johnson's General Preface to the 1753 Index to the *Gentleman's Magazine* and the sections uncongenial with Jacobitism or a kind of liking for it. Johnson says that there has been no more important period "in which Publick Events have been more numerous or more important" than during the last twenty-three years. Two major wars affecting much of the known world have begun and ended, and different nations' views and powers have been revealed.

Johnson then adds sentences that demonstrate his dramatic change from opposition in the late 1730s.

> With respect to our own Country, the deep and extensive designs of a Statesman, as able perhaps as any that ever existed, and the Opposition that was inflexibly maintained against him, have produced such a Series of Argumentation as has comprised all Political Science, and ascertained the Right of the Crown and the Privileges of the People, so as for ever to prevent their being confounded in the Cause either of Tyranny or of Faction. The Period will be render'd still more remarkable in English History by a Rebellion, which was not less contemptible in its beginning than threatening in its progress and consequences; but which, through the Favour of Providence, was crushed at once, when our Enemies abroad had the highest expectations from it, and has contributed to our greater security.[39]

This demonstrably Johnsonian text is neither Linear B nor Babylonian cuneiform. Its plain sense is that the vigorous and often crude public debates—in parliament, the press, and the coffee houses for example—were models of free political exercise that confirmed the powers of different constituencies. Tyranny and faction cannot shake this mature state—that is, not the Jacobites and their arbitrary monarchical government, nor the extreme Patriot Whigs who would have made governance impossible by the excesses of opposition. God Himself joins the fight to secure this "ascertained" nation against a powerful foreign enemy. The threatened French invasion, the '45 and its danger to the nation and to Lichfield, the disillusion with Walpole's hypocritical successors and of course the Parliamentary Debates all contributed to Johnson's rejection of earlier opposition. Johnson later joked that he did not let the Whig dogs get the better arguments in those debates; but his protagonists were Whigs and Johnson imaginatively engaged in a series of extensive exchanges. These indeed were lessons in political science that educated him regarding the strengths of noisy constitutional government.[40]

Dr. Clark and Professor Erskine-Hill misread these passages in order to explain them away. In *Samuel Johnson* Dr. Clark thus says that we here see "a cool appraisal of Walpole," of whom Johnson disapproves. The issues have been only "publicly aired, not constitutionally secured." As for providential design, that is "a cool comment on the extent of Hanoverian loyalty or the prowess of Hanoverian arms" (*SJ*, p. 184). Besides, Johnson probably did not write the compromising parts. There is no reason not to affirm my original judgment—this was and is Dr. Clark's "crudest exercise in denial" (*AJ* 7, p. 188). It also is a crude exercise in misreading: it regards God's approbation as a code for God's blame; and it regards Johnson's approbation of Walpole, who was "as able [a Statesman] perhaps as any that every existed," as a "cool appraisal."

5. The Pleasure of Parting

What, then, may we conclude after all these pages of ink, murdered trees, and well-exercised micro-chips? Johnson's politics were situational, but were based upon respect for the different rights and obligations of the subject, parliament, and crown. Invasion of any part of that constitutional balance was dangerous. Johnson himself was not a non-juror; he was not a Jacobite in any serious sense; he left Oxford because of poverty not because he refused to take oaths; upon matriculation he certainly subscribed loyalty to the Hanoverian government in the Oath of Supremacy and its implied oaths of Allegiance and Abjuration; he almost certainly took these as the law required in order to be an usher and teacher; he never believed in the divine right of kings; he certainly did not fight at Culloden or think of joining the French-led Jacobites if they invaded London; he did not refuse service in the Trained Bands to avoid oaths, but because he was half blind, half deaf, in poor health, probably at or near the maximum age, and was among the large majority of educated previous oath-takers who also hired surrogates. By the early 1740s he modified his earlier opposition politics; he became favorable to Walpole and hostile to the alien Stuart invader who offered bloody civil war and subservience to a foreign power. Given Johnson's comprehensive mind and respect for order and legitimacy, he nonetheless was aware that the Stuarts had a case, however modest, and not worth more than the heat of the intellectual political debate that essentially had settled the rights of the crown and the people by 1742 and Walpole's defeat. J. C. D.

Clark is entirely correct: "We have no evidence that Johnson ever said 'I am a Jacobite'" (*AJ* 8, p. 20).

I hope that this exchange is over and that all those involved will keep their piece, and peace, for nine years. Each side is likely to believe that it smote the other while taking slight blows on sturdy shields. The decision is not up to the combatants but to the scholarly audience who will evaluate evidence, read or reread Johnson and other relevant sources, and perhaps profit from the heat, and possibly light, generated from these pages. If that happens we can again agree upon God's strange but wonderful ways. Hope and Fear may indeed find their object.

APPENDIX: SOME RELEVANT DOCUMENTS CONCERNING JOHNSON AND JACOBITISM

THE OATH OF ALLEGIANCE

I A. B. do sincerely Promise and Swear, That I will be faithful, and bear true Allegiance to His Majesty King *George*. *So help me God*, &c.

II. THE OATH OF SUPREMACY

I A. B. do Swear, That I do from my Heart Abhor, Detest, and Abjure, as Impious and Heretical, that Damnable Doctrine and Position, *That Princes Excommunicated or Deprived by the Pope, or any Authority of the See of* Rome, *may be Deposed or Murthered by their Subjects, or any other whatsoever.*

And I do declare, That no Foreign Prince, Person, Prelate, State or Potentate, hath, or ought to have any Jurisdiction, Power, Superiority, Preeminence, or Authority Ecclesiastical or Spiritual within this Realm.

So help me God, &c.

III. THE OATH OF ABJURATION
(italics and roman type reversed)

I *A.B.* do truly and sincerely acknowledge, profess, testify, and declare in my Conscience, before God and all the World, That our Sovereign Lord King *George* is lawful and rightful King of this Realm, and all his Majesty's Dominions thereunto belonging. And I do solemnly and sincerely declare, That I do believe in my Conscience, that the Person pretended to be the Prince of *Wales*, during the Life of the late King *James*, and since his Decease, pretending to be, and taking upon himself the Stile and Title of King of *England*, by the Name of *James* the Third, or of *Scotland*, by the Name of *James* the Eight, or the Stile and title of King of *Great Britain*, hath not any right or Title whatsoever to the Crown of this realm, or any other the

Dominions thereto belong: And I do renounce, refuse, and abjure any Allegiance or Obedience to him. And I do swear, that I will bear true Faith and true Allegiance to his Majesty King *George*, and him will defend, to the utmost of my Power, against all Traiterous conspiracies and Attempts whatsoever, which shall be made against his Person, Crown, or Dignity. And I will do my utmost Endeavour to disclose and make known to his Majesty, and his Successors, all Treasons and Traiterous Conspiracies which I shall know to be against him, or any of them. And I do faithfully promise, to the utmost of my Power, to support, maintain, and defend the Succession, by an Act intituled, *An Act for the further Limitation of the Crown, and better securing the Rights and Liberties of the Subject*, is and stands limited to the Princess *Sophia*, Electress and Dutchess Dowager of *Hanover*, and the Heirs of her Body, being *Protestants*. And all these Things, I do plainly and sincerely acknowledge and swear, according to these express Words by me spoken, and according to the plain and common Sense and Understanding of the same Words, without any Equivocation, mental Evasion, or secret Reservation whatsoever. And I do make this Recognition, Acknowledgment, Abjuration, Renunciation, and Promise, heartily, willingly, and truly, upon the true Faith of a Christian.

So help me God.

IV. FROM *A Summary of the Penal Laws Relating to Nonjurors, Papists, Popish Recusants, and Nonconformists. And of the late Statutes Concerning the Succeeding Riots, and Imprisonment of Suspected Persons* (London, 1716)

94. To the Intent, all the Acts heretofore made for Securing the Succession of the Crown in the Protestant Line, and extinguishing the Hopes of the Pretender, might be for ever inviolably preserved, it was Enacted as follows:

That all Persons, as well Peers as Commoners, who shall bear Office Civil or Military, or receive any Pay, Salary, Fee, or Wages, by reason of any Patent or Grant from his Majesty, or shall have command or Place of Trust, from or under his Majesty, or his Predecessors, or by

Authority deriv'd from him, her, or them, within the Kingdom of *Great Britain*, in the Navy, or in the Islands of *Jersey* and *Guernsey*, or shall be of the Household, or in the Service [95] or Employment of his Majesty, the Prince or Princess of *Wales*, or their Issue; and all Ecclesiastical Persons, Heads, or Governours, and all other Members of Colleges and Halls in Either University, being of the Foundation, or who enjoy any Exhibition (being of the Age of 18.) And all Persons teaching or reading to Pupils in the Universities, or elsewhere; all School-masters, Ushers, Preachers and Teachers of Separate Congregations; every Constable, Serjeant at Law, Counsellor, Advocate, Attorney, Sollicitor, Writer in *Scotland*, Proctor, Clerk or Notary, who shall be within 30 Miles of *London* or *Westminster* the first Day of *Michaelmas* Term 1715. shall appear before the End of the said Term in the Courts of Chancery, King's Bench, Common Pleas or Exchequer, and take the several Oaths enjoin'd them by this Act....

102 The Oaths prescribed to be taken by this Act, are the Oaths of Allegiance and Supremacy, and the Abjuration Oath, for that Part of *Great Britain* called *England*; and in *Scotland*, only the Oaths of Allegiance and Abjuration, with the Assurance, which is to be subscribed by them in the Room of the Oath of Supremacy.....

V. Johnson's signature in subscribing
the Oath of Supremacy on 16 December 1728

"Term 1[t] Mick 1728": "Samuel Johnson è Coll: Pembr: Gen: Fil:" See Oxford University Archives (OUA) SP 12, from 8 October 1714-23 December 1740.

NOTES

1. For the relevant works in this exchange, see J. C. D. Clark, *Samuel Johnson: Literature, Religion and English Cultural Politics from the Restoration to Romanticism* (Cambridge: Cambridge University Press, 1994), hereafter referred to parenthetically in the text as *SJ* with following page numbers. The other works are Howard Erskine-Hill, "Johnson the Jacobite? A Response to the New Introduction to Donald Greene's *The Politics of Samuel Johnson*," *Age of Johnson*, 7 (1995), 3-26; J. C. D. Clark, "The Politics of Samuel Johnson," *AJ* 7, 27-56; Donald Greene, *AJ* 7, 57-135; Thomas Curley, "Johnson No Jacobite: Or, Treason Not Yet Unmasked," *AJ* 7, 137-61; and Howard D. Weinbrot, "Johnson, Jacobitism and the Historiography of Nostalgia," *AJ* 7, 163-211, from which I have recycled some information. All references are cited parenthetically in the text. I wrote the earlier review-essay without having seen any of the other contributions. Students of Johnson and of eighteenth-century intellectual and political history also should see John Cannon's *Samuel Johnson and the Politics of Hanoverian England* (Oxford: Clarendon Press, 1994). His Johnson, though still misunderstood and maligned by Americans (p. 6), emerges as the practical and moderate political thinker and a founder "of mainstream conservative thought" (p. 7).

   In the current exchange Dr. Clark warns: "I shall publish elsewhere an extended reply to Weinbrot on Johnson's classicism" (*AJ* 8, p. 60, n. 12). Caveat emptor. Dr. Clark also twice insists that Johnson's politics have "been established beyond a reasonable doubt" (*AJ* 8, pp. 15, 43) and cites his own discredited work as proof. That assertion has the evidentiary value of Henry Ford on the success of the Edsel and Saddam Hussein on the success of the Iraqi army.
2. *Gustavus Vasa, The Deliverer of his Country. A Tragedy* (London, 1739), pp. iv, [85]). For some discussion of the political force of the German connection, see Johnson's *Dictionary* (1755) definition of "King" and its use of Richard Verstegen's *Restitution of Decayed Intelligence* (1605) at the end of section 3 within. See also my "Politics, Taste, and National Identity: Some Uses of Tacitism in Eighteenth-Century Britain," in *Tacitus and the Tacitean Tradition*, ed. T. J. Luce and A. J. Woodman (Princeton: Princeton Univ. Press, 1993), pp. 168-84.
3. This originally was part of a Clark Library seminar by J. Douglas Stewart and H. W. Liebert, *English Portraits of the Seventeenth and Eighteenth Centuries* (Los Angeles: William Andrews Clark Memorial Library, 1974). I here quote from Liebert's expanded reprint, *Lifetime Likenesses of Samuel Johnson Reissued With Additional Plates . . . To Commemorate The Two Hundred and Sixty-fifth Birthday of Dr. Samuel Johnson* (New

York: Privately Printed for the Johnsonians, 1974), p. 52. See plates 4, 9, 10, pp. 76, 81, 82, for the other paintings. Anecdotes regarding Johnson's inability to see the stage are in Arthur Murphy's *An Essay on the Life and Genius of Samuel Johnson, LL.D.* (London, 1792) and in George Steevens' anecdotes. See *Johnsonian Miscellanies*, ed. George Birkbeck Hill, 2 vols. (Oxford: Clarendon Press, 1897), I, 157, and II, 318 respectively. Murphy's comment on Johnson's perennial ill health is in I, 199. For other comments on Johnson's poor eyesight or hearing, see *Johnsonian Miscellanies*, I, 319 (Mrs. Piozzi), II, 209-10 (Bishop Percy), II, 275-76, 292, 298 (Frances Reynolds), II, 343-44 (Thomas Tyers). Some time between 1778 and 1780 Johnson tells Edmund Hector that "My Health has been from my twentieth year such as has seldom afforded me a single day of ease." See *The Letters of Samuel Johnson*, ed. Bruce Redford, 5 vols. (Princeton: Princeton Univ. Press, 1992-94), V, 7. I suspect that the officer Benjamin Rackstrow would gladly have paid not to have such a man in his unit. Johnson may allude to his poor hearing and its role in oaths. See his "Remarks on the Militia Bill" (1756) at n. 18 within.

4. Linda Colley, *Britons: Forging the Nation 1707-1837* (New Haven: Yale Univ. Press, 1992), p. [377] for the map. Colley stresses that these actions before Culloden entailed risk and "should be taken seriously" (p. [376]).
5. Hawkins, *The Life of Samuel Johnson, LL.D.* (Dublin, 1787), p. 146.
6. See the unsigned *The Servitour: A Poem. Written by a Servitour of the University of Oxford, and Faithfully taken from his Own Original Copy, &c.* (London, 1709), p. 9, quoted, pp. 12, 14, 15 for the other references. See also *The Commoner, A Poem* (London, 1710) for the other references cited in the text. According to David Foxon this may have been "a piracy of an untraced edition" and was reissued in *A Collection of the Best English Poetry* (1717)—which I have not seen. The putative author is "Incognito." See Foxon's *English Verse 1701-1750* (Cambridge: Cambridge Univ. Press, 1975), I, 135.
7. *Boswell's Life of Samuel Johnson*, ed. George Birkbeck Hill, rev. L. F. Powell, 6 vols. (Oxford: Clarendon Press, 1934-50), I, 76-77. Hawkins contributes supporting details and notes the "pity" that Johnson's appearance evoked: "a gentleman of his college, the father of an eminent clergyman, now living, directed a servitor one morning to place a new pair at the door of Johnson's chamber." Johnson's "indignation" may also have stemmed from the gentleman's implicit insult by means of the servitor. See *The Life of Samuel Johnson*, pp. 9-10.

8. For these, see Hearne, *Remarks and Collections of Thomas Hearne*, ed. Oxford Historical Society (Oxford: Clarendon Press, 1902), VI, 300, and Hearne's manuscript *Collections*, XII, 192 as quoted in Anthony à Wood, *Athenae Oxoniensis*, ed. Philip Bliss, 3rd ed. (London, 1812), I, ii; Amhurst, *Terrae Filius* (1721), 3rd ed. (London, 1754), p. 14; Blacow, *A Letter to William King LL. D. Principal of St. Mary Hall in Oxford. Containing a particular Account of the Treasonable Riot at Oxford, in Feb. 1747*, 2nd ed. (London, 1755), p. 17; [Kennicott] *A Letter to Doctor King Occasion'd by his Late Apology* (London, 1755), p. 25. I have discussed these at greater length in "Johnson, Jacobitism, and the Historiography of Nostalgia," *AJ* 7, pp. 191-99.
9. *A Summary of the Penal Laws Relating to Nonjurors, Papists, Popish Recusants, and Nonconformists. And of the late Statutes Concerning the Succeeding Riots, and Imprisonment of Suspected Persons* (London, 1716), pp. 94, 102. See the Appendix for fuller references.
10. I am grateful to Simon Bailey, Archivist of the Bodleian Library, for his courtesy and efficient help with the Oxford University subscription registers. See the Appendix for fuller information.
11. Amhurst, *Terrae Filius*, p. 88. This is from a letter by "Philalethes," a Truthlover who may or may not be Amhurst himself.
12. I am indebted to Dr. Clark for identifying George Cooper as Hearne's examiner: *AJ* 8, p. 32.
13. *Oxford University Statutes*, trans. G. R. M. Ward (London, 1845), I, 10.
14. Allix, *An Examination of the Scruples of Those who refuse to Take the Oath of Allegiance. By a Divine of the Church of England* (London, 1689), p. 18.
15. See *Undergraduate Subscription. Extracts From A Collection of Papers Published in Oxford 1772 On the Subject of Subscriptions to the XXXIX Articles, Required from Young Persons at their Matriculation* (Oxford, 1835), p. 30.
16. *Terrae Filius*, p. 12. This is one of the references that suggest oath-taking as a communal and often casual experience.
17. This response was part of an extended debate regarding the efficacy of oath taking. One side argued that an oath taken with God as witness was a matter of the highest seriousness, the violation of which was virtual atheism and attempted deicide. The other side argued that frequent oath-taking so cheapened the process that it was unreliable and infamous on the face of it. Johnson appears to have recognized the validity of each argument, but stressed that oaths were largely deceitful and futile.

18. The Yale Edition of the Works of Samuel Johnson, vol. 10, *Samuel Johnson: Political Writings*, ed. Donald J. Greene (New Haven: Yale Univ. Press, 1977), pp. 158-59.
19. Johnson, *Lives of the English Poets* (1779-81), ed. G. B. Hill, 3 vols. (Oxford: Clarendon Press, 1905), II, 557.
20. *A Summary of the Penal Laws*, p. 94.
21. For whatever reason Dr. Clark cites his authority on school-masters' oaths from the second edition of Donald Greene's *The Politics of Samuel Johnson* (Athens: Univ. of Georgia Press, 1990); Greene, p. xxxi; Clark, *SJ*, p. 118, n. 57. In the text of his essay on Johnson's "cultural identity" he merely says that the statute included unnamed "others," without making plain that such a category included Johnson as teacher: *AJ* 8, p. 24.
22. I quote Johnson's poetry from *Samuel Johnson: Poems*, ed. E. L. McAdam, Jr., with George Milne (New Haven: Yale Univ. Press, 1964), *YE* 6, and cite future references in the text.
23. For another work hostile to France in a "French" voice, see *The Manifesto of a Certain Power* (London, 1744), in which "*Charles Edward Stuart* (Son of James Francis Stuart)" (p. 7) makes plain that he is France's surrogate and will destroy British culture as Britons know it: "no *British* Bosom shall presume to swell with the generous Glow of Freedom" (p. 18). See also *The Liberty of Britons, The Noblest Motive to Loyalty and Courage* (London, 1744), which includes the Manifesto and two other "French" speeches.
24. See his "The Political Character of Samuel Johnson" in *Samuel Johnson: New Critical Essays*, ed. Isobel Grundy (London: Vision Press, 1984), pp. 107-36, and "The Political Character of Samuel Johnson: *The Lives of the Poets* and a Further Report on *The Vanity of Human Wishes*," in *The Jacobite Challenge*, ed. Eveline Cruickshanks and Jeremy Black (Edinburgh: John Donald, 1988), pp. 161-76. Professor Erskine-Hill there draws our attention to Edward Young's "Some Thoughts, Occasioned by the Present Juncture: Humbly Inscribed to His Grace the Duke of Newcastle." This poem was appended to *The Consolation* (1745), the eighth of Young's *Night Thoughts* and refers to Charles Edward as "A *Gothic* Hero rising from the *Dead*." Young observes: "How must *This* Strike a Horror thro' the Breast,/Thro' every *generous* Breast, where *Honour* reigns?" (p. 136). If Young's Charles XII "reminds" one of Charles Edward the prompt is wholly negative. Murray Pittock supports the Jacobite Johnson. See his *Poetry and Jacobite Politics in Eighteenth-Century Britain and Ireland* (Cambridge: Cambridge Univ. Press, 1994), pp. 126-32, especially 130-31: Johnson is "a nonjuror of robust and

provocatively pro-Stuart views" (p. 128); Charles Edward's fall is "among the vain wishes uppermost in his reader's mind" (p. 129). For other relevant discussions, see Herbert G. Wright, "Some English Writers and Charles XII," *Studia Neophilologica*, 15 (1942-43), 105-31, and James Gray, "Johnson's Portraits of Charles XII of Sweden," in *Domestick Privacies: Samuel Johnson and the Art of Biography*, ed. David Wheeler (Lexington: Univ. Press of Kentucky, 1987), pp. 70-84, 173-77.

25. Pope, The Twickenham Edition of the Poems of Alexander Pope, *An Essay on Man*, ed. Maynard Mack (London and New Haven: Methuen and Yale Univ. Press, 1950), lines 220-22, p. 148; *Samuel Johnson: The Idler and The Adventurer* ed. W. J. Bate, John M. Bullitt, and L. F. Powell (New Haven: Yale Univ. Press, 1963), *YE* II, 433.
26. See the review of the *Memoirs* in the *Literary Magazine*, 1 (1756), as in *The Works of Samuel Johnson, LL.D* (Oxford, 1825), VI, 12. For comparable references, see my "Johnson, Jacobitism, and the Historiography of Nostalgia," *AJ* 7, pp. 187-88.
27. Gerrard, *The Patriot Opposition to Walpole: Politics, Poetry, and National Myth, 1725-1742* (Oxford: Clarendon Press, 1994), pp. 231-32. Dr. Clark cites this book in support of an enduring Jacobite-Johnson (*AJ* 8, p. 68, n. 107), but as with my own work and with Robert DeMaria, Jr.'s, again misreads a source he claims to support his position. He argues that Brooke's *Gustavus Vasa* and Paterson's *Arminius* are part of the Anglo-Latin Tory tradition (*SJ*, pp. 166-67, 156); Dr. Gerrard more accurately sees them as a "flagrantly political version of anti-Augustan Gothic liberty" (p. 114).
28. See, if you can, *Samuel Johnson's Preface to the 1753 General Index to the First Twenty Volumes of the Gentleman's Magazine. Presented as a Keepsake for the Johnson Society of the Central Region* (East Lansing: Michigan State Univ. Press, 1987), p. [1]. See Nichols' Preface to the *General Index to The Gentleman's Magazine, From the Year 1787 to 1818*, 3 (1821), xlviii. The first putative "partial" attribution is on p. [xlv]. As late as 1988 Professor Erskine-Hill referred to this Preface as "only recently republished" in the Sherbo keepsake and was unaware that it had been in the Johnson canon since 1807, and had been reprinted in 1943. See the following discussion of that history, and Howard Erskine-Hill, "The Political Character of Samuel Johnson," p. 162.
29. Boswell's *The Life of Samuel Johnson*, ed. Edward [*sic*] Malone (New York, 1807), I, xii; Chalmers, Boswell's *The Life of Samuel Johnson LL.D.* (London, 1822), I, xxxv. The R. B. Adam library included Chalmers undated unaddressed autograph note saying:

from internal evidence—etc." See *The R. B. Adam Library Relating to Dr. Samuel Johnson and His Era* (Printed for the Author in Buffalo, N.Y.; Oxford: Oxford Univ. Press, 1929), III, 56. Chalmers' knowledge of Johnson may have been fertilized by his friendship with John Nichols. I thank Dr. Bonnie Ferrero for advice regarding Chalmers.

30. Nichols, *Literary Anecdotes of the Eighteenth Century* (London, 1821), V, 15. This volume had long been in process and in press. See I, ix-x.

31. James M. Kuist, *The Nichols File of "The Gentleman's Magazine": Attributions of Authorship and Other Documentation in Editorial Papers at the Folger Library* (Madison: Univ. of Wisconsin Press, 1982), p. 82.

32. See L. F. Powell, "An Addition to the Canon of Johnson's Writings," *Essays and Studies by Members of the English Association*, 28 (1943), 38-41; Powell, *Boswell's Life of Johnson*, VI, 189; E. L. McAdam, Jr., "Johnson, Walpole, and Public Order," in *Johnson, Boswell and Their Circle: Essays Presented to Lawrence Fitzroy Powell In Honour of His Eighty-Fourth Birthday* (Oxford: Clarendon Press, 1965), pp. 93-98; Arthur Sherbo, "Samuel Johnson and the *Gentleman's Magazine*, 1750-1755," in *Johnsonian Studies*, ed. Magdi Wahba (Cairo: Distributed Outside the U.A.R. by the Oxford Univ. Press, 1962), pp. 143, 144 (a tentative reservation). The Powell attribution is starred as an item of special importance in *Samuel Johnson: A Survey and Bibliography of Critical Studies*, ed. James L. Clifford and Donald J. Greene (Minneapolis: Univ. of Minnesota Press, 1970), p. 286, no. 25:44. Gough also was part ofthe Nichols circle. I continue to suspect that the Johnson-Nichols friendship is the ultimate source of the attribution. See *Literary Anecdeotes* I, xii for Gough's importance to Nichols. See also Nichols' praise of Gough and Gough's bequest of books to Nichols in *A Catalogue of the Entire and Valuable Library . . . of Richard Gough* (London, 1810), pp. iv, vi, xv, xviii. I am indedebted to Professor O M Brack, Jr., for his examination of the late, and sorely missed, J. D. Fleeman's bibliographical entry regarding the Preface: "attributed to SJ by [?Richard Gough], cf. Kuist, Nichols File (1982), 82a, and adopted by O M Brack, Shorter Prose."

33. For contemporary comment see [Richard Rolt?], *Historical Memoirs of His Late Royal Highness William Augustus, Duke of Cumberland. Including the Military and Political History of Great-Britain, during that Period* (London, 1767), pp. 304-306.

34. We should remember James M. Kuist's observation regarding the importance of *The Gentleman's Magazine*: it was "one of the most widely circulating publications of the English-speaking world." See his *The Nichols File*, p. vii. I cite references to the *GM* parenthetically in the text by volume and page.

35. *The History of Tom Jones. A Foundling*, ed. Martin C. Battestin and Fredson Bowers (Middletown, Conn: Wesleyan Univ. Pres, 1987), pp. 578 (11:2) and 593 (11:6).

36. J. Colin, *Louis XV et les Jacobites: le projet de débarquement en Angleterre de 1743-1744* (Paris: Librarie Militaire R. Chapelot, 1901). All references are cited in the text. The French continued to contemplate, or appear to contemplate, an invasion, the proclamation for which in 1746 had been written by the compliant Voltaire. This was printed in Voltaire's unsigned *Commentaire historique sur les oeuvres de l'auteur de la Henriade* (Basle [=Geneva], 1776), and translated as *Historical Memoirs of the Author of the Henriade* (London, 1777). This is reprinted in the Appendix to Theodore Besterman's *Voltaire* (New York: Harcourt, Brace and World, 1969). See Voltaire's "MANIFESTO *Of the King of France in Favour of Prince Charles Edward*": Louis XV "sends the Duke of Richelieu at the head of his army, only because the English, of the purest intentions, have solicited this assistance; and he sends only the precise number of troops which have been demanded, because he will be ready to withdraw them as soon as the English nation shall require it" (Besterman, p. 566). Presumably, this generous act was to have taken place before Culloden, when "a revolution was expected" (p. 565). Richelieu promised all Britons the protection of Louis XV, and was "to shed his blood for their service" (p. 566). He reconsidered.

The *DNB* article on Charles Edward, records this exchange between Cardinal de Tencin and Le Prince de Galles: "He was informed by Cardinal Tencin that Louis might be induced to grant help [in subjecting Britain on the Stuarts' behalf] on one condition. 'And that condition?' eagerly asked the prince: 'That Ireland be ceded to France,' replied the cardinal, 'as a compensation for the expense the court at Versailles must necessarily be put to.' The prince rose angrily from his seat and cried out, "Non, Monsieur le cardinal, tout ou rien! point de partage! point de partage!" (IV, 10).

37. Modern Jacobite scholars argue that one cannot legitimiately expect eighteenth-century Jacobites to have died or to have suffered property damage to show their zeal for revolution. Apparently it is appropriate for them to expect 10,000 Frenchmen to die for them. The latest such arguments on the proper pacifism of revolutionaries are in Murray G. H. Pittock, *Poetry and Jacobite Politics*, p. 3, and Eveline Cruickshanks, *TLS* 8 September 1995, p. 17. These commentators may be puzzled by the final words of the American Declaration of Independence: "with a firm reliance on the protection of divine Providence, we mutually pledge to each other our Lives, our Fortunes and our sacred Honor." Compare

Colin's observation: "les jacobites étaient de singuliers conspirateurs, à en juger . . . par la répugnance de bon nombre d'entre eux à quitter leur foyer et monter à cheval en hiver pour leur Roi légitime. Le dévouement, le courage, et la discipline semblaient n'exister que chez quelques membres du parti" (p. 122).

38. Clark, *English Society 1688-1832: Ideology, Social Structure and Political Practice During the Ancien Regime* (Cambridge: Cambridge Univ. Press, 1985), p. 173. Dr. Clark may imply that such an invasion should have taken place. See his "On Moving the Middle Ground: The Significance of Jacobitism in Historical Studies," in *The Jacobite Challenge*, pp. 180-81, and his discussion of "the penalties of an opportunity missed" to overthrow "Hanoverian repression."

39. For easier further examination, I quote from the reprinted Preface in L. F. Powell's "An Addition to the Canon of Johnson's Writings" in *Essays and Studies*, 28 (1943), here 39-40.

40. Thomas Kaminski has an excellent discussion of the debates and Johnson's supposed bias. See *The Early Career of Samuel Johnson* (New York: Oxford Univ. Press, 1987), pp. 123-43, especially pp. 132-39, which argues that the "fairness of the debates can hardly be disputed" (p. 139).

JOHNSON NO JACOBITE; OR, TREASON NOT YET UNMASKED: PART II, A QUOTABLE REJOINDER FROM A TO C

THOMAS M. CURLEY

A. Two Epigraphs

They are both men of integrity where no factious interest is to be promoted, and both lovers of truth, when they are not heated with political debate. Samuel Johnson, *Idler* 10 (1758)

The editor, or author, never could shew the original; nor can it be shewn by any other; to revenge reasonable incredulity, by refusing evidence, is a degree of insolence, with which the world is not yet acquainted; and stubborn audacity is the last refuge of guilt. . . . To be ignorant is painful; but it is dangerous to quiet our uneasiness by the delusive opiate of hasty persuasion.

Samuel Johnson, on James Macpherson's non-existent primary evidence of the manuscript of an ancient Scottish epic poem in *A Journey to the Western Islands of Scotland* (1773)

B. A Quotable Rejoinder

Thomas M. Curley: "Johnson No Jacobite; or, Treason Not Yet Unmasked" (pp. 151-52): "There appear to be two overarching weaknesses in the argumentation of Clark and Erskine-Hill. [1] The first is their failure to describe and differentiate adequately their political and religious terminology. . . . [2] As I have already stressed, the second major weakness of both essays is, I think, fatal to their case. They skimp on evidence taken from Johnson's own writings and often skew their interpretations of the few primary sources consulted therein to fit a preconceived thesis or idée fixée about Johnson's alleged Jacobitism."

[1] Howard Erskine-Hill, "Johnson the Jacobite?" (pp. 10-11): "The topics of the poem [Johnson's *London*] are not 'the merest commonplaces' of opposition propaganda. What was commonplace was to attack Walpole as the corrupt manager of a corrupt system, but not *to attack the king on the throne. . . . The latter is crypto-Jacobite* [italics mine]. . . ."

[1] Howard Erskine-Hill, "A Kind of *Liking* for Jacobitism" (p. 10): "A Jacobite, one may suggest, was a person who sought to serve some major religious, political or personal interest *through the restoration* [italics mine] of the *de jure* royal line [of the Stuarts].

[1] J. C. D. Clark: "The Politics of Samuel Johnson" (pp. 44-45): "Hostility to Walpole is not in itself sufficient proof of Jacobitism: this would otherwise convict members of the Whig opposition like Pitt, Pulteney, Carteret, and Sandys, whose Hanoverianism is not in doubt. *Yet part of the uneasy coalition between Tories and Opposition Whigs did take their opposition to the length of hostility to the dynasty* [of Hanoverian George II], *and some Whigs, from Marlborough to Chesterfield, could always be found to treat a restoration* [of the Stuart dynasty] *as a legitimate move in the political game* [italics mine]."

[2] J. C. D. Clark, "The Cultural Identity of Samuel Johnson" (p. 3): "The politics of Samuel Johnson have now been *established, beyond reasonable doubt,* [italics mine], by the normal processes of literary and historical scholarship. . . ."

[2] J. C. D. Clark, "The Cultural Identity of Samuel Johnson" (p. 20): "We have no evidence that Johnson ever said 'I am a Jacobite': such

a statement might meet Greene's and Curley's incantatory demands for 'incontrovertible evidence' and for 'irrefutable proof. . . .'"

[2] J. C. D. Clark, Introduction (p. 9) to *Literature, Religion and English Cultural Politics from the Restoration to Romanticism* (1994): "But we have no records of his conversation on this point, and *our estimate of* [Johnson's Jacobitism] *must always be circumstantial* [italics mine]."

[2] Howard Erskine-Hill, "A Kind of *Liking* for Jacobitism" (pp. 4): "I think *The Vanity of Human Wishes* may mark the end of *serious* [italics mine] Jacobite inclination in Johnson; but on some future occasions he would still write and talk like a sometime-Jacobite sympathizer."

[2] Howard Erskine-Hill, "A Kind of *Liking* for Jacobitism" (p. 10): "Johnson was not an all-or-nothing Jacobite—*nobody has argued that he was* [italics mine]."

[2] J. C. D. Clark, Introduction (p. 7) to *Literature, Religion and English Cultural Politics from the Restoration to Romanticism* (1994): "Johnson was a Jacobite."

C. Johnson Without Boswell or Hawkins: Some Primary Evidence for Johnson's Cool Opinion of Jacobites and Stuarts

[1] Samuel Johnson, *The Vanity of Human Wishes* (1749), ll. 29-34, on the harsh British treatment of aristocratic pro-Stuart *traitors* for supporting the failed uprising of 1745-46:

> Let Hist'ry tell where rival Kings command,
> And dubious Title shakes the madded Land,
> When Statutes glean the Refuse of the Sword,
> How much more safe the Vassal than the Lord,
> Low skulks the Hind beneath the Rage of Pow'r,
> And leaves the *bonny Traytor* [italics mine; MS 1749] in the
> *Tow'r. . . .*

[2] Samuel Johnson on James II, the last Stuart king of England, in *Introduction to the Political State of Great-Britain. 1756*: "He was not

ignorant of the real interest of his country; he desired its power and its happiness, and thought rightly, that there is no happiness without religion; but he thought very erroneously and absurdly, that there is no religion without popery. When *the necessity of self-preservation had impelled the subjects of James to drive him from the throne* [italics mine], there came a time in which the passions, as well as interest of the government [under William III], acted against the French [England's great national rival, with whom Charles II had felt inclined to be friendly and with whom James II had no time to quarrel because of his unhappy British subjects]. . . ."

[**3**] Johnson on foolish political partisanship in *Idler* 10 (1758): "Of all kinds of credulity [defined in the first paragraph as 'confidence of opinion too great for the evidence from which opinion is derived'], the most obstinate and wonderful is that of political zealots; of men, who, being numbered, they know not how nor why, in any of the parties that divide a state, resign the use of their own eyes and ears, and resolve to believe nothing that does not favour those whom they profess to follow. . . .

"Yet so it is that an Idler meets every hour of his life with men who have different opinions upon every thing past, present, and future; who deny the most notorious facts, contradict the most cogent truths, and persist in asserting to-day what they asserted yesterday, in defiance of evidence, and contempt of confutation.

"Two of my companions, who are grown old in idleness, are Tom Tempest and Jack Sneaker. . . .

"Tom Tempest is a steady friend to the House of Stuart. He can recount the prodigies that have appeared in the sky, and the calamities that have afflicted the nation every year from the Revolution, and is of opinion, that if the exiled family had continued to reign, there would have neither been worms in our ships nor caterpillars on our trees. He wonders that the nation was not awakened by the hard frost to a revocation of the true king, and is hourly afraid that the whole island will be lost in the sea. . . . In the succeeding reigns [after those of William III and Queen Anne] all has been corruption, malice, and design. . . . Many and many a time has Tom told me, in a corner, that our miseries were almost at an end, and that we should see, in a month, another monarch on the throne; the time elapses without a revolution; Tom meets me again with new intelligence, the whole scheme is settled, and we shall see great events in another month.

"Jack Sneaker is a hearty adherent to the present establishment. . . . He considers a standing army as the bulwark of liberty, thinks us secured from corruption by septennial parliaments, relates how we are enriched and strengthened by the electoral dominions [of Hanover], and declares that the public debt is a blessing to the nation. . . . He cannot believe that the nonjurors are so quiet for nothing, they must certainly be forming some plot for the establishment of Popery; he does not think the present oaths sufficiently binding, and wishes that some better security could be found for the succession of Hanover."

[4] Samuel Johnson in Scotland to Mrs. Thrale, 24 September 1773, on Jacobite Highlanders supportive of Prince Charles Edward Stuart, called the (Young) Pretender: "You may guess at the opinions that prevail in this country, they are however content with fighting for *their* king, they do not drink for him, we had no *foolish* healths [italics mine]."

[5] Samuel Johnson to Sir Robert Chambers, 19 April 1783, more than two decades after the accession of Hanoverian George III: "The state of the Publick, and the operations of government have little influence upon the private happiness of private men, nor can I pretend that much of the national calamities [e.g., the successful American Revolution] is felt by me; yet I cannot but suffer some pain when I compare the state of this kingdom, with that in which we triumphed twenty years ago [on the eve of Johnson's secret collaboration with Chambers on *A Course of Lectures on the English Law* loyally surveying the British constitution]. *I have at least endeavoured to preserve order and support Monarchy* [under the current Hanoverian dynasty; italics mine]."

DR. CLARK AND MR. HOLMES: SPECULATION IN JOHNSONIAN BIOGRAPHY

MICHAEL CALDWELL

Without question, Jonathan Clark is one of the most controversial British historians working today. One can trace a flurry of exchanges between Clark and other members of the historical profession following the publication of almost every one of his works.[1] Clark continues to provoke interest and controversy with his latest book on Samuel Johnson's politics.[2] On this occasion, however, the controversy takes place on literary, or (to be more precise) literary-historical ground, and the combatants include literary scholars as well as historians.

The latest quarrel—between Clark and Howard Erskine-Hill on the one side, and Donald Greene and much of the Johnsonian literary establishment on the other—has produced, for this reader at least, equal parts heat and light. Greene's fear that his opponents' views "might be harmful to the welfare of Johnson studies" notwithstanding, the argument has benefited students of Johnson by raising questions of evidence, by debating historiographical issues of great moment, and by refining and clarifying our notions of Johnson.[3] The question of Johnson's alleged Nonjuring Jacobitism has injected an element of vigor and urgency into Johnsonian studies. And the ensuing discussions carried forth in journals, conferences and reviews have certainly added to the stock of knowledge we possess regarding Johnson and his time.

One could wish however, that the debate were less acrimonious. The most recent exchanges in the previous number of this journal extend the argument in a tone of (what seems to me) increasingly personal hostility. While the issues discussed in this debate are surely important, lying at the very heart of the eighteenth century's political character, one would like to see more attempts at conciliation and friendly exchange such as occur in the first paragraphs of Erskine-Hill's article.[4]

Perhaps one way to bypass some of the acrimony permeating this debate would be to shift the terms of the discussion slightly by asking a new question of Clark's work. Regardless of the relative merits and deficiencies of Clark's larger argument, can we still attach value and importance to his speculations that Johnson was a Nonjuring Jacobite? Note that asking such a question says nothing about whether such speculations are accurate. For it may be the case that Clark's hypotheses regarding Johnson's political character have enduring value *even should they be proven completely false*. In what follows, I will argue that how Clark carries out his speculations—secretly or configured as positive declarations, in an absence of engagement with previous scholarship, without due caution or reserve, based too often on negative evidence and insensitive readings of literature—vitiates whatever force they may have once had. But even in this failure to speculate well, Clark performs a valuable service by giving Johnsonian scholars new ideas and perspectives to which they may react. If only in that sense, Clark's book has both value and importance.

As anyone who has taught a literature course knows, speculation can be a valuable pedagogical tool. Teachers can place students on the horns of a real dilemma—why *does* Hamlet delay?—in order to elicit discussions of problems that continue to exercise the imaginations of literary scholars. And what can serve the pedagogue's need to excite and engage students, can also serve the scholar's desire to fill or otherwise account for historical *lacunæ*. But just as a teacher recognizes that many student guesses will not bear fruit, so also does the scholar know in advance that speculation does not always solve problems. For this reason, while they do not expect every guess or hypothesis to be verified, many scholars work (and sometimes publish) in the hope that tentative hypotheses can attract the kind of support or refutation which in itself produces new knowledge. One need think only of W. W. Greg's and John Dover Wilson's classic exchanges over the text of *Hamlet* or (closer to home) Greene's numerous skirmishes with various Johnson scholars over the years to realize that this is so.

The reason I raise this link between pedagogy and scholarly practice is because I believe we need to be reminded at times that all new knowledge begins in query and hypothesis, trial and error, and that the mere act of conjecturing is not, by itself, a bad thing.

By the same token, idle speculation, guesswork hazarded in the absence of an awareness of and respect for previous labor in a field on the part of professional historians and scholars is a bad thing. For while the rash assumptions of overzealous students can be excused and gently corrected, the insufficiently researched and hastily assembled hypotheses of our colleagues usually draw stronger reactions. We rightly demand cognizance of and engagement with the "giants" of the field for even the tamest emendation to the most feeble and tottering tenet. And in a case where one argues for incendiary revisions to long-held and seemingly unassailable orthodoxies, we demand a good deal more. So, if one should, say, propose new theories about Johnson's membership in the Nonjuring community or his allegiance to deposed Stuart monarchs, one ignores Clifford, Greene, Bate, Kaminski and De Maria (among others) at one's peril.

One way to measure how well Clark's speculations engage with existing Johnson scholarship is to contrast his efforts with those of another recent writer on Johnson—Richard Holmes. Holmes' work, *Dr. Johnson and Mr. Savage*, appeared in 1993, one year before Clark's book.[5] In his account, Holmes rereads Johnson's early life in light of Johnson's friendship with the poet Richard Savage. Holmes subjects Johnson to a fair amount of psychologizing, finding in Johnson's relationship with the cantankerous, self-declared bastard of Earl Rivers and Lady Macclesfield a potential key to the young Johnson. Whatever one's misgivings about projects which seek to explain Johnson by reference to his psyche, it is difficult to deny that Holmes has startlingly reinterpreted some old data. I believe that the chief reason one hears Holmes out and (at times) grants him to have compellingly altered our understanding of Johnson is that he acknowledges the speculative nature of his enterprise and because he triangulates his new theories with respect to the existing orthodoxies of current Johnsonian scholarship. In what follows, I am less concerned with vindicating Holmes's individual readings than I am with showing how his speculative activity sets a kind of standard that Clark is either unwilling or unable to approach.

Take for example Holmes's opening gambit. In the early pages of his text he imagines Johnson to have died in the summer of 1749, and produces a fictional obituary of Johnson:

> Mr Samuel Johnson: poet, playwright and sometime political journalist. News has just reached us in the London mails, of the untimely death of Mr Samuel Johnson, late of the city of Lichfield, at his house at Gough Square, Fleet Street, London, of a sudden Apoplexy. (p. 10)

The exercise goes on for some time, describing what would have been the noteworthy facts of Johnson's life had he not lived long enough to produce his major works:

> Disappointed in his literary Prospects, disheartened in his political Hopes, and perhaps embittered by his personal Affairs (but we make no Windows into the human Heart), Mr Johnson threw himself at last into the harmless drudgery of his English Dictionary, for which he undertook a contract in 1746. . . . (p. 11)

This fictional obituary, designed to suggest to the reader how unfamiliar Boswell's paternal moralist would be to those who knew only the young Johnson, is frankly acknowledged as an imaginative exercise (Holmes calls it a "fiction"). While Holmes's method may seem a little unorthodox to more traditional Johnsonians, his aim is one shared by scholars like Greene. What concerns me here is not so much the effectiveness of this experiment (though I believe it has merit, particularly paired as it is with another fictional obituary of Savage) as the gesture's honesty: Holmes does not pretend to be doing "history" when he compiles his obituary. He sees himself engaged in a process of taking the reader back to a pre-Boswellian time, to a Johnson "embittered by his personal Affairs" and thrown into "harmless drudgery," to a Johnson disturbingly similar to Savage.

The openness with which Holmes acknowledges his fictional obituary is typical of his work generally. In his final chapter, titled "Charon," Holmes talks about his motives and hopes in embarking on yet another excursion into the murky waters of the Johnson-Savage relationship. He makes it clear that he sees the biographer in this instance as a necessary medium, who conjures forth the traces of what he calls "an invisible friendship":

> This book is not a complete biography of either writer, but the fragment of two lives brought together for a period of some two years. I have tried to approach this central period of intimacy—that so puzzled Hawkins and Boswell—from a number of different angles, to throw as much light as possible on what is really an astonishing lack of solid evidence or definite fact. (p. 228)

Holmes modestly admits that he has written a "fragment"—as if his book were a kind of extended entry in *Notes and Queries*. Moreover, this admission is accompanied by the only possible justification for speculation a biographer or scholar could offer: a "lack of solid evidence." In addition, Holmes confesses his abdication of a linear narrative here, in favor of a number of different temporal and historiographical perspectives on Johnson and Savage. Such admissions only enhance a reader's impression of Holmes's trustworthiness. Like other more traditional readers of the Johnson-Savage relationship, Holmes may guess from time to time; like the best of them, he confesses where he extrapolates from slender evidence.

A good example of how Holmes limits his speculations, particularly when they are based on negative evidence, appears in his discussion of the strange fact that though it seems almost certain that Johnson and Savage spent time together in Richmond in the spring of 1738, there are no recorded instances of their meeting. Reading the appointment book of Thomas Birch, an antiquary who knew and socialized with all the members of the literary circle who were in Richmond at that time —Johnson, Savage, James Thomson, Eliza Carter, and David Mallet—Holmes notes that Johnson and Savage freely mixed with various persons within this circle, but never with each other. On the basis of this negative evidence, Holmes wonders whether Savage was "deliberately keeping Johnson to himself" (p. 190). If so, it might explain how Johnson alone among the *literati* was privy to Savage's late-night, carousing, cursing-the-Walpole-establishment side. Yet Holmes is careful not to speculate on the basis of the fact that no one can show that Johnson and Savage did meet, admitting that "it is difficult to gauge, on such slight evidence, the exact shifting nuances of their social relation in 1738" (p. 190).

More importantly, Holmes acknowledges his use of scholars who have previously worked on Savage and Johnson. He freely admits his debts to Clarence Tracy's work on Savage; the monographs of Clifford, Bate, Kaminski and Wain on Johnson; and a pair of books that usefully describe the early eighteenth-century milieu of Savage and Johnson: Pat Rogers's study of Grub Street and Maynard Mack's biography of Pope. His identification of Savage as the model for Thales neatly demonstrates his willingness to engage the work of previous writers and politely dissent when he differs from them. The key for Holmes, as it has been for every writer who has attempted to treat the Johnson-Savage relationship, is the date of these two writers' first encounter. Building on the work of Clifford's *Young Sam Johnson* and A. D. McKillop's edition of James Thomson's letters, Holmes pins down July 1737 as the earliest possible date for a meeting. A quick reference to Johnson's letters places Johnson in Greenwich in July 1737. Holmes then uses Savage's own poetry to demonstrate that Savage preferred to spend the warm, humid months out of London, in Greenwich. This resourceful use of existing scholarship allows Holmes to speculate that Savage and Johnson probably first met in Greenwich at this time. Holmes finds corroborating evidence for this speculation in *London*, when the poem's speaker recalls how "On Thames's Banks, in silent Thought we stood,/Where Greenwich smiles upon the silver Flood."

A further measure of his respect for these scholars appears in Holmes's cautious dissent from their positions from time to time. For example, Holmes takes issue with Kaminski's defense of Boswell's dismissal of Savage as the inspiration for Thales. This dissent, framed largely as a rejection of Boswell, is offered in a spirit of moderation and respect. Holmes is not as concerned with crushing two hundred years' worth of energetic biographical denial (beginning with Boswell) that Savage was Thales, or with scoring points off a respected, contemporary biographer of Johnson, as he is with reinstating the possibility that Savage was the model for Thales and with demonstrating how that possibility makes clear a number of other unclear matters in the Johnson-Savage relationship.

His evidence for Savage as the inspiration for Thales comes primarily from a sensitive reading of *London*. Like Clifford before him, Holmes finds echoes of Savage's work within the poem.[6] The speaker's opening salute to "the Hermit" is taken as a reference to the Hermit figure in Canto I of *The Wanderer.* The same passage, according to Holmes, contains an allusion to Savage's fourth "volunteer laureate" poem, of 1735. Perhaps the strongest evidence produced by Holmes, however, is the striking moment in *London* (ll. 65-72) where Thales expatiates against a number of London evils, evils which Holmes rightly points out concerned Savage:

> the perjured witnesses at his trial; the plucking of his title "Volunteer Laureate" from Cibber; the mockery of Sir Robert Walpole's policies; the farce of the royal pension; and even perhaps the social agonies of dining obsequiously with friends of Lord Tyrconnel. (p. 179)[7]

The syntax of the couplets which give rise to this allusion is rather difficult, but Holmes's previous experience as a reader of Shelley's poetry (the Romantic poet who arguably worked in more poetic genres than any other) stands him in good stead here.

Lest it be thought that Holmes's hypotheses are merely the amateurish guesses of an unwanted interloper in Johnsonian studies, a quick glance back at Clifford's book will show how closely Holmes's speculations resemble more traditional Johnson scholarship. Recall that in his book Clifford discusses the possibility that Savage was the model for Thales. Despite the fact that he sees many similarities, Clifford cites Savage's departure for Wales in 1739—a year after *London*'s publication—as the main problem with seeing Savage as a model for Thales. Nevertheless, Clifford "hazard[s] a guess at what happened," finding it possible that Johnson "either consciously or unconsciously identified the hero of his poem with another romantic but unhappy patriot, whose story he had often heard." Clifford even speculates that Savage may have discussed a country retirement before 1739.[8] By speculating as to the identity of Thales through a sensitive and contextual reading of *London*, all Holmes has done is follow in the footsteps of more traditional scholars like Clifford.

Taken together, Holmes's forthrightness about his speculative activity, his skeptical use of negative evidence, his willingness to engage with the scholarly tradition, his measured dissent from that tradition, and his sensitivity as a reader of literature, can be taken to constitute a kind of poetics of speculation. That is to say, one who wishes to speculate needs to do so openly, on the basis of sustained engagement with knowledgeable scholars, with a cautious regard for the possibility that one is or will be proven wrong and a good ear for literary discourse. Holmes satisfies these conditions. Moreover, the graciousness with which Holmes presents his arguments makes it all the easier to take seriously his speculations—for a more sexually appetitive Johnson than some are perhaps accustomed to, for Savage as the inspiration for Thales, for Savage and Johnson as distorted mirror images of each other—even if we choose finally to disagree with him. In short, I think Holmes offers us an excellent model for how speculation can operate usefully in a crowded field such as Johnsonian biography.

Turning to Clark, we see speculation in biography used rather differently. For while Holmes is forthright in admitting where he is guessing, Clark rarely acknowledges conjectures. When he does so, he tends to configure his speculations as positive assertions. Perhaps it is Clark's unwillingness to admit the speculative nature of his book that disturbs readers as much as (and perhaps even more than) the book's thesis that Johnson was a Nonjuring Jacobite. Add to this the facts that Clark nearly always takes negative evidence as a confirmation of his theories and that he reads literature for historical revelation, without any sensitivity to diction, irony, or genre, and one has pinpointed, I believe, precisely what so irritates Johnsonians about *Samuel Johnson*.

To begin with, Clark almost never acknowledges that his pronouncements on Johnson are speculative. Take, for example, his argument that an Anglo-Latin tradition existed which was Tory and Jacobite in character. This early phase of Clark's larger argument is important because it allows him to claim *London*, *The Vanity of Human Wishes* and even the *Lives of the Poets* as essentially forms of Stuart propaganda. In order to make these claims, Clark posits an unbroken chain of Stuart Anglo-Latinists which progresses from Dryden to Pope and William King, and which ends with Johnson. Glossing over the fact that his chain contains only a few links, Clark goes on to state that "Both Whigs and Tories sought to master and appropriate the classical tradition, but Tories did so in larger numbers and normally with greater success than their opponents" (p. 34). Such a breathtaking assertion would seem to call for at least a footnote, perhaps tabulating numbers of copies sold or contrasting the number of editions of Tory Anglo-Latinists with the Whig Anglo-Latinists. Clark's note however merely admits that "Dr. Bentley . . . is the exception to this rule" (p. 34, n. 83). What of Addison, Brooke, Dennis, Garth, Glover, Hervey, Hill, Montagu, Oldmixon, Philips, Pitt, Rowe, Steele, Tickell, Thomson or Young? Or Swift for that matter, who identified himself politically as a Whig? Perhaps it is because literary Tories—a label I think we must be more careful in applying than Clark is—ultimately won many of the canon wars in English literary history, that Clark sees no need to prove that the Anglo-Latin tradition he describes is fundamentally Stuart and Tory in character. In fact, there is a good deal of evidence—ably synthesized in Christine Gerrard's *The Patriot Opposition to Walpole*—that the existence of Clark's Anglo-Latin tradition owed perhaps as much to Whigs as it did to Tories.[9] The main point to remember here is that Clark's dubious and unacknowledged speculation about an Anglo-Latin Tory tradition is merely asserted, not proved, and is never acknowledged as a conjecture.

Where Clark does tacitly admit the speculative nature of some of his arguments—and here I refer to the startling number of "seems likely"s, "evidently"s, "probably"s and "it was not unreasonable"s that appear in his chapter on Johnson and oaths—he manages frequently to reshape the speculative nature of his claims as positive assertions by rhetorical sleight-of-hand. So, for example, the series of tenuous and hedged claims in chapter four for Johnson's avoidance of oaths because of his allegiance to the Stuart cause, begets the following conclusion:

> Some of this [Johnson's reticence about his Oxford years] *may* have to do with poverty; more, *it seems*, with the acute melancholia, *perhaps* verging on insanity, which gripped him from that time. How much of this distress was caused by the prospect that his labours were futile because he would in any event be unable to qualify himself for a degree *is unknown*. What is certain is that no evidence has been presented that he ever did take the oath of allegiance or the oath of abjuration at Oxford, or anywhere else. (p. 122, italics mine)

Note how the uncertainty over the cause of Johnson's silence (signified by the italicized expressions), slowly gives way to certainty. But what is revealed here is not "the cause of Johnson's silence," but rather, the fact that no one else has shown that Johnson took an oath. The innuendo Clark hopes to plant in the reader's mind is that the cause for Johnson's silence was, in fact, his near-maniacal despair over not being able to qualify for a post. But no hard evidence is ever adduced.

Clark hopes to use the negative evidence that no one has shown that Johnson took an oath to establish that Johnson must have been silent about Oxford for that reason. But there is no necessary or causal relation between Johnson's silence and the supposed lack of evidence for his taking an oath. All Clark has done here is to offer a series of guesses as to why Johnson might have been silent about Oxford (poverty, depression) along with a *non sequitur* showing that no one has demonstrated that Johnson took an oath at Oxford. In other words, the chapter's tissue of speculations (never proved, by the way, only asserted by assorted references to Michael Johnson and post-Civil War Lichfield) about why Johnson might not have taken oaths, is here framed as a positive explanation casting definitive light on why Johnson is so silent about his Oxford years. Clark's burden—to prove Johnson took no oaths at Oxford because he was a Nonjuring Jacobite—has been subtly shifted to the skeptics: they have offered no evidence that Johnson took oaths.[10] Yet I believe we have the right to ask someone who claims that we can "say with confidence that Johnson was a Tory, a Nonjuror, and a Jacobite" (p. 7) not only to prove that others have not shown Johnson took oaths, but also to show that Johnson abstained from oaths precisely because he was a Nonjuring Jacobite. In short, the reader can reasonably expect that Clark will either prove Johnson was a Nonjuror through the use of positive evidence or frankly admit that his account, based as it is on negative evidence, is only one of several possible explanations.

Clark does open the following chapter with some acknowledgment that his proof regarding Johnson and oaths is inconclusive. He confesses, "The negative evidence that Johnson never took oaths of allegiance and abjuration and sought ways of securing a degree while evading the obligation to do so does not in itself establish his professed position on the Nonjurors" (p. 126). Having offered this acknowledgment, Clark seems poised to embark on an exposition of the positive evidence which will clinch his case for Johnson as a Nonjuror. One waits in vain. For while the fifth chapter of Clark's book contains a wealth of interesting information, suggesting that Johnson was familiar with the writings of many Nonjurors and admired some Nonjurors greatly, even to the point of including quotations from four of their number in his Dictionary, it does not provide evidence showing that Johnson agreed with their arguments and avoided oaths to the Hanoverian regime as a result (pp. 130-31). Clark usefully immerses his reader in the milieu of Nonjuring theological argument over and meditation on political questions, a milieu in which Johnson certainly moved; he also creates, convincingly I think, a case for Johnson's awareness of and empathy for some Nonjurors. He does not, however, show that Johnson was one of their number. In the end, the only evidence for Johnson as a Nonjuror is negative or circumstantial.

One might expect Clark to turn to Johnsonian scholarship to support his notions here. The reason Clark has difficulty in engaging previous scholarship that might provide some positive evidence for his claims has to do with the fact that, in the end, Clark is not interested in Johnson *per se*. Where Holmes, with his introduction of deftly-handled psychoanalytic techniques, intends to write a book *about* Johnson, Clark, by his own admission, sees Johnson as a means to an end, as the Nonjuring test case who proves the *ancien régime* rule.[11] This explains why he is less concerned to engage in any sustained way with previous Johnson biographers (with the important exception of Hawkins). Since they do not share his twin programs of rewriting eighteenth-century historiography by reinstating *ancien régime* values at the center of eighteenth-century society and of banishing the materialist social theories of an earlier historical generation, their observations about Johnson's political and religious commitments are easily passed over.

Clark's willful and careless disregard of literary scholarship appears over and over again in his book. I have already provided one example of Clark's recklessness by referring to his "Anglo-Latin, Tory tradition"; another appears in his opening claims regarding English literature. Clark identifies Dryden's translation of Virgil as England's national epic and Pope's translation of Homer as her "military epic" (p. 2). Epic by definition, however, *is* military; Clark creates two categories where really only one exists.[12] Further, he conveniently overlooks *Paradise Lost—the* English epic poem if one counts numbers of editions. (Perhaps Milton's Whiggish politics have something to do with Clark's omission.) Strangely then, Clark identifies Addison's *Cato* as the most popular vernacular play in Britain, ignoring Gay's *The Beggar's Opera*, which had a longer initial run and enjoyed many more revivals than *Cato*.[13] What leads Clark into these mistakes is his desire to prove how powerful the classical strain was in England, in order to lay the groundwork for his notions regarding an Anglo-Latin Tory tradition. To be sure, Englishmen's love of the classics should be a key feature of any complete account of early eighteenth-century English literature. But by focusing too narrowly on how the Anglo-Latin strain might support his other hypotheses, Clark misjudges that literature's popularity.

Clark's single-mindedness not only leads him to misread or ignore literary scholarship, it induces him to misread literature as well. His selective quotation of Boswell's narration of Johnson's burning of his papers shortly before his death, allows him to project a crypto-Jacobite motive on Boswell's text (p. 114). But a careful look at the entire context suggests that Boswell at that moment was busily engaged in his constant practice of denigrating Hawkins and his biography. The passage in question occurs immediately after a narration of Hawkins's success in finally getting Johnson to commit a will to paper.[14] Only then did "the consideration of numerous papers of which he was possessed" strike Johnson as an urgent matter (p. 403). The implication appears to be that drawing a will put Johnson in mind of death and the potentially revealing papers he would leave behind him. Boswell then notes that "it is much to be lamented that [Johnson] had not entrusted some faithful and discreet person with the care and selection" of his papers (p. 404). The text continues with the revelation that Boswell had read portions of the "two quarto volumes" which contained many details of Johnson's early life.[15] It is in this context that Boswell relates Hawkins's theft of one of the quartos, along with Johnson's approbation of Hawkins's letter of apology following the incident.[16] The approbation contrasts strongly with Boswell's reflection on the incident, which Clark quotes as circumstantial evidence that the quartos contained dangerous Jacobite implications: "The agitation into which Johnson was thrown by this incident, probably made him hastily burn those precious records which must ever be regretted" (p. 406, n. 1). In fact, the more likely reading of this sentence, particularly in light of Johnson's jocular approval of Hawkins's penitence, is that Boswell hoped to further tarnish Hawkins's qualifications as a biographer by showing how he was responsible for a dearth of information regarding Johnson's early life. At the very least then, it seems that Clark has been insufficiently attentive to speaker and context here.

So if we were to measure Clark's speculative activity against the standard exemplified in Holmes's work, I believe we have strong reason to take issue with Clark's book. He does not admit that he is guessing much of the time. Further, he bases many of his conjectures on negative evidence and, through the clever use of rhetoric, transforms then into positive assertions. Because his conjectures are based on a partisan historical reading, for which he wishes to make Johnson a kind of emblem, Clark tends to ignore the more experienced voices in Johnsonian studies, denigrating them as historiographically backward. While the charge is no doubt true in some instances, it does not invalidate the expertise of formidable Johnson scholars like Greene. Indeed, Clark would have done well to rely more heavily on the abilities of literary scholars to examine and cross-check his readings of literature and literary history.

The possibility that his conjectures are based on the slenderest of evidence, involving massive massaging of source material, does not seem to cross Clark's mind. Even here, Holmes manages to set the standard, since, as an astute critic of Holmes's work has noted, he "has always been as much interested in the nature of his craft as in his subject."[17] Doubtless, the Johnson-Savage nexus attracts Holmes as much for its potential to illuminate the mysterious arts of biography as for its own strange and twisted logic. But Holmes never allows his interest in matters biographical to overcome his allegiance to his materials. Clark does, often.

What I find most regrettable about the unacknowledged speculations in Clark's book is that they not only enable Johnsonians to dismiss his ideas about Johnson out of hand, but they also open the way for skeptics to ignore or devalue Clark's other, more important contributions to theories of party and ideology in Hanoverian Britain. Clark's participation in the ongoing project of revising Namier, for example, is significant and, I believe, essential. He has also done truly valuable work in charting the Whig party's evolution under Walpole by distinguishing the theories of providential divine right from theories of indefeasible divine right.[18] Though I cannot share the insistence of Clark, Erskine-Hill, and Eveline Cruickshanks that the Tory party between the rebellions was comprised entirely or even mostly of Jacobites, I believe their assertions of party conflict during those years is fundamentally sound. So I can only regret that Clark's larger theses about an *ancien régime* England have allowed him to produce a book which, among literary scholars, may ultimately obscure his other, more valuable contributions to eighteenth-century history.

Despite these serious difficulties with the speculative nature of Clark's book, does it have enduring value? I think the answer here must be yes, and for the following reason. Speculations such as Clark's —whether refuted or confirmed, offered in a spirit of generous conjecture or a mood of self-righteous certainty—sometimes offer the best chance to advance our knowledge of a writer like Johnson. The exchanges which this journal has already printed on this matter are ample repayment for any discomfort that Clark's incursions into Johnsonian studies may have caused. If for no other reason than that we learn Johnson did take oaths at Oxford, Clark's provocative book will have served a useful purpose. Moreover, asking whether the opposition sentiments of Johnson's early political writings echo Tory ideology, or whether Johnson's association with particular individuals known to be Jacobites and Nonjurors in turn qualifies him as a Nonjuring Jacobite, are not unfruitful questions. In the words of another Johnson scholar, "We must ask such questions, without fearing that there is anything impious in them, or that we are somehow keeping imperfect faith with a beloved biographical representation."[19] I would only add that if we cannot give convincing evidence for our answers to such questions, we would surely be better off grounding our conjectures in familiarity with the best scholars, and acknowledging where we know and where we merely guess.

NOTES

1. For a review of the reactions to Clark's first three books, see G. S. Rousseau, "Revisionist Polemics: J. C. D. Clark and the Collapse of Modernity in the Age of Johnson," *Age of Johnson*, 2 (1989), 421-50.
2. J. C. D. Clark, *Samuel Johnson: Literature, Religion and English Cultural Politics from the Restoration to Romanticism* (Cambridge: Cambridge Univ. Press, 1994). Pages are cited parenthetically in the text.
3. Donald Greene, "Johnson: The Jacobite Legend Exhumed. A Rejoinder to Howard Erskine-Hill and J. C. D. Clark," *Age of Johnson*, 7 (1995), n. 1.
4. Howard Erskine-Hill, "Johnson the Jacobite? A Response to the New Introduction to Donald Greene's *The Politics of Samuel Johnson*," *Age of Johnson*, 7 (1995), pp. 3-4.
5. Richard Holmes, *Dr. Johnson and Mr. Savage* (London: Hodder and Stoughton, 1993). Pages are cited parenthetically in the text.
6. James Clifford, *Young Sam Johnson* (New York: McGraw-Hill, 1955), p. 207.
7. Thomas Kaminski has a very different reading of this allusion, which he notes was first discovered by John Gough Nichols (a fact of which Holmes is unaware). It seems to me that the alternate reading of the lines referring to Colley Cibber which Kaminski provides, does not eliminate the possibility that the lines in question refer to Savage. See Thomas Kaminski, "Was Savage 'Thales'?: Johnson's *London* and Biographical Speculation," *Bulletin of Research in the Humanities*, 85, pp. 329-30.
8. Clifford, pp. 207-208.
9. Christine Gerrard, *The Patriot Opposition to Walpole: Politics, Poetry, and National Myth, 1725-1742* (Oxford: Clarendon Press, 1994). See also Howard D. Weinbrot, "Politics, Taste, and National Identity: Some Uses of Tacitism in Eighteenth-Century Britain," *Tacitus and the Tacitean Tradition*, ed. T. J. Luce and A. J. Woodman (Princeton: Princeton Univ. Press, 1993), pp. 168-84.
10. In a paper presented at the Northeast American Society of Eighteenth-Century Studies in fall 1995 in Ottawa, Howard Weinbrot produced evidence which demonstrates that Johnson did, indeed, take oaths.
11. "After 1760, the whole disposition in favour of order, rank and subordination, which Johnson classically represents, found a single and undivided object of loyalty." *English Society*, p. 189.
12. The turn to "peaceful" epics was, as Michael Murrin has shown, largely a response to the realities of gunpowder warfare, which made obsolete the old heroic hand-to-hand combats of the *Iliad* and the *Aeneid*. Nevertheless, epic poets like Milton still managed to incorporate warfare

into their poems, as in books 5 and 6 of *Paradise Lost*. See Michael Murrin, *History and Warfare in Renaissance Epic* (Chicago: Univ. of Chicago Press, 1994), especially ch. 12.

13. My informal count suggests that Gay's play was performed three times as often as Addison's through 1751. Moreover, from 1751 on, Addison's play fell out of the repertory periodically. Gay's play, of course, held the stage through the eighteenth century. See Ben Ross Schneider, Jr., *Index to the London Stage 1660-1800* (Carbondale: Southern Illinois Univ. Press, 1979), pp. 4-5, 344-45.
14. James Boswell, *Life of Johnson*, ed. George Birkbeck Hill (Oxford: Clarendon Press, 1934), IV, 401-407. Pages cited parenthetically in text.
15. Nowhere in this material does Boswell suggest there were potentially damaging political or dynastic affiliations revealed in the quartos. The implication seems to be that there was information relating to Johnson's psychologically-troubled young adulthood.
16. Johnson quipped "Melius est sic penituisse quam non errâsse" or "Better to have repented thus than to have not sinned." Clearly, Johnson's initial reaction to Hawkins's theft and justification was anything but that of a frantic Jacobite hoping to secure damaging documents.
17. Anne Barton, "Over the Dark River," *New York Review of Books*, 16 Feb. 1995, p. 6.
18. See *English Society*, especially ch.3.
19. Philip Edward Baruth, "Recognizing the Author-Function: Alternatives to Greene's Black-and-Red Book of Johnson *Logia*," *Age of Johnson*, 5 (1992), p. 54.

NEW LIGHT ON JOHNSON'S DUCK

LAWRENCE LIPKING

In all the Johnson canon, it seems safe to say that no work is more prodigious, or more strangely neglected, than what might be called his fledgling effort, the Epitaph on a Duck. From the beginning biographers have recognized its importance. Within a few months of Johnson's death, Anna Seward supplied the text to the *Gentleman's Magazine*, pointing out that "the great infant" had "lisped 'in numbers, for the numbers came.'"[1] The following year Mrs. Piozzi quoted the verses as "a striking example of early expansion of mind, and knowledge of language."[2] A year after that Hawkins' *Life* of Johnson confirmed that this "dawn of his poetical faculty" was "well attested, and therefore makes part of his history."[3] And even though Boswell, in his *Life*, accepts Johnson's assurance that he remembered his father's making the verses (or half of them), Boswell said something else to his *Note Book*: "I trust to his mother's relation of what happened in his childhood rather than to his own recollection; and Miss Porter assured him in my presence upon his mother's authority that he had made this epitaph himself"[4] (as we shall see, Johnson had his own reasons for denying the attribution). No anecdote of Johnson's youth is better warranted by those who knew him best. Indeed, the epitaph is the only one of the author's works that we can be sure his father and mother liked.

Nor did the Lichfield circle fail to appreciate the significance of this seminal work. Seward, the Swan of Lichfield, told Boswell that "there can be no doubt of the authenticity of that little anecdote of Johnson's infancy; the verses he made at three years old, on having killed, by treading upon it, his eleventh duck,"[5] and it was her version he used:

> Here lies good master Duck,
>   Whom Samuel Johnson trod on,
> If 't had liv'd 't had been good luck:
>   For then there'd been an *odd one*.[6]

But more important, she knew the value of what she had found. "It is curious to remark, in these little verses, the poetic seed which afterwards bore plenteous fruits, of so rich a lustre and flavour." Who better than a Swan to understand the possibilities of even the oddest duckling? And Seward saw still more: an epitome of Johnson's whole career. "We observe, also, in those infant verses, the seeds of that superstition which grew with his growth, and operated so strongly through his future life."[7] No later biographer has picked up the hint. But already the lines predict the author to come: his fascination with epitaphs, his fondness for calculation, and most of all his fear of death, combined not only with superstition but with a sense of guilt. Who, after all, had trod on the duck? and who else was compelled to make the funeral arrangements?

All this is obvious. Nevertheless, much more remains to be explored in these provocative verses. With the help of recent scholarly advances, I think it is not too much to say that the lines can furnish a key to unlocking the deepest secrets of Johnson. All that we need is a willingness to use the insights of some of the leading psychoanalytical and historical thinkers of our time, those virtuosi who have lately opened up whole discourses and startling new ways of reading. Faced with such an embarrassment of riches, I shall confine myself to two.

The first is Jacques Lacan. If Lacanian analysis of the mirror stage can tell us much about Johnson, then Johnson can also do something for Lacan. For one problem of verifying the mirror stage has been a scarcity of written records: very few infants have put down their thoughts, let alone finished works of art. Hence the epitaph supplies a unique opportunity for observing the process at work, a document *about* the mirror stage by someone *at* the mirror stage. Johnson provides a classic of self-recognition. Nothing in the epitaph is more striking than its third-person naming, as SAMUEL JOHNSON becomes both subject and object. Never again, in more than seventy years, would he refer to himself in a poem. It is as if he were the master spirit of the verses. As is well known, Johnson detested epitaphs that omitted names: "The end of an epitaph is to convey some account of the dead; and to what purpose is any thing told of him whose name is concealed?"[8] Why then does he name himself and not the ill-starred duck? Evidently some transference has occurred: the duck is a mirror in which the killer detects his own identity or even his own memorial. Henceforth he will always tread the same set of murderous letters.

The process as a whole has been described with characteristic lucidity by Lacan himself.

> The *mirror stage* is a drama whose internal thrust is precipitated from insufficiency to anticipation—and which manufactures for the subject, caught up in the lure of spatial identification, the succession of phantasies that extends from a fragmented body-image to a form of its totality that I shall call orthopaedic—and, lastly, to the assumption of the armour of an alienating identity, which will mark with its rigid structure the subject's entire mental development. Thus, to break out of the circle of the *Innenwelt* into the *Umwelt* generates the inexhaustible quadrature of the ego's verifications.[9]

It would be hopeless to try to improve upon that. Nonetheless, a few vulgar glosses might spell out the implications for Johnson's poem. One key is surely the "odd one." The phrase itself insinuates something distinctly odd. If the duckling had lived, it would evidently have needed some orthopedic correction. Exactly what part of its body-image has been fragmented, at least in the imaginary order of the poet's fantasy? The answer is almost embarrassingly clear. The mark of castration anxiety stamps the dismembered duck as well as its executioner, and in the unconscious such misfortunes can never be accidental. In this case everything points to castration; it is overdetermined by a family romance involving both parents. "Immediately after the accident, he came to his mother, and desired she would *write*. 'Write, *what* am I to write?' 'Write upon poor Duck.' 'Well then, Sam, tell me what to say.'"[10] And so he dictated, her hand becoming his scribe, as if for the moment he were the lord of the manor, wielding the potent phallus of language. But a rival account exists, in which Johnson "was walking by his father's side" when he trod on the duck and was told "he must make an epitaph for it."[11] Here the infant cock of the walk turns into an underling, and his seminal verses shrivel into his first piece of hackwork, commissioned by an oedipal patron. Just who wore the pants in this family? Johnson would spend the rest of his life attempting to put them on.

The plot thickens, moreover, when one more detail is added, the poet's claim "that his father made one half of this epitaph." If true, this is highly significant; if untrue, even more significant, exposing a desperate psychic struggle to convert the father to collaborator rather than rival. But castration is not that easy to avoid. Which half of the epitaph might belong to the father, which half to the son? Unfortunately no record has survived. But the unlikelihood that a three-year-old could count to eleven, or tell the difference between even and odd, suggests that the senior partner took responsibility for completing the rhymes.[12] Momentous consequences follow from this surmise. In the mind of the infant Johnson, already identified with poor master Duck, he was labeled forever an *odd one*; his own father had said so. What could he do but accept it, assuming "the armour of an alienating identity, which [would] mark with its rigid structure the subject's entire mental development"? When he grew up, his favorite seat would be a three-legged chair.[13] Henceforth he would always be haunted by a sense of incompleteness, of something missing that his mastery of language both repaired and confirmed. The more he dictated, the more he served as the slave of a patron and public, and the less satisfaction he found. The great Sam Johnson would always be what one calls an odd duck.

Indeed, the duck would never really leave him. Everyone knows that Johnson often talked to himself, or "fell into a reverie accompanied with strange antic gesticulations."[14] What did he talk about? We do not know, except that he seemed to be expressing guilt and making arithmetical calculations, sometimes on his fingers. By this time he certainly knew how to tell the odd from the even. But the combination of guilt and counting suggests that some primal scene, or prime numbers, continued to obsess him. At times he murmured "too, too, too"—perhaps a mistranscription of the number two, whose repetition always comes out even? I never like to speculate. Nevertheless, it seems inescapable that Johnson was reliving some early calamity, and trying this time to get the numbers right. And one more piece of evidence may be added. In recent years a controversy has flared around Johnson's "secret," the mysterious letter "M" that appears in his diaries and might stand for masturbation or defecation (bowel *m*ovements).[15] I cannot hope to settle this weighty issue. But another possibility might well be considered. The Latin name for the *m*allard or drake, the male duck with which Johnson was so deeply involved (a *m*aster, never a mistress), is *anas mas* or *masculus*—again the letter "M" (though it is probably only a coincidence that "mas" is "Sam" spelled backward). Was Johnson once more treading the memory of his infant castration, compulsively recalculating on paper where he might have gone wrong, and trying to propitiate the ghosts of duck and father? Intelligent scholars will draw their own conclusions.

Yet our reading ought not to end there. For up to this point I have to confess that my interpretations of the epitaph, however plausible, do have a single grave flaw: they are too superficial. No matter how many improvements Lacan and his followers have brought to the task of deciphering the mechanisms of psyches and texts, they do not exhaust the findings of recent scholarship. In particular, my reading of the family romance has paid too little attention to the larger political and ideological contexts that dominate current studies of Johnson. The work of Howard Erskine-Hill and J. C. D. Clark has opened our eyes to the covert messages concealed in many Johnsonian texts.[16] We need to look again at our "innocent" verses.

The year 1712-13, when the young poet was three, was probably the most exciting of his life. A recent book has claimed that "The point at which Johnson seems to have wished to make his stand was 1714, when he was 5 years old: on the throne, a queen devoted to the Church; a huge Tory majority in the House of Commons; an act against occasional conformity in force; the Schism Act to curtail the teaching power of dissent just about to come into effect."[17] But this seems postdated. Aside from the fact that Johnson did not turn five until September of 1714, a month after the queen had died, he had long since taken his stand. The high point occurred just before the treaty of Utrecht, in April 1713. Just at that moment the champion of the High Church, Henry Sacheverell, came to Lichfield, and Johnson was observed "at the Cathedral perch'd upon his Fathers Shoulder listning and gazing at the much celebrated Preacher. Mr. Hammond ask'd Mr. Johnson how he cou'd possibly think of bringing Such an

Infant to Church and in the Midst of so Great a Croud. He answer'd because it was impossible to keep him at home for young as he was he believ'd he had caught the Public Spirit and zeal for Sacheverel and woud have staid forever in the Church Satisfied with beholding him."[18] Evidently "the infant Hercules of toryism" already knew where he stood. The previous year he had met Queen Anne in London; and though no one kept minutes of their conversation, Johnson could hardly have been unaware of the hope springing up around him, the prospect of a restoration of the Stuarts. The death of Queen Anne, the fall of the Tories, and finally the doomed '15, would thoroughly damp those hopes within a few years. But the state of affairs had been quite different in early spring of 1713. Bliss was it in that dawn to be alive, and to be a young Jacobite was very heaven.

What do such attitudes have to do with the Epitaph on a Duck? A short but accurate answer might be, *everything*. Even the form and genre of the verse proclaim the Stuarts. With unusual boldness, the young author alludes to a poem familiar to every three-year-old of his time, the anticipatory epitaph for Charles II:

> Here lies our sovereign lord the king,
>   Whose word no man relies on,
> Who never said a foolish thing,
>   And never did a wise one.

The resemblance is plain as day. Not only does Johnson borrow structure, phrasing, and rhyme scheme, he borrows the rhyme itself, the whimsical "on" and "one." No one could miss the allusion. Indeed, the Jacobite reference appears so blatant that one wonders at its daring. It is hardly surprising that Johnson later wanted to disavow the verses, especially after the '45 and his pension. "My father was a foolish old man," he told Boswell; "that is to say, foolish in talking of his children."[19] But Boswell seems to have missed the point, which must be that the elder Johnson, so zealous in spreading the word of his son, might well have put him in danger. Sacheverell had been impeached for much less.

Yet the epitaph did have one means of protection: its message was coded. Young Johnson artfully cloaks his Jacobite politics in barnyard images. He had probably learned the technique from Dryden. In the

Dictionary, Johnson's first illustration of "duck" refers to Dryden's fable of the Cock and the Fox (from Chaucer's *Nun's Priest's Tale*):

> The *ducks* that heard the proclamation cry'd
> And fear'd a persecution might betide,
> Full twenty mile from town their voyage take,
> Obscure in rushes of the liquid lake.[20]

These are clearly very cautious political fowl. In fact the main point of the fable, which Johnson must have known well, is the way that allegorical meanings hide in plain sight: "The Truth is moral, though the Tale a Lie." Ducks, like loyalists, keep alive by learning to lie low and stay obscure.

The Epitaph on a Duck is a perfect example. Speaking in parables, it cleverly alleges the reverse of what it means, revealing its sentiments only to those of the same persuasion. Is the good master dead, would Johnson tread upon him (for instance, by taking an oath), is Britain so unlucky as to have no legitimate Stuart heir, or was James Francis Edward really an odd one (as his enemies had rumored), smuggled into the royal bed in a warming-pan? Surely this cannot be. True believers will understand the radical subtext, a message of hope: the Stuarts will rise again. For good luck is on the way, even though Charles II and James II have died. The Pretender is only pretending to lie underfoot, and when he appears he will certainly be an *odd one*: James *III*, and behind him Charles *III*.

One question about the epitaph remains, however, and it is vital. Those whom my argument has not yet convinced are likely to harbor the same doubt that Leonard Marx expressed so cogently sixty years ago. The question is, Why a duck? Indeed, *why* a duck? No scholar has ever given an adequate answer to Marx. But once the solution is seen, I believe, any reader will take it to be mere child's play, or to use the vulgar expression, duck soup. As Jonathan Clark has lately taught us, one dependable sign of Jacobite principles was adherence to the Anglo-Latin tradition: "Latinity, episcopalianism and dynastic legitimacy were united in the minds of such men."[21] Hence knowledge of Latin offered an initiation rite for Stuart sympathizers, as well as a coded language that Hanoverians could not comprehend. When Jacobite babies prattled, they lisped their numbers in Latin. It is there we must seek the answer to our question.

As soon as we grasp this clue, we shall pick up the final trail of Johnson's duck. What is the meaning of "Dux"? Even a well-schooled three-year-old would know that the Latin word stands for a leader or *ruler*—il duce, the root of duke. This simple pun provides the key to Johnson's epitaph. The young poet conveys his point with oral elegance, perceptible even to those who, like him, have not yet learned how to read and write or to care about the orthography of ducks and dux. Here, in the bosom of dynastic legitimacy, a good master holds sway. The poem deals with rulers and those who are born to rule: their apparent demise, the chance of their future return. Let Hanoverians tremble or misconstrue. The king is dead, long live the king, the dux will never die.

I hope it is not immodest to say that, in this new light, Marx's question has been answered once and for all. But I am far from thinking that all the mysteries of "Epitaph on a Duck" have yet been solved. Indeed, this piece should serve as a new beginning. In his updated guide to Johnson, no less an authority than Donald Greene remarks that "A somewhat jaundiced reviewer recently wrote, 'Why do people go on writing books and articles about Johnson?'" and continues, "An answer to the reviewer's question might be 'For the same reason they go on writing books and articles about Shakespeare.'" Quite so; and anyone who has read much current criticism of Shakespeare will realize that the present essay is in their spirit. The epitaph demands the same sort of critique. A short but urgent list for further research might include a study of textual problems ("Johnson's Two Ducks"), a post-colonial inquiry ("How to Read Johnson's Duck: The Master and the Subaltern"), a deeper look at the culture of sexuality ("'My Dainty Duck, My Dear': Homosocial Desire across Species"), and of course full discussions of the epitaph's place in the movement for animal rights and the development of chaos theory. But this important work I leave to the next generation, in confidence that they, like Jonathan Clark, will fulfill Greene's prediction: "There is still much more of value to be learned about Johnson."[22]

NOTES

1. *Gentleman's Magazine*, 57 (February 1785), 100. The quotation comes, of course, from Pope's "Epistle to Dr Arbuthnot," 128.
2. *Johnsonian Miscellanies*, I, 153.
3. Sir John Hawkins, *The Life of Samuel Johnson, LL.D.*, ed. Bertram H. Davis (New York: Macmillan, 1961), p. 3.
4. *The Correspondence and Other Papers of James Boswell Relating to the Making of the "Life of Johnson,"* ed. Marshall Waingrow (New York: McGraw-Hill, 1969), p. 64n. Cf. *BLJ*, I, 40; *Works*, VI, 354; and Waingrow's two appendices on "Anecdote of the Duck" in his edition of the original manuscript of Boswell's *Life of Johnson* (Edinburgh: Edinburgh Univ. Press, 1994), I, 351-54.
5. *Correspondence*, p. 79.
6. *Gentleman's Magazine*, 57 (February 1785), 100. This text seems to have served as the basis for both Hawkins and Boswell, though each "improved" it slightly for publication. Quite different versions are given in Mrs. Piozzi's *Anecdotes* and Boswell's *Note Book* (see Johnson's *Works*, VI, 354).
7. *Correspondence*, pp. 79-80. Boswell alters this passage substantially when quoting it in the *Life* (*BLJ*, I, 40-41n.).
8. *Lives of the Poets*, III, 257.
9. Jacques Lacan, *Écrits*, tr. Alan Sheridan (New York: Norton, 1977), p. 4.
10. *Gentleman's Magazine*, 57 (February 1785), 100.
11. *Works*, VI, 354.
12. This is the conclusion of O M Brack, Jr., in his authoritative study, "Samuel Johnson and the Epitaph on a Duckling," *Books at Iowa*, 45 (November 1986), 62-79.
13. Fredric V. Bogel has plumbed the relation of the chair to guilt, authority, and Lacan in *The Dream of My Brother: An Essay on Johnson's Authority*, *ELS* 47 (Victoria: Univ. of Victoria, 1990).
14. Sir Joshua Reynolds, *Portraits*, ed. F. W. Hilles (London: Heinemann, 1952), p. 70.
15. See *The Age of Johnson*, 4 (1992), 1-40; and 6 (1994), 131-49.
16. Howard Erskine-Hill, "The Political Character of Samuel Johnson," in *Samuel Johnson: New Critical Essays*, ed. Isobel Grundy (London: Vision Press, 1984), pp. 107-36; J. C. D. Clark, *Samuel Johnson: Literature, Religion and English Cultural Politics from the Restoration to Romanticism* (Cambridge: Cambridge Univ. Press, 1994).
17. John Cannon, *Samuel Johnson and the Politics of Hanoverian England* (Oxford: Clarendon Press, 1994), pp. 20-21.

18. Letter to Boswell from Mary Adey, *Correspondence*, p. 64. As Hill points out, the likely time for this sermon would have been March or April of 1713 (*BLJ*, I, 39n.).
19. *BLJ*, I, 40.
20. "The Cock and the Fox," 736-39; Dryden's *Fables Ancient and Modern* (1700).
21. Clark, *Samuel Johnson*, p. 37. The quotation paraphrases Douglas Duncan's views of Thomas Ruddiman. For the theme in general, see Clark, pp. 20-21, 33-41, 59-66.
22. Donald Greene, *Samuel Johnson* (Boston: G. K. Hall, 1989), p. ix.

CONVULSION OF THE LUNG: AN HISTORICAL ANALYSIS OF THE CAUSE OF DR. JOHNSON'S FATAL EMPHYSEMA

JEROME M. REICH, M.D.

Cough: a convulsion of the lungs vellicated by some sharp serosity; pronounced coff.[1]

INTRODUCTION

That Samuel Johnson, the Colossus of Literature,[2] died of emphysema complicated by cor pulmonale will not be doubted by anyone familiar with the observations at his necropsy. It has not however been sufficiently considered why the Great Lexicographer, who neither smoked nor engaged in an industrial occupation, developed this lethal affliction. In undertaking an analysis of the cause of Johnson's emphysema, it should be understood that diagnosis of pulmonary disease was primitive in the eighteenth century, and his necropsy, perfunctory by modern standards: one is therefore left with the compensatory expedient of reconstructing the nature and cause of his illness from a historical record that is both imperfect and incomplete, and in which the signification of terms differs from contemporary usage.

That this re-examination of the historical record is merited will be questioned; votaries of Johnson will doubtless find its conclusions of interest, and among non-devotees, one hopes to stir gainful regard. Raleigh's assessment goes far in explaining this enduring reverence: "This is the greatness of Johnson, that he is greater than his works. He thought of himself as a man, not as an author; and of literature as a means, not as an end in itself. Duties and friendships and charities were more to him than fame and honour. . . . 'Books without the knowledge of life are useless; for what should books teach but the art of living'";[3] and "He has almost become the tutelary genius of the English people. He embodies all that we most admire in ourselves. When we pretend to laugh at our national character, we call it John Bull; when we wish to glorify it, we call it Samuel Johnson."[4] An indication of the remarkable length of duration and continuance of esteem in this extraordinary personage is the existence of a periodically updated bibliography[5] which cites 4,282 critical studies in the 1970 edition, and acknowledges the incomplete state of this exhaustive compendium because of the absence of records of many eighteenth-century publications.

HISTORICAL REVIEW

In 1849, Dr. Squibb provided the first exclusively medical account of Dr. Johnson's life in a paper read before the Harveian Society of London,[6] in which he acknowledged the "kind permission and liberality" of Dr. James Arthur Wilson in permitting him to see the manuscript of the post-mortem examination that his father, James Wilson, had performed on Samuel Johnson in 1784.[7] This account Sir Humphrey Rolleston received from Lee Dickinson, the prosector's great-grandson by marriage in 1891, and quoted from it in the presidential address given before the Royal Medico-Chirurgical Society of Glasgow in 1924:[8]

> Wednesday, Dec. 15th, 1784, opened the body of Dr. Samuel Johnson for Mr. Cruikshank, in the presence of Dr. Heberden, Brocklesby, Butter. . . . On opening into the cavity of the chest, the lungs did not collapse as they usually do when the air is admitted, but remained distended, as if they had lost the power of contraction; the air-cells on the surface of the lungs were also very much enlarged; the right lobe adhered very

> strongly to the diaphragm; the internal surface of the trachea was somewhat inflamed; no water was found in the cavity of the thorax. The heart was exceedingly large and strong. . . .

Dr. Wilson, nineteen years of age at the time of the dissection, classified Johnson's disorder under the heading "asthma." It has however the pathological features of emphysema complicated by cor pulmonale; and it appears to provide the basis of the first pathological report of this disease,[9] having preceeded Laennec's description[10] by sixteen years. Baillie's account of this condition,[9] given below, was based on the lung he obtained from the person who had requested the necropsy, Dr. William Cruickshank, referred to above, one of four physicians attending Dr. Johnson, without charge, during his last illness.[11] "The lungs are sometimes, although I believe very rarely, formed into pretty large cells, so as to resemble somewhat the lungs of an amphibious animal. Of this I have now seen three instances. The enlargement of the cells cannot well be supposed to arise from any other cause, than the air not being allowed the common free egress from the lungs, and therefore accumulating in them. It is not improbable also that this accumulation may sometimes break down two or three contiguous cells into one and thereby form a cell of very large size."

Regrettably, dissection of the lung was not undertaken and the specimen no longer exists [written communications March 1992: E. Allen, H. Attwood]. Moreover, Attwood[12] argues persuasively that the engraving in Baillie's *Morbid Anatomy* (fasc. 2, plate 6, fig. 1), taken to represent Dr. Johnson's lung, does not correspond to the description given above and in fact represents a lung from an individual who suffered from fibrosing alveolitis. One is therefore left with the sole recourse of attempting to reconstruct, principally from epistolary sources, the cause of Johnson's emphysema, which, as Larson points out, cannot be ascribed to tobacco smoke.[13]

That Dr. Johnson suffered from ill health since infancy is attested to in his autobiographical sketches and recounted in some detail by Rolleston[8], who indicates that he had unilateral blindness due to a corneal leukoma resulting from tuberculous keratitis. He was one of two-hundred individuals suffering from The King's Evil touched by Queen Anne at St. James's Palace in 1712. Profound near-sightedness of the remaining functional eye is said to have accounted for his disinterest in the visual and performing arts. He suffered from periods of melancholy so profound that he feared that madness would ensue.[14] In 1778, Johnson wrote his friend Dr. Edmund Hector, "My health has been, since my twentieth year, such as has seldom afforded me a single day of ease." His respiratory health until the winter of 1755 appeared excellent, as McHenry reports.

His ultimately fatal ailment, of which a minute account will be naturally desired, began shortly after the publication of his prodigious work, the two-volume *Dictionary of the English Language*. "Meanwhile, during the lonely Christmas season he became ill with a severe case of bronchitis, possibly even pneumonia. Johnson described the illness to a friend, Robert Paul, on Dec. 29, 1755, 'I have never been out of doors since you saw me. On the day after I had been with you I was sore with a hoarseness which still continues; I had a cough so violent that I once fainted under its convulsions. I was afraid of my lungs. My physician bled me yesterday and the day before, first against his will, but the next day without any dissent. I had been bled once before, so that I have lost 54 ounces. I live on broaths and my cough, thank God, is much abated, so that I can sleep.'"[14] Succeeding correspondence frequently refers to episodes of severe and persistant winter bronchitis, remitting in the spring with the coming of warmer weather.

In 1762, Johnson, then fifty-three, summered at the estate of his good friend Sir Joshua Reynolds, the same to whom Boswell dedicated his justly famous biography in 1791. Johnson's bronchitis had remitted and his respiratory status was excellent as the following anecdote, related by Miss Reynolds[15] attests:

> Before the house was a spacious lawn. When someone remarked that the lawn was ideal for racing, a young lady who was present boasted that she could outrun anyone there. At this, Johnson arose and said "Madam, you cannot outrun me."
>
> The lady at first had the advantage; but Dr. Johnson, happening to have slippers on much too small for his feet, kicked them off up into the air, and ran . . . leaving the lady far behind him and...returned, leading her by the hand, with looks of high exultation and delight.

In the early 1760s Johnson wrote that he had been suffering from "convulsions of the breast" since 1759; ten years later, "The last year has been wholly spent in a slow process of recovery. My days are easier, but the pertubation of my nights is very distressful. . . . My lungs seem encumbered, and my breath fails me, if my strength is any unusual degree exerted, or my motion accelerated. I seem to bear exercise with more difficulty than in the last winter."[16] On the return from his annual trip to Lichfield in December 1773, he caught a cold. ". . . He then developed a 'violent' fever which lasted six weeks. His cough too, grew more violent, harassing him almost incessantly day and night, finally becoming productive of sputum."[14] In February 1774 he wrote Boswell: "I have been troubled for many weeks with a vexatious catarrh which is sometimes sufficently distressful." In March of the same year: "Of myself I can only add that having been afflicted many weeks with a troublesome cough, I am now recovered." In July, to Langton:[17] "I have never recovered from the last dreadful illness, but flatter myself that I grow gradually better: much however, yet remains to mend." In January 1776: "I have been afflicted through the whole Christmas with the general disorder of which the worst effect was a cough. . . ."[14] McHenry observes "This bout of pulmonary disease lasted nearly four months. By March of 1776, with the help of warm weather, he was better. Johnson had had an attack of acute bronchitis probably complicated by pneumonia in view of high persistant fever. This illness was more severe than is at first apparent, for later he wrote that he had never cared for much walking since this illness in the winter of 1773-1774."[14]

Three years later nocturnal dyspnea and orthopnea ensued, for which he received phlebotomies, physic, and opium. In 1780, he noted that "by spasms in the breast I was condemned to the torture of sleepyness without the power to sleep. These spasms after enduring more than twenty years I ease by three powerful remedies, abstinence, opium and mercury."[18] The same year he wrote Boswell, "I have this summer sometimes amended, and sometimes relapsed, but, upon the whole, have lost ground very much. My legs are extremely weak, and my breath is very short, and the water is now increasing upon me."[18] In 1783, two years before the publication of Withering's monograph on the foxglove, and one year before his death, he received squill for treatment of progressive dropsy; he passed twenty pints of urine, allowing him to leave his home after four months of confinement. After a brief respite, edema reappeared, which, added to his dyspnea, caused great suffering; sleep became possible only by the employment of opiates to relieve the latter. In December of 1784, during an exacerbation of his cough and dyspnea, he asked Dr. Brocklesby for a candid opinion of his case, and, being advised that he could not recover without a miracle, replied, "Then . . . I will take no more physic, not even my opiates; for I have prayed that I may render up my soul to God unclouded."[19] On 13 December 1784, at age seventy-four, an effusion of blood from a self-inflicted scalpel wound in his calf, performed in a heroic but misguided effort to palliate his dyspnea, precipitated his death from cardio-respiratory failure.

DISCUSSION

Dr. Johnson's death occurred nearly one-half century prior to Laennec's recognition and description of bronchiectasis and introduction of diagnostic auscultation, some two centuries prior to the development of imaging procedures capable of confirming his diagnosis during life, his post-mortem examination limited to superficies, and the specimen no longer existing, it remains to inquire what anamnestic criteria would prompt an experienced observer to a provisional diagnosis of bronchiectasis. A history of protracted and severe lower respiratory tract infection, followed by episodic recurrences of a similar nature, accompanied by expectoration of copious quantities of purulent sputum, terminating in respiratory failure and cor pulmonale, would suffice, since these these were the characteristic features described by Laennec in 1819:[20]

Chapter VI. OF DILATATION OF THE BRONCHIA

> The organic lesion which I am now to notice, seems to have been hitherto entirely overlooked, both by the anatomist and the practitioner. This oversight is easily accounted for by the circumstance that, as it generally occurs in a small portion of a bronchial tube, when observed, it has been mistaken for a larger branch. It can only be detected by tracing the individual bronchial tubes to their ultimate ramifications,—a thing which is rarely done in our examination of the lungs.

> This species of dilatation is generally met with only in subjects that have died, after being affected by chronic catarrh. It is sometimes so considerable that the bronchial ramifications, which in their natural condition, would scarcely admit the point of a very fine probe, acquire the diameter of a goose-quill, or even of the finger. These dilated portions terminate in culs-de-sac capable of containing a hemp-seed, a cherry-stone, a filbert, or even an almond. . . .
>
> Case 18. A woman, 62 years of age, had been affected, ever since she was sixteen, with a disease of the chest, which exhibited most of the usual symptoms of consumption. The principal of these were—hemoptysis, very frequent, and renewed by the slightest causes; constant cough, with expectoration of opaque, yellow sputa, having sometimes the character of pus, sometimes of a puriform mucus; and respiration more or less impeded. These symptoms varied much; they had very marked remissions, but scarcely ever any positive intermission: they never prevented her from following her occupation of teaching the piano-forte. Her chest was well formed, and she had nothing of the consumptive configuration. Without any material aggravation of the pectoral affection, this woman became universally anasarcous, and died shortly after entering the hospital.

The history of Dr. Johnson's ultimately fatal illness conforms closely to the clinical characteristics of this disease provided by Laennec, with one omission: the character of the phlegm or catarrh is never described.

Free of unusual respiratory problems until the age of forty-six, Dr. Johnson developed a severe and prolonged pulmonary ailment in 1755, followed by episodes of protracted winter bronchitis. A decade later he developed persistant dyspnea, which, after severe exacerbation in 1773-74, became both more pronounced and persistent. Two years before his death in 1784, he developed profound peripheral edema. Postmortem examination showed advanced emphysema, cor pulmonale, and a dense adhesion of one lung to the diaphragmatic pleural surface. The lungs were not dissected.

The heralding and constant feature of Johnson's illness was a violent, prolonged, and pernicious cough, exacerbating in winter, remitting in summer, sometimes accompanied by fever and phlegm. His choice of term, "convulsion," in defining "cough," considering his remarkable linguistic precision, is revealing; nor is it likely that its severity was exaggerated, for he regarded the display of personal accounts of suffering distasteful. "Dear Doctor (said he one day to an acquaintance, who lamented the tender state of his inside), do not be like the spider, man; and spin conversation thus incessantly out of thy own bowels."[21]

In pursuing the inquiry that Johnson's emphysema may have been the consequence of bronchiectasis, it is instructive to inquire into coeval accounts of these disorders, which, since they precede the era of cigarette smoking to which nearly all instances of emphysema are presently imputed, may permit a clearer understanding of the genesis of the subject's illness.

Wilson, as noted above, classified Johnson's lungs as asthmatic. But Laennec, who undertook to define emphysema as a clinical-pathological entity, also recognized its association with bronchitis:[22]

> Before terminating this chapter, it may be useful to make a few observations on one symptom which many nosologists have erected into a distinct disease—I mean Asthma. This word, which properly signifies difficulty of breathing, has been as much misused, and has been made the cognomen of as many different diseases as any work in medicine. It has been proved by Corvisart that a great part of the diseases so denominated, are, in fact, diseases of the heart and large vessels; and every person accustomed to morbid dissection is now aware that the cases denominated humid or humoral asthma are simply examples of chronic catarrh.

Laennec's cases 19 through 23, which provide an account of emphysema, are instructive: three of the five patients had a history of a severe respiratory infection early in life, commonly in infancy or childhood; the same proportion had cor pulmonale. Two evidenced an underlying infectious, as opposed to inflammatory process: localized pneumonitis, pleural adhesions, and/or bronchial dilatation. Case 23 may have had cystic bronchiectasis. The average age at death of the five reported cases was forty-two.[23] It should be well considered—since these clinical features are not those one expect to encounter in emphysema—that cases which, in the absence of prominent findings of bronchial dilatation, Laennec denominated emphysema, we should currently style emphysema consequent to cylindrical bronchiectasis.

It is not commonly appreciated that emphysema can arise as a consequence of bronchiectasis. Ogilvie[24] examined pulmonary specimens of thirty-one individuals who underwent resection for this disorder and found that "emphysema was noted in all specimens and varied from small patchy areas to extensive areas, even in specimens in which collapse appeared complete on gross inspection. This last observation was interesting and rather unexpected, as little or no mention of emphysematous areas was made in the literature and from accounts of 'atalectasis' one gained the impression that collapse is complete." In a more detailed analysis, Reid[25] concluded that obliterative airways changes, and not dilatation, were the the structural abnormality of greatest significance in bronchiectasis and bronchiolectasis. McLean[26] observed that the earliest lesions in centrilobular emphysema were an obliterative bronchiolitis and alveolar ductitis and hypothesized that the distending forces distal to this obstruction resulted in alveolar wall disruption. In contrast, Leopold and Gough,[27] in their classic article on the pathology of emphysema, published the same year, observed patent communications between the bronchioles and the emphysematous spaces in all lesions of centilobular emphysema, and suggested an inflammatory mechanism for the alveolar loss. The "supplying" bronchioles invariably demonstrated inflammatory changes but none was obliterative; narrowing was common but severe in only 12 percent nor was it invariably present. Thurlbeck,[28] referring to bronchiectasis, echoes Reid's conclusions: "It should be reiterated at this point that it is not the dilated airways which produce airway obstruction in these patients. The site of airway obstruction is in the obliterated distal bronchi and bronchioles, and this is compounded by mucus plugs." He concludes, "Bronchiectasis (or, more precisely, its associated feature of obliterative airway disease) is occasionally the major source of chronic airway obstruction, and these patients are often misdiagnosed today."[29] In a physiological assessment of the same issue, Cherniack and Carton[30] concluded, "The pattern of physiologic dysfunction observed in bronchiectasis is quite similar to that occurring in chronic bronchitis and emphysema."

Two observations at Johnson's necropsy which help to define the nature of the process underlying his emphysema are deserving of comment: pleural changes, and right heart failure. Focal pleural thickening is a frequent finding in bronchiectasis: Ogilvie noted "gross pleural thickening," in twenty-five of thirty-five excised lobes, and Goodman[31] stated that "The lobes in all cases were markedly adherent to surrounding structures in the chest." One may reasonably infer one or more past episodes of suppurative infections from this finding. Cor pulmonale, attributable to the pulmonary hypertension resulting from numerous bronchial artery-pulmonary artery shunts that evolve in this disorder, frequently attends bronchiectasis; it was noted to be the cause of death in from 0 to 47 percent in five series summarized by Konietzko.[32]

That Johnson's emphysema was caused by chronic bronchitis merits consideration. The poet Robert Southey vividly characterized London's atmosphere in 1808 as ". . . a compound of fen-fog, chimney-smoke, smuts and pulverised horse-dung!"[33]

Brimblecombe's evaluation of London air pollution[34]—based on a mathematical model employing geologic estimates of coal sulphur content, the records of London coal imports, and maps indicating the dimensions of the city—indicates that sulphur dioxide levels rose to a plateau between 1690 and 1880, reaching an annual average value of $150\mu g\ m^{-3}$, twice current United States National Ambient Air Quality Standard of $80\mu g\ m^{-3}$. Eighteenth-century London ambient H_2SO_4 levels are not readily derivable from this figure because the ratio of the two is highly variable, depending as it does on atmospheric conditions, height of discharge, and presence of neutralizing levels of NH_4. H_2SO_4 comes into existence when SO_2 undergoes sequential oxidation and hydration, and is ten to twenty times more potent than SO_2 as a respiratory irritant.[35] Also unknown are levels of black smoke and suspended particulate matter, believed to act synergistically with H_2SO_4 as respiratory irritants, exacerbating asthma and chronic bronchitis. In contrast to the acute adverse health effects of SO_2 and other pollutants derived from fossil fuel combustion, a causal role in chronic bronchitis has been difficult to establish; epidemiologic studies of workers chronically exposed to high levels of SO_2 have shown no clearcut evidence for chronic effects.[36] Oswald and Medvei[37] evaluated symptoms of respiratory disease in London clerical workers by means of a questionnaire in the early post-war period. They observed a surprisingly high prevalence (9.4 percent) of bronchitis, defined as habitual productive cough associated with disability from exacerbations and/or breathlessness, among nonsmokers compared to 15.9 percent among smokers. The clinical import of these observations cannot be assessed as neither radiographic nor spirometric analysis was undertaken; nor was any attempt made to assess the role of London air-pollution in the genesis of the bronchitis, although this was suggested as a possible cause. In summary, there is no credible historical or epidemiologic evidence to indicate that Johnson's emphysema arose as a consequence of urban air-pollution. Furthermore, one might anticipate that, were this its cause, the condition he suffered from would be commonly seen and therefore easily recognized. Boswell however makes no mention of others among Johnson's circle of close friends who had this disorder, and Johnson in his own inquiries found no parallel example.

SUMMARY AND CONCLUSIONS

Of Johnson's fatal emphysema, it appears probable, on available historical and anatomic evidence, that it resulted from bronchiectasis, a diagnosis favored by the pattern of illness: a protracted and severe respiratory infection succeeded by annual episodes of severe winter bronchitis, remitting in summer, and culminating in respiratory insufficiency; and by the findings of pleural adhesion and cor pulmonale at necropsy. That it resulted from chronic bronchitis is a proposition both plausible and irrefutable without the specimen. "To be ignorant is painful; but it is dangerous to quiet our uneasiness by the delusive opiate of hasty persuasion."[38]

ACKNOWLEDGEMENTS

I wish to express my gratitude for the help offered by B. Corrin, department of pathology, NHLI, and The Royal Brompton Hospital; E. Allen, Curator, Hunterian Museum, Royal College of Surgeons of England; and H. Attwood, Medical History Unit, The University of

Melbourne in ascertaining that the specimen of Dr. Johnson's lung no longer existed. I wish to thank Dr. G. Snider for his many helpful observations and suggestions, and to add that he is of the opinion that Dr. Johnson's emphysema was more likely the consequence of obstructive bronchitis than bronchiectasis.

GLOSSARY OF MEDICAL TERMS

Alveolus: An "air sac"; the basic unit of gas exchange in the lung.

Anamnesis: The complete medical history of a patient.

Auscultation: Listening to the sounds made by the body, employed as a diagnostic method. René Laennec invented the stethescope and applied it to the diagnosis of diseases of the lung by correlating the auscultatory findings in the lungs of his patients with the pulmonary pathologic observations at postmortem examination. He died at age thirty-five of tuberculosis contracted as a result of close and frequent exposure to the causative agent as it affected his patients and persons on whom he conducted postmortem examinations. His 1819 treatise on this subject (see note 10) is a key advance in the scientific practice of medicine.

Bronchiectasis: A destructive condition of the walls of the conducting airways (bronchi) caused, most often, by a protracted and severe infection of the lower respiratory tract. As a consequence of the loss of structural integrity, the bronchi dilate and bronchial secretions pool in the airways. Persons with this condition typically experience repeated lower respiratory infections as a consequence of the stasis of their secretions.

Bronchiole: Diminutive of *bronchi*—a small conducting airway not supported by cartilege.

Centrilobular emphysema: A form of emphysema in which the destructive changes to the air spaces (alveoli) predominate in the central region of the lobule. Emphysema refers to dilatation and coalescence of alveoli ("air sacs"). A lobule is a primary lung unit; the normal human lung is estimated to contain twenty-three million lobules.

Consumption: Tuberculosis.

Cor pulmonale: Heart disease consequent to disease of the lung, most commonly, emphysema. The principal sign of cor pulmonale is swelling of the legs due to fluid retention ("dropsy"), the result of elevated back pressure from the failing heart.

Cylindrical bronchiectasis: A commoner form of bronchiectasis than the cystic variety (*vide infra*); on section, the bronchi retain the expected cylindrical contour but the calibre is greater than normal. Cylindrical bronchiectasis is usually the consequence of a destructive lower respiratory infection.

Cystic bronchiectasis: A type of bronchiectasis in which the dilatated bronchi end in large, spherical, and infected grapelike clusters. This form of bronchiectasis is thought to be the result of congenital malformation of the bronchial tree. See bronchiectasis.

Dropsy: Swelling of the legs (and, frequently, the abdomen) due to fluid retention, most often resulting from heart failure. Peripheral edema is a mild, and anasarca a severe, manifestation of dropsy.

Dyspnoea: A sense of shortness of breath that is felt to be disproportionate to the provoking physical effort or which may occur at rest. The usual causes are diseases of the heart or lungs.

Emphysema: Pathological dilatation and coalescence of alveoli.

Fibrosing alveolitis: Inflammation of the alveolar walls of the lung which results in scarring, and, ultimately, in respiratory failure. The cause is not determinable in most instances.

Foxglove: William Withering found *Digitalis purpurea*, the common foxglove, to be the active ingredient in a folk remedy for the treatment of dropsy. The active agent is the cardiotonic glycoside, digitalis. His monograph on the use of digitalis is a classic in medical literature. See Squill.

Haemoptysis: Bloody expectoration.

King's Evil: Tuberculosis involving the lymph glands of the neck. In the eighteenth century, it was commonly believed that it could be cured by the king's touch.

Orthopnoea: Dyspnoea in the supine position that is relieved by adopting a semi-upright position. It is most often caused by heart failure.

Phlebotomy: Drawing off blood from a vein.

Pleural adhesion: The pleura is a thin elastic membrane that covers the lung and lines the interior of the chest cavity. Bands of fibrous tissue connecting the two surfaces are called adhesions; they are most often the result of a prior lung infection.

Pneumonitis: Inflammation of the lung(s), most often infectious in cause.

Pulmonary hypertension: Elevation of the pressure in the arteries of the lung due either to failure of the left side of the heart, or, in the case of Johnson, to intrinsic lung disease.

Squill: Urginea maritima, a plant containing the cardiotonic.

Tuberculosis keratitis: Tuberculous inflamation of the cornea.

NOTES

1. *A Dictionary of the English Language* (London: Longman, Hurst, Rees, Orme and Brown. 1818).
2. James Boswell, *The Life of Samuel Johnson, LL.D.* (Chicago: William Benton, 1971), p. ix.
3. Walter Raleigh, "Samuel Johnson," in *Six Essays on Johnson* (London: Oxford Univ. Press, 1919), p. 31.
4. Walter Raleigh, "On the Two-Hundredth Anniversary of Johnson's Birth," in *Six Essays on Johnson*, p. 32.
5. James L. Clifford and Donald J. Greene, *Samuel Johnson, a Survey and Bibliography of Critical Studies* (Minnesota: Univ. of Minnesota Press, 1970).
6. G. J. Squibb, "Last Illness and Post-Mortem Examination of Samuel Johnson, the Lexicographer and Moralist, with Remarks," *London Journal of Medicine*, 1 (1849), 615-23.
7. J. P. Bishop, "Samuel Johnson's Lung," *Tubercle*, 40 (1959), 478-81.
8. H. Rolleston, "Medical Aspects of Samuel Johnson," *Glasgow Medical Journal*, 101 (1924), 173-91.
9. M. Baillie, *The Morbid Anatomy of Some of the Most Important Parts of the Human Body* (London: W. Bulmer and Co, 1803).
10. R. T. H. Laennec, *De l'auscultation mediaté; ou, traité du diagnostic des maladies des poumons et du couer, fondé principalemente sur ce nouveau moyen d'exploration* (Paris: Brosson et Chaude, 1819).
11. Boswell, *Life*, p. 575.
12. H. D. Attwood, "A Dissertation upon the Lung of the Late Dr. Samuel Johnson, the Great Lexicographer," *Lancet* (21-28 December 1985), 1411-13.
13. R. K. Larson, "Historical Note on Emphysema" [Editorial], *American Review of Respiratory Disorders*, 91 (1965), 277-78.
14. L. C. McHenry, "Dr. Samuel Johnson's Emphysema," *Archives of Internal Medicine*, 119 (1967), 98-105.
15. *Johnsonian Miscellanies*, ed. G. B. Hill (Oxford: Oxford Univ. Press, 1897), II, 278.
16. Samuel Johnson, *Diaries, Prayers, and Annals*, ed. E. L. McAdam, Jr. (New Haven: Yale Univ. Press, 1958).
17. Boswell, *Life*, p. 232.
18. Samuel Johnson, *The Letters of Samuel Johnson, with Mrs. Thrale's Genuine Letters to Him*, ed. R. W. Chapman (Oxford: Clarendon Press, 1952).
19. Boswell, *Life*, p. 581.
20. R. T. H. Laennec, *A Treatise on the Diseases of the Chest*, tr. John Forbes (New York: Hafner, 1962), pp. 76-80.

21. *Johnsonian Miscellanies*, I, 276.
22. Laennec, *A Treatise*, pp. 81-95.
23. Laennec, *A Treatise*, pp. 76-78.
24. A. G. Ogilvie, "The Natural History of Bronchiectasis," *Archives of Internal Medicine*, 695 (1941), 395-465.
25. L. M. Reid, "The Pathology of Obstructive and Inflammatory Airway Disease," *European Journal of Respiratory Disease*, 69 (Suppl. 147) (1986), 26-37.
26. K. H. McLean, "The Histology of Generalized Pulmonary Emphysema," *Australasian Annals of Medicine*, 6 (1957), 124-40.
27. J. G. Leopold and J. Gough, "The Centrilobular Form of Hypertophic Emphysema and its Relation to Chronic Bronchitis," *Thorax*, 12 (1957), 219-35.
28. W. M. Thurlbeck, *Chronic Airflow Obstruction in Lung Disease* (Philadelphia: W. B. Saunders, 1976), p. 392.
29. Thurlbeck, p. 21.
30. N. S. Cherniack and R. W. Carton, "Factors Associated with Respiratory Insufficiency in Bronchiectasis," *American Journal of Medicine*, 41 (1966), 562-71.
31. H. I. Goodman, "Suppurative Bronchiectasis," *American Journal of Surgery*, 26 (1934), 543-49.
32. N. F. J. Konietzko, R. W. Carton, and E. P. Leroy, "Causes of Death in Patients with Bronchiectasis," *American Review of Respiratory Disease*, 100 (1969), 852-58.
33. R. Southey, letter in *The London Anthology*, ed. H. Massingham and P. Massingham (London: Phoenix House, 1950), p. 218.
34. P. Brimblecombe, "London Air Pollution, 1500-1900," *Atmospheric Environment*, 11 (1977), 1157-62.
35. M. Lippmann, "Sulfur Oxides—Acidic Aerosols and SO_2," in *Environmental Toxicants: Human Exposures and Their Health Effects*, ed. M. Lippmann (New York: Van Nostrand Reinhold, 1992), pp. 543-74.
36. C. F. Federspiel, J. T. Layne, C. Auer, and J. Bruce, "Lung Function among Employees of a Copper Mine Smelter: Lack of Effect of Chronic Sulfur Dioxide Exposure," *Journal of Occupational Medicine*, 22 (1980), 438-44.
37. N. C. Oswald and V. C. Medvei, "Chronic Bronchitis: The Effect of Cigarette Smoking," *Lancet*, 2 (1955), 843-44.
38. Samuel Johnson, *Journey to the Western Islands of Scotland*, in *Samuel Johnson: Rasselas, Poems, and Selected Prose*, ed. Bertrand H. Bronson (New York: Holt, Rhinehart and Winston, 1958), p. 211.

MYTH UPON MYTH: JOHNSON, GENDER, AND THE MISOGYNY QUESTION

JAMES G. BASKER

In a recent book review, Marie McAllister raises important questions about the topic of "Johnson and Gender," and about the implications —some welcome, some (in her phrase) "troubling"—of recent scholarship on the subject. The book under review is a collection of six essays edited by Charles Hinnant and published in the *South Central Review* as a *Special Issue: Johnson and Gender.*[1] In addressing the volume, Professor McAllister immediately raises the essential question —"What precisely is meant by 'Johnson and Gender'?"—and then lists the "three basic answers" she finds in its contents: first, "the exploration of Johnson's attitudes toward women"; second, "the study of Johnson's female contemporaries"; and third, "the application to Johnson's own work of current theoretical approaches to gender studies." But she also discerns another answer beneath the surface: "Several of the essays here hint at a fourth, more troubling, explanation of what it means to think about 'Johnson and Gender,' one which I would like to consider before going on to the individual essays in this volume" (p. 397). In her analysis of this covert and "more troubling" rationale for some of the essays, Professor McAllister offers a thoughtful challenge to those of us who have written on the subject, a challenge that forces us to reexamine our critical positions and their underpinnings. Her analysis deserves careful consideration, and so I would like to examine her arguments in some detail before trying to respond with some additional evidence that might point to conclusions very different from hers.

Professor McAllister's first concern is with the scholars whose impetus, at least in part, is to refute Johnson's reputation as a misogynist and to show him to be "fundamentally sympathetic to the situation of women." "The battle they fight," she says, "may well be an invented one," and she goes on to question "to what extent 'the myth of Johnson's misogyny'[2] has ever really prevailed." The problem, she contends, is evidence—or the lack of it. Duly noting the infamous remark attributed to Johnson by Boswell—"A woman's preaching is like a dog's walking on its hinder legs. It is not done well, but you are surprised to find it done at all"—Professor McAllister doubts that "such remarks [have] really served readers as proof of Johnson's attitudes towards women" (p. 398). She also scrutinizes, and ultimately remains unpersuaded by, other evidence offered by the essayists, including reports of their students' preconceptions about Johnson, anecdotal evidence about scholars who view Johnson as sexist, and "the press's frequent choice of Johnson as an ideal 'quotable conservative to pillory.'" Professor McAllister goes on to report that her "own quick review of the scholarly record turns up many recent works demonstrating Johnson's respect for women . . . but next to nothing arguing the opposite." She is careful to mention a few recent works that criticize some aspect of Johnson on women, but notes that "none of these treatments labels Johnson a misogynist." After all this she concludes that, "at least in the scholarly literature . . . the 'myth of Johnson's misogyny' seems itself to be a myth" (p. 399).

She is quick to say that this may not in itself be a bad thing:

> The energy put into refuting a mythical myth, let me say, may not be entirely misapplied. After all, scholars who seek to prove that Johnson respected women, advocated their education, and thought them men's potential equals reveal something about themselves: that they believe those things themselves and enjoy knowing that the author whose works they love shares some of their perceptions. (p. 399)

In an additional but perhaps double-edged concession to such scholars, she also credits the value of the information they turn up: "Even if no critical tradition, scholarly or popular, claimed the opposite, evidence that Johnson liked and supported women is inherently valuable as a contribution to our knowledge of the man and his works." By this point, even as Professor McAllister has bent over backwards to acknowledge the redeeming qualities in such scholarship, her rhetoric has hardened in characterizing the degree of its authors' wrongheadedness. Doubt gives way to dismissal: Johnson's reputed misogyny has become a "mythical myth," behind which there seems to be "no critical tradition, *scholarly or popular*" (emphasis added) to substantiate it.

This leads to Professor McAllister's second main concern: motive. Justifiably alert to the possibility that for some scholars "Johnson the misogynist" serves as a straw man or convenient stalking horse, she pushes her conjectures to the extreme. "We need to ask whether the desire to believe in a mass of readers who dismiss Johnson as a misogynist might not stem from a less reasoned position: the fear that recent interest in gender studies somehow poses a threat to Johnson's position in the canon" (p. 399). It is the kind of question one might well ask, but one could wish it were in more measured terms. It is hardly fair to characterize the conclusions of scholars, however mistaken one thinks them, as springing from "the desire to believe [something]" or to transform the anti-Johnson opinions they report into a self-serving figment of their collective imagination.

As she expands on this idea of imaginary fears of Johnson being under threat from gender studies, it becomes unclear how much Professor McAllister is drawing her conclusions from the essayists in *Johnson and Gender* and how much from her sense of the climate of ideas in our time. Acknowledging that "the initial gesture of Anglo-American feminism was indeed oppositional [in the 1970s]," she describes a tremendous and dire reaction today:

> Now, two decades later, we are beseiged by jeremiads claiming that feminist proponents of "political correctness" wish to ban entirely any work perceived as misogynist (or in the most hysterical representations anyone identifiably white and male). Such jeremiads ignore contemporary feminist theory and practice. They also ignore practical evidence like the MLA's recent survey of members, which demonstrates what teachers of literature already knew: Johnson has not disappeared from the classroom, even if the syllabus may now include his female contemporaries. The perception that

> Johnson needs defending can easily lead to false feelings of embattlement. (p. 399)

This is a lot to contend about anyone, but I don't think Professor McAllister means to suggest that Donald Greene, Claudia Thomas, Bonnie Hain, Carole McAllister, or Catherine N. Parke—all of whom she goes on to praise later in the review—write such anti-feminist jeremiads. And perhaps out of courtesy or discretion, she does not name anyone who *has* written such jeremiads or cite any examples, though this might also be taken as a problem in a review so concerned with evidence up to now.

But the progress of her argument makes it easy to infer that those who write anti-feminist jeremiads and those who praise Johnson's views on women are, if not the same, then closely connected. Quickly and seamlessly her argument moves from the general—"we are beseiged by jeremiads," "such jeremiads ignore contemporary feminist theory and practice"—to the Johnsonian particular—"Johnson has not disappeared from the classroom," "the perception that Johnson needs defending." Distinctions are elided. It is not clear who exactly *does* "ignore practical evidence like the MLA's recent survey" or claim that "Johnson has . . . disappeared from the classroom," but readers would be forgiven for concluding that it was the Johnsonian scholars under review or others like them.

It is regrettable also that the language of conflict and warfare seems to be necessary to characterize differences of critical opinion. Professor McAllister describes the essayists' sense of purpose in terms of "warring factions," the need to "enter the battle," and "the battle they fight," and the tone is not alleviated by the vocabulary of fear, threat, seige, and embattlement that seasons the rest of her discussion. This is doubly unfortunate, as all the battles and feelings of embattlement eventually prove, in her view, to be "invented" or "false"—all except the sense that "we [feminists?] are beseiged by jeremiads," which is allowed to stand as true.

But even allowing for the rhetorical excesses (of a kind we all indulge in) and the dangers of trying to discern underlying motives (a procedure which can always be turned against us), Professor McAllister's fundamental question still stands: What about the evidence? *Is* the "myth of Johnson's misogyny" itself a myth? I would argue that it is not, and that the persistence of this myth, together with a widespread neglect of Johnson's writings on women, has shaped perceptions of Johnson over the past two decades and continues to shape them in various contexts today.

The first is the context in which Johnson is probably read more often than anywhere else these days: the classroom. Those who teach will have their own views of student preconceptions about Johnson, though I think a poll would discover that most of us have had experience similar to that so eloquently described by Catherine N. Parke in her essay on "Johnson, Women, and Gender in the Classroom."[3] But the books we use in the classroom offer more concrete evidence. Despite the fact that as early as 1789 some of Johnson's writings were included by Mary Wollstonecraft in what has been called the first feminist anthology (*The Female Reader*),[4] the twentieth century has witnessed an almost systematic suppression, or at least omission, of Johnson's writings on women. In a survey of some of the most widely available modern anthologies and textbooks, I found that the majority of them did not include in their contents even a single one of the more than forty *Rambler* and *Idler* essays Johnson had written in women's voices or on women's issues; only 25 percent of the anthologies printed more than one such essay. These included texts published by Yale, Oxford, the University of California, Penguin, Norton, and New American Library, among others, several of them still in print and available in paperback today.[5] Not only does the prevalence of such texts impede the teaching of Johnson's writings on women, it also makes it unlikely that students will accidentally discover that side of Johnson for themselves.

But to neglect or to be ignorant of Johnson's pro-feminist writings is not the same as to treat him as a misogynist. Professor McAllister is right that within the narrow compass of the "scholarly literature" most specialists know Johnson's views on women; she is also right that Virginia Woolf, "who helped to popularize the 'women preaching' remark in *A Room of One's Own*, there cites Johnson as her model of a man who believed in women's educational capabilities" (pp. 398-99). Why, then, does it seem that only the dancing dog remark gets remembered and repeated so persistently, even in the works of writers and scholars who, if they have not read Johnson, must have read Woolf? One hears it, for example, in the title of the essay Cynthia Ozick contributed to *Woman in Sexist Society* (1971), "Women and

Creativity: The Demise of the Dancing Dog." The specific source of her allusion is evident, as she prefaces her essay with a passage quoted from *A Room of One's Own*.[6] Adrienne Rich conjures up the same demon in her poem "Snapshots of a Daughter-in-Law":

> *Not that it is done well, but*
> *that it is done at all?* Yes, think
> of the odds! or shrug them off forever![7]

Some writers seem to treat the dancing dogs remark as a kind of touchstone, the one thing their readers would "know" in advance about Johnson and therefore necessary to be acknowledged before making any other kind of point about Johnson and women. Thus, in her book *The Brink of All We Hate: English Satires on Women 1660-1750* (1984), Felicity Nussbaum writes: "Samuel Johnson, in spite of his famous saying regarding women, preachers, and dogs, creates an intelligent woman in Rasselas's sister Nekayah."[8] It is exceptional, one gathers, that Johnson might imagine an intelligent woman.

One can only conjecture why people privilege one of Woolf's passages about Johnson over all others (Woolf was a great admirer of Johnson, and not just in *A Room of One's Own*[9]), but clearly such passages do constitute some evidence that there is a tendency to invoke Johnson's apparently misogynistic remark even (or especially) when little or nothing else is said about him and the topic of women. Moreover, writers like Cynthia Ozick and Adrienne Rich reach readerships far beyond the scope of our scholarly literature and thus, while one hesitates to call them "popular," they undermine somewhat Professor McAllister's suggestion that "no critical tradition, scholarly or popular" supports the "myth of Johnson's misogyny."

Perhaps in a soundbite culture given to "quotable quotes," instances such as these—however partial or misleading—are inevitable. What is more troubling are those more substantive cases in which Johnson is raised as the symbol, the totem of everything anti-feminist in eighteenth-century literary culture. I want to focus on two of these in particular because they both are the products of extensive and serious scholarship, and yet both address readerships far beyond academic specialists or the literary elite.

The first is Janet Todd's enormously valuable and pioneering work *A Dictionary of British and American Women Writers 1660-1800*.[10] This award-winning volume, compiled with contributions from more than 100 scholars all over the world, presents biographical information, much of it new, about more than 450 early women writers. First published in 1985, it was selected by the American Library Association as an "Outstanding Reference Source of 1986," and reissued in paperback in 1987. Useful to students and scholars alike, it has provided (as its publishers predicted) "a strong foundation for further scholarship in the burgeoning field of women's studies."

Inevitably Johnson is very "present" in this volume. At least thirty-four of these biographies of women writers mention Johnson, usually as friend or supporter, which accords completely with other scholarship about Johnson's extensive support for women writers throughout his life.[11] In fact Johnson is cited more than forty times in the index to this book, making him the most frequently cited author—male *or* female—in the entire *Dictionary of British and American Women Writers 1660-1800*. Apparently one cannot tell the story of women writers in the eighteenth century without taking Johnson into account.

Thus it seems a bit unfortunate that when Professor Todd sat down to write the "Introduction" to this important book, she set Johnson up as the apparent spokesman for male resistance to the emergence of women writers. He is invoked almost immediately, in paragraph two:

> Women's arrival in literature was frequently seen as a threat to male hegemony: "In former times," wrote Samuel Johnson in an essay of 1753, "the pen like the sword, was considered as consigned by nature to the hands of men. . . . The revolution of years has now produced a generation of Amazons of the pen, who with the spirit of their predecessors have set masculine tyranny at defiance."[12]

Presented thus, Johnson could seem at first glance the archetypal enemy to women's accomplishments. But in fact the real Johnson, rather than feeling any "threat" from women writers, welcomed and encouraged them, and even if one misses the jocular irony in his tone here (essentially applauding women for asserting themselves), a literal reading of the term "masculine tyranny" can scarcely suggest that his sympathies lay on the other side.

Of course Professor Todd has an historic point to make (and a valid one) and quotable quotes from Johnson can be found to support (with perhaps a little shading) almost any point. She is not responsible in this context for a full and balanced presentation about Johnson's views on women, and elsewhere she does mention him very briefly in more neutral tones.[13] But her readers are left to their own devices to discern that Johnson was *not* the leading opponent of women writers in the eighteenth century, that this paragraph goes *against* the grain of his real views and behavior, and even that, strictly speaking, the quotation may not actually support the generalization it is supposed to.

Johnson's presumed misogyny is presented unequivocally in the second example, a book of even greater reach and influence: *The Norton Anthology of Literature by Women*, edited by Sandra M. Gilbert and Susan Gubar. First published in 1985, it too is a pioneering work, "designed [according to its editors] to serve as a 'core-curriculum' text for the many courses in literature by women that have been developed over the past ten years."[14] Widely respected and continuously in print since 1985, it has been used in thousands of college courses and tens of thousands of copies have been sold. (A new edition is reportedly in preparation, to be published in 1996.)

Johnson appears in a section of the Introduction to the "Literature of the Seventeenth and Eighteenth Centuries," where this time, rather than being raised at the outset as a colossal symbol of opposition to women, he is instead positioned as the culmination of a two-century tradition of misogyny. Entitled "Images of Women," from its opening sentence ("Literary ladies, no less than other women, had to confront the misogynist images of women still prevalent in their culture . . . [since] medieval times") the section traces negative male images of women from the early 1600s down to 1800, drawing examples along the way from the works of Jonson, Donne, Marvell, Herrick, and especially the "misogynist poems" of Rochester, Swift, and Pope, among many others.[15] Attitudes hardened over time and the editors point especially to one key development:

> The emergence of a new image of woman, a satiric protrait of the female artist which constituted a reaction against the growing visibility of women. The late-seventeenth- and eighteenth-century attacks on literary ladies are important not only because they demonstrate why many women are unable to write or publish, but also they explain why those who did write

> and publish frequently felt it necessary to apologize or justify their efforts. (p. 52)

Again here the editors have a valid historical argument to make, but again here the use of Johnson is unfortunate. After going on to present a series of eighteenth-century satirists who "implied that reading and writing would derange weaker female minds," they draw the century (and the section) to a close this way:

> Toward the end of the century, when Dr. Johnson compared a woman preacher to a dancing dog, or when he condemned portrait-painting as improper in women because "public practice of any art . . . and staring in men's faces, is very indelicate in a female," he was assuming that creativity and femininity were contradictory terms. (p. 53)

Johnson's becomes the final pronouncement in a terrible two-hundred-year progress of misogyny. How ironic that the man who did perhaps as much as any other writer of his time to assist and encourage women writers should be rendered thus, but this is the Johnson students will find in *The Norton Anthology of Literature by Women*.

It's hard to measure what influence such books as Professor Todd's *Dictionary* or the *Norton* have actually had on people, or even to know whether students read introductions to reference works and anthologies (perhaps only if they've been assigned, if then?). But for the students who have read these remarks, one cannot imagine that they would be drawn to read more of Johnson any time soon, whether on women or any other topic. Their views, fleetingly formed, are likely to remain uncomplicated, unelaborated, uninformed. And how could we expect students to do otherwise? To ignore or resist the editorial authority of their texts, in order to go out and form, on their own, fuller ideas of Johnson than the scholars who compiled the books?

Within the scholarly world itself, I agree with Professor McAllister's perception of important developments in feminist criticism and gender studies. Second- and third-wave feminism seems to have evolved less oppositional approaches, subtler readings, and a greater variety of methodologies. Still, the soundbite invocation of Johnson as misogynist seems unlikely to die out soon. A recent article in the *Women's Review of Books* (September 1993) opens with a classic of its kind:

> In the eighteenth century, Samuel Johnson famously compared women preaching to dogs walking on their hind legs, concluding that "It is not well done [*sic*]; but you are surprised to find it done at all." Johnson's casual ridicule, of course, was the fruit of a patriarchal religion that rigidly prescribed and enforced women's submission within the church and interpreted Christian doctrine so as to efface the possibility that a woman could mediate God's word.[16]

Here again there is nothing further mentioned about Johnson that might temper readers' impressions of him as a sneering misogynist, although in this case the invocation of Johnson prompted a clarifying rejoinder from another periodical within a few months. In the summer 1994 issue of *The New Idler*, the editor quoted this passage from *The Women's Review of Books* and then conceded that "In some ways, you have to admit, this is fair enough. *If* Johnson did say these words, then his ridicule did indeed originate from Anglicanism's rigid subjection of women to the mediation of male priests."[17] *The New Idler* then goes on to provide important contextualizing detail and to identify a diversity of attitudes among modern feminists towards Johnson:

> We shall never know if it was Johnson or Boswell who shaped this casual snub; we do know that Johnson extended professional encouragement to women who *published* sermons, and Boswell needed and desired a patriarchal (that is, an antifeminist) Johnson. Plenty of today's feminists need and desire that same patriarchal figure against whom to rebel. Others, more and more, are glad to perceive a different Johnson—one who, in some contexts and at some moments, might be truthfully described as feminist.

To be sure, somewhat like Professor McAllister in her review, *The New Idler* attributes to part of one group a "desire" for a "figure against whom to rebel." But to his credit, he balances his observation by acknowledging and sympathizing with the historic wrongs done to women, and by pointing out the existence of other feminists, in growing numbers, who perceive Johnson in very different ways.

Two brief passages about Johnson appearing thus, which is likely to carry more weight? Published by the University of Chicago Press, *The Women's Review of Books* is an international journal that enjoys contributions from scores of distinguished scholars and an extensive readership in many different disciplines. By contrast, admirable and interesting as it is, *The New Idler* is a modest sheet, "issued occasionally by the Johnson Society of London" and circulated among a membership of perhaps a hundred Johnsonians. Which of them constitutes real evidence of the prevailing view of Johnson outside the relatively narrow circle of Johnson specialists?

How does one weigh, finally, the influence of Cynthia Ozick, Adrienne Rich, Janet Todd, Sandra M. Gilbert, Susan Gubar, and Elizabeth Langland, of Norton anthologies and biographical dictionaries and *The Women's Review of Books*, against all the Johnsonian scholars and periodicals in the world? In none of what I have said about individual scholars and publications do I mean to impute to any of them a real antipathy to Johnson. They touched on him or quoted him in ways that were immediately useful to them, that served other purposes, other priorities, in particular contexts. Indeed, I may well have done some of them a disservice, failing to cite other instances in their work of favorable opinions or more thorough treatment of Johnson on women. I meant only to suggest that there was more than enough evidence around to explain why scholars might feel that Johnson was perceived in some quarters as a misogynist and therefore that efforts to defend or praise him could not be attributed solely (if at all) to a groundless, unreasoned response to feeling threatened by gender studies. Such imputations can divide us needlessly. Perhaps in future discussions of the subject, we should all take our cue from the more balanced and generous ethos that emanates from *The New Idler* and from the rest of Professor McAllister's review of *Johnson and Gender*.

NOTES

1. Marie McAllister, review of Charles H. Hinnant, ed., *Special Issue: Johnson and Gender. South Central Review.* The Journal of the South Central Modern Language Association. Vol. 9, no. 4 (Winter 1992), in *The Age of Johnson*, 6 (1994), 397-404. All references hereafter cited parenthetically in the text.
2. The phrase comes from the title of my 1990 article, "Dancing Dogs, Women Preachers, and the Myth of Johnson's Misogyny," *The Age of Johnson*, 3 (1990), 63-90.
3. Catherine N. Parke, "Negotiating the Past, Examining Ourselves: Johnson, Women, and Gender," *Special Issue: Johnson and Gender. South Central Review*, vol. 9, no. 4 (Winter 1992), 71-80.
4. See recent editions of *The Female Reader* by Moira Ferguson (Delmar, N.Y.: Scholars Facsimiles & Reprints, 1980) and by Janet Todd and Marilyn Butler (vol. 4 of *The Works of Mary Wollstonecraft* [New York: New York Univ. Press, 1989]) and, for a fuller discussion of Wollstonecraft's selections from Johnson, see my "Radical Affinities: Mary Wollstonecraft and Samuel Johnson," in *Tradition in Transition: Women Writers, Marginal Texts, and the Eighteenth-Century Canon*, ed. Alvaro Ribeiro, S.J. and James G. Basker (Oxford: Oxford Univ. Press, 1996), pp. 47-48.
5. For details of the anthologies and other texts surveyed, see Basker, "Dancing Dogs," pp. 78-87.
6. Cynthia Ozick, "Women and Creativity: The Demise of the Dancing Dog," in *Woman in Sexist Society: Studies in Power and Powerlessness*, ed. Vivian Gornick and Barbara K. Moran (New York: Basic Books, 1971).
7. Adrienne Rich, "Snapshots of a Daughter-in-Law," *The Fact of a Doorframe: Poems Selected and New 1950-1984* (New York: Norton, 1984), p. 38. I am indebted to Catherine Parke's essay "Negotiating the Past" for pointing out this passage.
8. Felicity Nussbaum, *The Brink of All We Hate: English Satires on Women: 1660-1750* (Lexington: Univ. of Kentucky Press, 1984), p. 162.
9. For Woolf's view of Johnson, see, among many others, her expressions of esteem for him as a critic in "Coleridge as Critic," *TLS*, 7 February 1918, reprinted in Virginia Woolf, *Books and Portraits* (London: Triad/Granada, 1979), pp. 46-47; as an enemy to social pretense in "Dr. Burney's Dinner Party," *The Common Reader: Second Series* (London: Hogarth Press, 1932); and as a kind of oracle from whom "rolled out the most magnificent phrases that have ever left human lips" in *Orlando* (London, 1928; repr. London: Harcourt Brace Jovanovich, [n.d.]), p. 223.

10. Janet Todd, ed., *A Dictionary of British and American Women Writers 1660-1800* (Totawa, N.J.: Rowman & Allanheld, 1985; rev. ed., Rowman & Littlefield, 1987).
11. See especially Isobel Grundy, "Samuel Johnson as Patron of Women, *The Age of Johnson*, 1 (1987), 59-77, as well as other works listed by Parke (p. 79, n. 7) and Basker ("Dancing Dogs," p. 90, n. 47).
12. Janet Todd, "Introduction," *Dictionary of Women Writers*, p. 1.
13. Todd, "Introduction," pp. 8, 9, and 16.
14. Sandra M. Gilbert and Susan Gubar, "Preface," *The Norton Anthology of Literature by Women* (New York: W. W. Norton & Company, 1985), p. xxvii.
15. "Images of Women," in "The Seventeenth and Eighteenth Centuries," *The Norton Anthology of Literature by Women*, pp. 50-53.
16. Elizabeth Langland, "Sin and Sensibility" (a review of Christine L. Krueger, *The Reader's Repentance: Women Preachers, Women Writers, and Nineteenth-Century Discourse*), in *The Women's Review of Books*, 10 (Sept. 1993), p. 23.
17. *The New Idler*, 16 (Summer 1994), p. [1].

THE SENSIBILITY OF SIR HERBERT CROFT IN *LOVE AND MADNESS* AND THE "LIFE OF EDWARD YOUNG"

MAXIMILLIAN E. NOVAK

When Sir Herbert Croft visited Eleanor Butler on Wednesday 12 November 1788, he identified himself as the author of "Young's life in Johnson's Poets" and the compiler of a yet unpublished dictionary. Two month's later he dropped by again with the latest gossip about Pitt and the mental condition of George III. This time Butler identified him as the author of *Love and Madness* and concluded, "Like him most exceedingly, pretty manners, sociable, accomplished."[1]

Such a favorable opinion is hardly to be discovered in the notes and anecdotes about Croft encountered in Johnsoniana and Boswelliana. There he is described, at various times, as a duplicitous scoundrel and a "rascal" and his account of Edward Young criticized as a futile imitation of Johnson's style and manner.[2] Boswell quoted Burke to the effect that Croft had "all the pomp without his force . . . all the contortions of the Sybil without the inspiration."[3] His style, judged alongside Johnson's, has "all the nodosities of the oak without its strength." He was accused of having cruelly exploited the Chatterton family after the death of the poet. And this was supposed to be all the worse because he was both a baronet and a priest.[4]

Before turning to Croft's writings I want to make two points. First of all, the Croft who appears in the notes to most studies of Johnson or Boswell as a seemingly shadowy and morally ambiguous figure was much more substantial a writer than he is depicted. *Love and Madness* was a huge success and one of the truly important works of fiction written during the twenty years following Smollett's death and the appearance of *Humphrey Clinker*. The subject was unpleasant—the psychology of someone obsessed with the love of the mistress of a public figure. Martha Ray was no Charlotte and Hackman was no Werther. It was an exercise in abnormal psychology bordering on madness, and it was written with great insight. The themes of suicide and murder that dominate it merge with a quest after true and authentic emotion that make the lengthy section on Chatterton perfectly compatible with the rest. By creating a web of allusions to real and fictional lovers who have committed murder and suicide, Croft gives a psychological portrait of the times that we find nowhere else.

Secondly, Croft was an ingenious literary projector, and in his series of fictional epitaphs to be found in Kilkhampton Abbey and in Westminster Abbey by investigators in the years 1980 and 2001, he used (if he did not invent) a form of science fiction which is still popular. That it was certainly popular at the time there can be little doubt. *The Abbey of Kilkhampton* went into fourteen editions, according to the *DNB*. The form allowed for satirical commentary on his contemporaries (including Boswell), along with the occasional inscription with unstinting praise, such as that devoted to Johnson. It was mainly a device for scandalous commentary, but the projection into the future along with the antiquarian exploration of the past must have provided genuine amusement to those familiar with contemporary scandals. His other projects proved less fortunate, but they were not without merit. His journal, *The Literary Fly*, has some lively numbers, and had to be abandoned, chiefly because it did not appear to promise much financial reward. That he was impoverished, arrested for debt at the time of his second marriage and forced to live most of his life in France did not endear him to success-oriented nineteenth-century critics. But he was not without talent.

1

Told in a series of letters, mostly from the clergyman, James Hackman, *Love and Madness* focuses upon an event which, as Horace Walpole remarked, fascinated much of London for the month of April 1779, the murder of Martha Reay or Ray by a man who loved her to distraction. She was the mistress of Lord Sandwich, whose connection with the government led everyone to think initially that there was a political motive to the killing. That, at least, would have made sense to Walpole and his contemporaries. Hackman had waited for her after she left Covent Garden theater where she had attended a performance of *Love in a Village*. Carrying two loaded pistols, he stalked her as she went to her carriage with Galli, her singing instructor, and as she was being handed into the carriage by a John Macnamara, with whom, rumor had it, she had been flirting during the performance, shot her through the head and then attempted to do the same for himself with the second pistol. When the shot merely grazed his head, he attempted to kill himself by hitting his head against the pavement. He was apprehended, tried at the Old Bailey on 17 April and hanged two days later. At the trial, he asked for no mercy, describing himself as the most miserable of men. He stated that his intention had been to kill himself in front of her, but seized with a sudden "momentary phrenzy," he found himself shooting her.[5] Martha Ray was thirty-four years old and the mother of five children with Lord Sandwich.

Since, at the time, there was no legal plea equivalent to temporary insanity, there was never much doubt about his fate. A typical reaction was that of Lady Sarah Lennox, who wrote her friend, "Don't you pity that poor devil Ld Sandwich just now, & the poor girl too, & I think the poor man too; tho' I think it quite right he should be hanged. I pity him monstrously."

Love and Madness is a book about violence—violence toward others, toward the self, and violence in and towards texts. Robert de Maria has recently repeated the notion floated by Robert Southey and apparently current in his circle that *Love and Madness* had little value except as—an excuse for a biography of Thomas Chatterton.[6] But Croft saw Chatterton's life and death in larger terms—in terms of the suicide of the hero in *The Sorrows of Werther*, in terms of forgery and plagiarism of the kind committed by Macpherson in his *Ossian*, and, of course, in terms of one of the most sensational murders of the eighteenth century—James Hackman's shooting of Martha Ray, the mistress of the Earl of Sandwich. All of this was intended as a display of that psychological state of excess sensibility known, in George Cheyne's phrase, as "the English malady."[7] For if Werther's tale of woe was advertised as a "German Story," it was quickly seen as the kind of tale of emotional excess that was extraordinarily English.[8] What I want to suggest in this part of my essay is that Croft's depiction of James Hackman's mental and emotional disintegration is delineated through a series of psychologically related allusions and fixations—that so far from being a series of letters on random subjects, it attempts to force the reader to make some difficult but intriguing connections.

2

Like so many of his contemporaries, Horace Walpole found the entire series of events "strange." "Now Madame," he wrote to Lady Ossory, "can one believe such a tale? . . . To bear a hopeless passion for five years, and then murder one's mistress—I don't understand it." Indeed, he found it so difficult to fit into his concept of human nature that he thought it had hardly any historical precedent. "What a shocking instance of the little power reason has over a violent passion, has been manifest in Hackman's case," he wrote his friend Mann. "One has often heard of people's hanging and drowning themselves for love, but not of destroying the object of it and himself too." He wondered at that older concept of insanity—being possessed of a demon as a possible explanation. Although Hackman did not seem like a "madman" at the trial, it could only be explained as a momentary fit of lunacy. He wondered if there were not twenty mad persons wandering about to one confined in Bedlam. Walpole's observations were not very profound, but, reminded of the case by the publication of Croft's *Love and Madness*, he did wonder along with many of his contemporaries whether the legal system was not entirely inadequate in such a case, being unable to "define the shades of madness."[9]

It should not surprise modern observers of such a case that most of the sympathy went out to James Hackman rather than to Martha Ray.[10] Although everyone agreed that she had considerable charm and some talent as a musician, she had been older than Hackman, who was only twenty-six, and, of course, she was the mistress of a man associated with the despised governing class.[11] Besides, she had apparently become less enthusiastic about the relationship, while Hackman's passion had burned fiercely over a number of years. Boswell brooded over Hackman's fate with his usual fascination for those afflicted with what seemed like a possible madness and doomed to be hanged for it. He was, after all, that ideal of the age—a man of exquisite feeling, who had seemingly sacrificed all for love.[12] A poem, supposedly by Hackman, was published in *The Public Advertiser* on 5 May 1779, a few weeks after his execution. It summarized all that the audience wanted to see in the case:

> I murder'd in my Heart, an Angel there.
> O! wretched Sense of Feeling exquisite,
> Expos'd to keenest Misery, or Delight.
> Nerves that vibrate in Agony of Pain,
> Then stretch to Joy each Ligament again.
> When Reason left the Vessel in ful Sail,
> How was I wreck'd by Passion's maddest Gale,
> Wreck'd on a Shore unknown, without a Ray,
> To chace that dreadful Spectre, Fear away.

Such a display of sensibility came closer to understanding a Hackman than Walpole even could, but it took a Herbert Croft, who at the age of fifteen wrote a book in which he expressed all the right sentiments against the Earl of Chesterfield's politeness, against the terrible mother of the poet, Savage, along with a sentimentalist's longing to join the Tahitian, Omai, on his distant island to empathize fully with Hackman's irrational feelings and longings.

Love and Madness begins much in the manner of *La nouvelle Héloïse* with a passionate exchange of letters, but unlike Rousseau's lovers, Hackman and Ray do not develop into a philosophic family of platonic lovers. Instead, their love turns into a nightmare of misunderstood passions. Hackman is presented as a person who sees his world through examples, both literary and real, and it is easy enough to see the confusion in his mind through the works he admires. He borrows a copy of *The Sorrows of Werther*, the seemingly dominant text for the way he conceives of his world. But he also feels an intense identification with Othello, and while he does not make the explicit connection, Ray is clearly to be cast as the murdered Desdemona. There are the dramatic lovers of Jerningham's *Faldoni and Theresa*, of Sergeant's *The Mine*, and of Oroonoko and Imoinda, who are quoted on the title page. Writing of the passions that consume him, Hackman states,

> . . . men, real men, have never been wild enough for my admiration: it has wandered into the ideal world of fancy. Othello, . . . Zanag, are my heroes. Milk-and-water passions are like sentimental comedy. Give me (you see, how, like your friend Montaigne, I strip myself of my skin, and shew you all my veins and arteries, even the playing of my heart)—give me, I say, tragedy, affecting tragedy, in the world as well as in the theatre.—I would massacre all mankind sooner than lose you.[13]

Such literary allusions are buttressed by references to real characters who managed to kill themselves and/or those they loved: a woman named Mrs. Dixon poisoned herself for love; a servant named Empson who tried to murder a maid servant who had rejected him; and most fully, the case of the Italian, Ceppi, who succeeded in his murder of the woman he loved. Just as Hackman felt an identification with Othello, so he believed that Ceppi truly loved the woman he killed and was to be pitied, even though such "Italian" emotion would, supposedly, be beyond the worst excesses of an Englishman.[14] In actuality he follows Ceppi's scheme of bringing the two loaded pistols exactly.

But it was with Werther that Hackman identified most fully, the Werther who can say about a poor man who loved deeply and lost his mistress when he tried to rape her, that it was she who was to blame, while the man remained faithful:

> This love, this faithfulness, this passion is, then, not an invention of poets. it exists, and it is purest in that class of people whom we call uncultivated, rude—we who are formed by culture and deformed into nobodies! Please read this story with reverence. I am calm today as I write this down; you see from my handwriting that I am not boiling and bubbling as usual. (179)

In much the same manner, Hackman pours out his praise of love toward his mistress:

> Oh love, love, canst thou not be content to make fools of thy slaves, to make them miserable, to make them what thou pleasest! Must thou also goad them on to crimes, must thou convert them into devils, hell-hounds! (123)

Werther was an important text for Croft in several ways. First of all, there was all the emotionalism and sensibility expressed toward the object of the loves of Hackman and Werther along with a love object who was not particularly memorable. Werther's emotion over the scene of Charlotte cutting bread with butter for her brothers and sisters, which moves Werther so, has its parallel in Hackman's love for a woman who had apparently lost most of her beauty, who was in her mid-thirties with five children. Secondly, the irrationality and instability of both lovers was obvious to all. The introduction to the first translation of Werther from the German argued that "*the design . . . is to exhibit a picture of that disordered state of mind, too common in our own country.*"[15] And Walpole, writing of the authenticity of Croft's novel, was willing to admit that Hackman might be regarded as a "wretched lunatic."

Finally, the greatest sympathy among readers was for Werther and Hackman, while Charlotte was pilloried as "an ungenerous, vain, unfeeling character,"[16] and Martha Ray as an experienced woman who had lost interest in her overly passionate lover. Of course, Croft goes far beyond Goethe in suggesting an only partially concealed aggression against Martha Ray. He attempts to make what Mrs. Thrale called "the strangest thing that has happened these hundred years"[17] completely understandable to the reader if not to himself.

Of course the large section on Chatterton should have established the notion of Hackman's suicide as something that the reader might expect in the same way that Hackman's continual discussion of literary works involving extravagant and violent scenes caused by excessive love warns the reader of some violent end.[18]

But it does more. The defense of Chatterton's fraudulent effort at passing off his poems as the work of the Monk Rowley has to be seen not as an interpolated piece of biographical criticism within a very different text. Rather, it is Chatterton's actions as seen through the eyes of a man of extraordinary sensibility, and one for whom consistency was not a hallmark. After insisting that there can be no doubt that Chatterton wrote all of the poems that he ascribed to Rowley, Hackman argues the impossibility of understanding the motivation of someone who was really just a boy. He suggests the possibility of a youthful pleasure in deception as well as the state of the times, when everyone was writing some kind of forgery or other. In Hackman's mind, all may be excused on the basis of his youth and genius. It was not worse, he argues than the statue Michelangelo attempted to pass off as an antique work. There is much of Croft's naive enthusiasm in his creation of Hackman, but there is also some distance from this unquestioning admirer of Chatterton. On the whole, Hackman's criticism of Chatterton is part of the novel—an enthusiastic picture of a genius whom society sacrificed, someone who took the privilege of the artist to violate the laws of society, in much the same way that Hackman was to violate those laws in murdering the woman he loved.

Although contemporaries found Croft's novel sensational enough to go into six editions in 1780, it is not hard to understand why it never became canonical. The entire story of Martha Ray would have been considered improper reading by nineteenth-century critics, and the violence of a disturbed lover stalking and murdering his prey, while it may capture modern interest for a time, is hardly very pleasant. The sensibility exhibited by Hackman, though it was of considerable interest in the years around 1780, was hardly acceptable to most readers without some kind of moral. What that moral might be was as puzzling to Croft's contemporaries as it is to us. And while we might find Croft's attempt to recreate the murderous intentions of Hackman fascinating enough, that fascination resides, at least partly, in our bemusement over the excesses and oddities of human passion.

3

If Croft did not identify entirely with James Hackman, he certainly sympathized with his creation as a person of profound sensibility. Under these circumstances, it is hardly surprising that, when he came to publish the "Life of Young" in the year following the appearance of *Love and Madness*, he should have found the poet lacking in true sensibility. The main point of this section of my essay is to argue that despite the often repeated notion of Croft's biography as an imitation of Johnson's method, it was anything but that. For if Johnson was to demonstrate that a man of bad character, such as he took Milton to be, might yet write great poetry, Croft was to suggest that the poetry written by a man of suspect sensibility had also to be regarded as suspect.

It is fair enough to say that Johnson was not averse to seeking shortcuts in his work on the *Prefaces Biographical and Critical* for the *Works of the Most Eminent English Poets* he had agreed to edit in 1777. Croft, who was acquainted with Young's son, Frederick, and was eager to establish his literary reputation, seemed to have special access to information about Young that Johnson would have been only too eager to use. Aside from these considerations, Young would have posed a particular problem for Johnson. Though he admired parts of Young's major series of verse satires, *The Love of Fame*, Johnson found the poem upon which Young's reputation was based, *Night Thoughts*, to be extremely uneven with occasional flashes of brilliance. To one of Young's admirers, Mrs. Thrale, he remarked, "Young's Works . . . are like a miry Road, with here and there a Stepping Stone or so; but you must always so dirty your Feet before another clean Place appears, that nobody will often walk that way." Mrs. Thrale rushed to Young's defense by noting that much the same could be said about Johnson's favorite, John Dryden. She felt she had made a point, but next morning, by way of reestablishing his position, Johnson returned to the attack as they were "drawing Spirits over a Lamp, and the Liquor bubbled in the Glass Retort; there says Mr Johnson—Young bubbles & froths in his Descriptions like this Spirit; but Dryden foams like the Sea we saw in a Storm the other day at Brighthelmstone."[19] In another version, of this anecdote, Johnson contrasted Young's poetry to Dryden's as the noise of a "tea-kettle" compared here with "the roaring of the ocean."[20] Such remarks make Boswell's version of Johnson's opinion of Young's poetry open to doubt. Boswell announces that he considered *Night Thoughts* "as a mass of the grandest and richest Poetry that human genius has ever produced" and is happy to report what he considers to be Johnson's concurrence; but it is Boswell, not Johnson, who insists that anyone not moved by Young's poem has a "hard and obstinate frame."[21] All that Johnson seems to concede is that, given the subject matter, blank verse might not have been a bad choice, that the poem had some originality, and that the power of the work was to be discovered in the "whole." How much of a compliment may be found in Johnson's praise of Young's poem as "a wilderness of thought" is difficult to say.[22] It is because Johnson's negative comments are so trenchant and his silences so deafening that Boswell depends more on his own opinions than on Johnson's in telling his readers what they ought to think of Young, resorting to quoting Johnson's brief critical remarks appended to the end of Croft's mixture of biography and criticism.

Having little appreciation for blank verse in general, Johnson was clearly never going to satisfy the legion of those who considered Young a great British poet. How convenient to have someone else treat a poet whose major claim to fame seemed to be based upon popular taste and sentiment rather than real poetic skill. Boswell reported that Johnson disapproved of the informal style of Croft's discourses for servants, *Sunday Evenings*, but that would have been several years down the road in 1784. Is it possible that Johnson didn't like the final quote from Young's *Night Thoughts* with which the series concluded?[23]

Certainly Croft cultivated the image of himself as a possible heir rather than a pale reflection of Johnson, going so far as to have a painting done by Abbot in which he was depicted with busts of Johnson and Bishop Lowth, the two men upon whom he hoped to model his life. Croft's plan for a new dictionary was intended to be an improvement upon Johnson's work but nevertheless a kind of monument to the creator of the *Dictionary*.[24] But for all of this, Croft was a person with a very different sensibility from that of his master, and his contribution to what were soon to be called the *Lives of the Poets* reflected something closer to the attitude of a young man of the 1780s than anything else found in this collection of biographies.

Croft held what we might call a "romantic" notion of the career of the writer. As Saul Bellow remarked of society's attitude toward his doomed poet, Von Humboldt Fleischer (modeled on Delmore Schwarz), "frantic desperate doomed crazy writers and suicidal painters are dramatically and socially valuable."[25] Britain had reached that stage at the end of the eighteenth century. As has been mentioned, in one of his earliest works, *A Brother's Advice to His Sisters*, which first appeared in 1775, Croft sounded all the right notes for a young man of the time. He excoriates Swift because he pretended to be worse than he was; Churchill was good because he once gave all the money he had to a poor woman. Boswell, who already had a reputation for a fascination for executions, hardly wins more approbation than Hume. Croft takes a high moral tone throughout and writes more like the future clergyman than the student of the law, the profession he first practiced. But what is most significant is his stance toward Gray.

Cowper may have considered any criticism of Gray a desecration, but Croft did not hesitate to find fault with the earlier generation of the poets of this era of sensibility. Croft thought that although Gray was an admirable poet, the man who emerged from Mason's biography did not deserve "praise" and threw doubt upon the authenticity of the poetry.[26]

On the other hand, Croft loved to indulge in scenes of violent sensibility of a kind that eventually brought on the high tide of Gothic fiction in England. The Bodleian has a manuscript poem by him in which a slighted heroine cuts out the heart of the false lover she has murdered and grinds it beneath her feet.[27] And as I have shown, in his novel, *Love and Madness*, he turned the fictitious letters between James Hackman, the soldier turned clergyman, and Martha Ray, the aging mistress of the Earl of Sandwich, into an allusive theater of proper sensibility. She loves *Poor Robin Gray*, an art ballad by Lady Anne Lindsay, and both admire Jerningham's *Faldoni and Theresa* in which a couple perform a ritual suicide. The story of a woman who joined her lover in the mines of Austria to which he had been sentenced unjustly brings tears to their eyes.[28] And the blend of sensuality with potentially murderous intent is very much a part of the sensibility of the end of the century.[29] That Croft was well positioned to admire certain elements in Young's *Night Thoughts* is certain, but whether he thought Young fully qualified to enter the pantheon of poetic geniuses either as a poet or a man is doubtful.

In the most recent biography of Young, Croft is blamed for giving birth to a biographical image of Young that was far from flattering and thereby depriving him of a proper appreciation of his artistic achievements.[30] In fact, the process began shortly after the poet's death in 1765 with the biography included in *Biographia Britannica* (1766). Its contemptuous evaluation of most of his poetry, along with implications of an unforgiving attitude toward his son, continued in the second edition of that work published in 1782, after Croft's life appeared. "He fell unwept by the Muses," wrote Isaac Reed in that work, "and passed as silent to the grave as piety or modesty could wish."[31] Croft was far more appreciative of Young's talents than Reed, but he knew Frederick Young quite well when they were both in Lincoln's Inn together. Frederick apparently resented his father's treatment of himself and his mother. The view of a son toward his parents is hardly to be accepted as the final word, but Frederick Young clearly felt neglected by his father and thought that his mother was both "overlooked" and "the object of his contempt." Johnson's one excision from Croft's life were these remarks on Young's treatment of his wife, either because he had better information or because he felt it too scandalous.[32] Croft, who was incapable of managing his own money, may have looked upon Frederick's complaints without sufficient distance, but it appears true enough that Frederick Young died miserably in a debtor's prison. As a result, whatever claim Young's poetry made for being placed in the canon of the geniuses of sensibility, his failures as a man had to raise perpetual questions about the authenticity of his feelings. There was some contemporary belief that Young's poem had a degree of vulgarity about it—that it was more verse for the popular consumption than a poetry of refined sensibility.[33] We should approach Croft's "Life of Young," with these contemporary attitudes in mind. Although Croft refers to Young at the beginning as "the great Author of *Night Thoughts*," there is a strain of irony throughout the biography.[34] Despite making apologies for Young's attempts to seek patronage, Croft leaves the implication that such self-interest is beneath the actions of a true poet. He cannot help remarking that the encomiums Young wrote for his great patrons, whether Queen Anne or George I, whether Lansdowne or Wharton, were often transferred to new patrons with only a few verbal adjustments. Young sought a position in Parliament but found his oratory failed him. He was clearly a poet who wrote out of a sense of isolation and loneliness, but he also exploited those qualities in order to gain attention. And there is an accusation of hypocrisy. In noting that Young advocated laughing at the world in the preface to his *Love of Fame*, Croft reminds his readers that Young was apparently entirely incapable of doing any such thing. He defends Young for writing in praise of his patrons and for accepting a pension with the off hand remark that at least he showed his gratitude. But such a defense, along with its attack upon "modern patriotism," begs the question that hovers over the entire biography: Is this the kind of behavior to be expected from a poet of sensibility? Given Young's perpetually seeking official position (He became Chaplain to George II in 1728), a poem such as "Ocean; An Ode," with its concluding "Wish" for humble retirement becomes in Croft's eyes "a curious page in the life of Young." And after quoting lines such as "Prophetic schemes,/And golden dreams,/May I, unsanguine, cast away!/Have what I *have*,/And live, not *leave*,/Enamoured of the present day," and "My chief revenue in content" the reader feels that Croft's remark on the poetry having a degree of "self burlesque" applies equally to the relationship between the poetic sentiments and the real career of the poet (III, 373-74).

The anecdote concerning Young's reading Aquinas on the advice of Pope after he took orders in 1728 and ending up close to "*an irretrievable derangement*" is told with a certain wryness, considering that, as Croft observes, Young should have sought the advice of a clergyman. And his account of the tragedies in Young's life—the deaths of a step-daughter and his wife—is treated with the awareness that Young exploited his grief to compose a popular poem. Croft notes the lapse of considerable time between the deaths in his family and the adventitious foreshortening in the poem as proof of a certain inauthenticity and suggests that by the fifth book of *Night Thoughts* Young had run out of all emotional energy. Croft remarks, Young was always ready to produce a melancholy line, implying that Young appeared to be merely biding his time, waiting for the occasion that could produced his major work. One is reminded of Johnson's strictures on Milton's emotions in writing "Lycidas," "Where there is leisure for fiction there is little grief" (I, 163)

Croft argues that the admonition about never speaking ill of the dead "savours more of female weakness than of manly reason," (III, 379), and he speaks out in defense of Young's son, Frederick, as a person who has been falsely vilified, even being identified with Young's arch libertine, Lorenzo. Although Croft shows that Frederick would have been too young to serve as a model, he nevertheless leaves floating an image of Young as the crucifier of his own son by describing Young's poem on Michelangelo's painting of the crucifixion. Croft attempts to excuse Young as a bad parent on grounds that appear almost deliberately weak:

> Young was a poet; poets, with reverence be it spoken, do not make the best parents. Fancy and imagination seldom deign to stoop from their heights; always stoop unwillingly to the low level of common duties. Aloof from vulgar life, they pursue their rapid flight beyond the ken of mortals, and descend not

> to earth when obliged by necessity. The prose of ordinary occurences is beneath the dignity of poetry. (III, 382).

At a time when Rousseau's *Emile* had made the education of children a high duty for parents, such an attempt at exculpation was surely not intended to carry much weight.[35]

Young the man is judged harshly even under the often repeated title, "the author of *Night Thoughts*. Croft notes that Young scattered compliments to "Earls and Dukes, from Speakers of the House of Commons, Lords Commissioners of the Treasury, and chancellors of the Exchequer." The poet is frequently betrayed by the "politician" (III, 382). He sees Young as continually desirous of rewards by those in power but, ironically enough, trapped by his glorification of retirement from the world that fills his poetry. By living quietly at Welwyn in Hertfordshire, Young managed to approach that "absolute retreat" that Winchelsea and so many other poets of the age glorified. Croft suggests that his success in achieving this state had to be disastrous for someone ambitious of preferment. "No hand," writes Croft, "was reached out to pull him self from that retirement of which he declared himself enamoured" (III, 391).

Croft admits that he was able to find out little about Young's domestic life, but after quoting a favorable comment by a foreign nobleman, Croft immediately suggests that this picture of Young, at Welwyn, tasting "all the ease and pleasure mankind can desire," (III, 391) was the result of Young's deliberate posturing. Knowing the man was coming, Young posed for the expected portrait.[36]

Croft was the product of the age's sensibility. His interest in Martha Ray, James Hackman and Chatterton involved a fascination for violent crimes of passion, suicide, and genius; not, as with Dr. Johnson, a concern for matters of right and wrong. And Croft could not forgive Young his failure to fulfill his expectation of what a true poet's character ought to be. He was ten years younger than Boswell and therefore far less forgiving of Young's foibles. For Croft, Young was a hypocrite, and hypocrites do not write great poetry. Croft's heroes were Savage, Ossian, and Chatterton, poets around whom there gradually grew the myth of the creative artist destroyed by the insensitivity and cruelty of society. Johnson concludes his brief critical analysis by stating that Young was "a man of genius and a poet" (III, 399). In light of the severity of his criticism but even more thoroughly in light of Croft's portrait, it comes as a surprise. Johnson seemed more willing to forgive Young his self-preoccupation, his ambition, and the errors of his private life. And Johnson always gave due note to the popular appeal of a work. Croft was not half so generous. With the exception of her analysis of Young's imagery and her attack upon the moral worth of his belief in an afterlife, almost every point made by George Eliot in the essay on Young that was thought to have given "a fatal blow to . . . [Young's] reputation" is based on Croft's doubts about Young's character.[37] For in Croft's eyes as well as Eliot's, Young was not a poet "true to his own sensibilities or inward vision,"[38] and while he remained a canonical poet through most of the nineteenth century, around 1780 there developed a canon of poets who met the test of true sensibility; Young was not among them.

NOTES

1. *The Hamwood Papers of the Ladies of Llangollen and Caroline Hamilton*, ed. Mrs. G. H. Bell (London: Macmillan, 1930), pp. 148, 159-61.
2. See Samuel Johnson, *Lives of the Poets*, ed. George Birkbeck Hill, 3 vols. (Oxford: Clarendon, 1905) III, 362.
3. See Joseph W. Reed and Frederick A. Pottle, *Boswell Laird of Auchinleck 1778-1782* (New York: McGraw-Hill, 1977), p. 447, n. 4; and *Letters of Samuel Johnson*, ed. Bruce Redford, 5 vols. (Princeton: Princeton Univ. Press, 1992), III, 322.
4. See the accusations by Robert Southey, *Monthly Magazine*, 8 (November 1799), pp. 777-78; and Joseph Cottle, *Early Recollections Chiefly Relating to the Late Samuel Taylor Coleridge* (London: Longman, 1837), I, 253-68. Southey painted Croft's role in obtaining letters and papers from Chatteron's mother and sister as hypocritical and vicious because he appeared before them as a clergyman, but Croft had not become a priest at that time and was short of money throughout his life. Although Croft's attempt to clear his reputation shows him paying for the use of the Chatterton material, the DNB article on Croft sides with Southey and Cottle.
5. *The Case and Memoirs of the late Rev. Mr. James Hackman, and of His Acquaintance with the Late Miss Martha Reay* (London, 1779), p. 24.
6. Talk presented at the annual meeting of the Johnson Society of Southern California, 12 December 1994. Joseph Cottle wrote that Croft's novel, "rhapsodical as it was, would, in all probability, have attracted little notice, but for the exhibition it made of Chatterton's letters" (*Early Recollections*, p. 259).
7. See George Cheyne, *The English Malady* (1733; rpt. Delmar, New York: Scholar's Facsimiles & Reprints, 1976). See especially p. 147 where Cheyne discusses the results of various nervous diseases such as hysteria, hypochondria, and the spleen.
8. See particularly the discussion by Charles Moore, *A Full Inquiry into the Subject of Suicide* (London: J. F. and C. Rivington, 1790), pp. 122-58. If Southey was to read Croft's character in terms of his later turn to becoming a clergyman, Moore, taking a more favorable view of that profession, insists on reading a moral view into *Love and Madness* (compared to Werter) which the text hardly sustains.
9. As Ida Macalpine and Richard Hunter demonstrated, not until the discussions rising from the apparent madness of George III, was there a broad public discussion about the nature of insanity and its treatment. See *George III and the Mad Business* (New York: Random House, 1969), pp. 99-110. For the relationship between legal decisions and definitions of madness, see pp. 310-20.

10. Unlike almost everyone else, Johnson believed that Hackman was a person who could not control his anger and that he intended to kill Martha Ray. He quarreled with Beauclerk over this matter, contending that Hackman had revealed his intentions by taking two pistols with him. Beauclerk argued that Hackman merely wanted to be sure. Werther borrowed two pistols for his suicide, and Croft, at least, would have known this. See James Boswell, *Life of Johnson*, ed. George Birkbeck Hill and L. F. Powell, 6 vols. (Oxford: Clarendon, 1934-64), III, 384-85.
11. Lord Sandwich, First Lord of the Admiralty, had been forced to flee with Martha Ray from their apartment at the Admiralty on 11 February 1799 when a mob rioted over attempts to censure Admiral Keppel after an inconclusive battle with the French fleet. See J. Hineage Jesse, *Memoirs of the Life and Reign of King George the Third* (London: Tinsley, 1867), II, 235-38.
12. In his column, "The Hypochondriack," Boswell tended to view violent passion and madness over love as more literary than real, pointing to the tendency in nature for the male to be dominant and distinctly unsentimental. But he went to see Hackman in prison and was deeply disturbed by the murder and Hackman's subsequent execution. See James Boswell, *The Hypochondriack*, ed. Margery Bailey (Stanford: Stanford Univ. Press, 1928), I, 178-93; and *Boswell Laird of Auchinleck, 1778-1782*, ed. Joseph Reed (New Haven: Yale Univ. Press, 1977), pp. 72-73, 86-94.
13. Herbert Croft, *Love and Madness. A Story Too True* (London, 1780), p. 26. The reference to Zanag is somewhat bizarre. A character in Edward Young's tragedy, *The Revenge*, he is a combination of Iago and Othello. A moor, he bitterly resents the fact that his master, Alonzo, struck him. Like Iago, he succeeds in making his master jealous of the woman he loves and is stabbed, much like Iago, at the end.
14. The final letter, written by Hackman's friend, Charles, adds the name of a man named Stirn to those who killed the woman they loved in a fit of madness. A more deliberate example of casting a woman in a sacrificial role may be found in the libretto of Handel's *Jephtah*, a work for which Martha Ray expresses admiration.
15. Johan Wolfgang von Goethe, *The Sorrows of Werter: A German Story* (London, 1784), p. v.
16. See for example *A Letter to a Friend, with a Poem Called the Ghost of Werter* (London, 1782), p. 1. Supposedly by an anonymous member of the aristocracy, the writer states (p. 14) that "the character of Charlotte should be pronounced ungenerous, selfish, cold, and actuated by vanity under the borrowed name of feeling, and rendered not less unworthy, but

more dangerous, by the false parade of sentiments, which have not generosity, true delicacy and feeling for their basis."

17. Hester Lynch Piozzi, *Thraliana*, ed. Katherine Balderston, 2nd ed. (Oxford: Clarendon Press, 1951), p. 385.
18. In his fascination for the death of the Duke of Monmouth, who had to be struck repeatedly by the executioner before he succumbed, Hackman also suggests a submerged feeling that his own death will not be a romantic suicide but a botched attempt followed by his execution. Of this scene, he writes, "The remainder of the scene is too shocking of the eye, almost for the ear.—But, I know not how, whenever I am away from you, nothing is too shocking for *me*." See Herbert Croft, *Love and Madness* (London, 1780), p. 113.
19. Piozzi, *Thraliana*, I, 174.
20. Quoted in *Lives*, III, 399. See also the notes to pp. 393-94.
21. Johnson told Boswell that he considered Young to have a mind filled with commonplace thoughts. See James Boswell, *Journal of a Tour to the Hebrides*, ed. Frederick Pottle and Charles Bennett (London: Heinemann, 1936), pp. 234-35.
22. Boswell, *Life of Johnson*, IV, 60.
23. Boswell remarked on Johnson's criticism of Croft's style in the "*Family Discourses*" as being "in too familiar a style to be approved of by so manly a mind." Croft intended the series as material to be read to servants. See Boswell, *Life of Johnson*, p. 554. Quoted in a note from *The Lives of the Poets*, III, 361.
24. For a discussion of Croft's dictionary and his expressions of admiration for Johnson's work, see J. E. Congleton, "Sir Herbert Croft on Revising Johnson's *Dictionary*," *Tennessee Studies in Literature*, 13 (1968), pp. 49-52. Croft emphasizes Johnson's poverty as the reason for errors in the Dictionary, but he was also aware of the way Johnson's political attitudes influenced the *Dictionary*, remarking (p. 58) that there were no quotations from Thomson's Whiggish poem *Liberty*.
25. Saul Bellow, *Humboldt's Gift* (New York: Viking, 1976), p. 51.
26. Sir Herbert Croft, *A Brother's Advice to His Sisters*, 2nd ed. (London: J. Ridley, 1776), p. 214. Similarly, in the thirteen issues of his journal, *The Literary Fly*, published between 18 January and 8 May 1779, Croft moralizes over contemporary evils and promises, in the third number, to devote future issues to "the unfortunate Chatterton."
27. Bodleian, Dep. c. 644, fol. 63-64.
28. This story was made into a popular contemporary drama by John Sargeant called *The Mine*.

29. The first edition of Croft's novel contained a highly suggestive poem by Thornton called "The Rose." Croft removed it from subsequent editions, but the blend of violence and sexuality appears to have been attractive to him.
30. Harold Foster, *Edward Young: The Poet of the Night Thoughts* (Harleston, Norfolk: Erskine Press, 1986), pp. 378-79.
31. Isaac Reed, "Edward Young," *Biographia Britannica* (London: J. Walthoe, 1782), VI, 259.
32. See Isaac Disraeli, *The Literary Character* (New York: Widdleton, 1868), p. 245. Disraeli states that he had this information from Croft's manuscripts. Another rumor had Johnson cutting away more than half of Croft's work. See the quote from *Johnson's Miscellany* in the notes to Johnson's *Lives of the Poets*, III, 361.
33. See Forster, *Edward Young*, pp. 385-94.
34. *Lives of the Poets*, III, 362.
35. Croft real opinion was clear enough. He told Lady Eleanor Butler that "Young the poet a bad man, thought of nothing but his poetry and drinking, and entirely neglected the education of his children." *Hamwood Papers*, p. 160.
36. The letter from de Tarschner to Haller written in March 1751 was reprinted by W. Thomas in his *Le poète Edward Young* (Paris: Hachette, 1901), pp. 599-600. In fairness to Croft, Young does seem to be playing his role as the retired poet with special emphasis for this foreign visitor who came to see him to clear up some points in *Night Thoughts* for himself and for the Swiss *philosophe* Albrecht von Haller.
37. See "Worldiness and Other-Worldliness: The Poet Young," in *Essays*, ed. Thomas Pinney (London: Routledge, 1963), pp. 335-85. The quotation (p. 335) about the decline in Young's reputation is taken from *Studies of English Poets*, by J. W. Mackail (New York, 1926), pp. 115-16. Eliot's comments on Young's tendcncy to rework old panegyrics for new patrons, his fictionalizing his grief, and the disconnection between his writing sublimely sad lines on any occasion are all drawn from Croft's criticisms.
38. Eliot, p. 367. Eliot was responding to the biographies of Young that presented him in an heroic light. They, in turn, had been called forth by the high romantic view sounded by Joseph Cottle, who argued that the genius of Young, however marred by poor "taste" in poetry, deserved a properly worshipful biography. Cottle argued that Croft "neither understood . . . [Young's] character, nor possessed the grasp of mind sufficient to comprehend his genius." See *Early Recollections*, I, 268.

WILLIAM COLLINS'S ODE ON THE DEATH OF CHARLES ROSS: THE SEARCH FOR AUDIENCE AND PATRONAGE

MARY MARGARET STEWART

William Collins's "Ode to a Lady, on the Death of Col. Charles Ross in the Action of Fontenoy" was one of Collins's poems which enjoyed some popularity during the poet's life. First published in *The Museum* on 7 June 1746, it appeared that same year in a revised version within his volume of *Odes*, published on 20 December 1746. Under a slightly altered title, it was reprinted in the July 1747 issue of *The British Magazine; or, The London and Edinburgh Intelligencer.* It then appeared in a third version within Robert Dodsley's *Collection of Poems. By Several Hands*, the second edition, published in late 1748 but not announced until 14 January 1749. The ode is mentioned and one stanza is recited by the main character in *The Life and Memoirs of Mr. Ephraim Tristram Bates, commonly called Corporal Bates, a Broken-Hearted Soldier*, a novel published anonymously in 1756.

In recent years this ode, although not highly admired by scholars, has received some attention because it is generally conceded to be Collins's earliest ode. Richard Wendorf, for instance, notes that this ode "surely marks a new direction in Collins's poetry." In *William Collins and Eighteenth-Century Poetry*, Wendorf primarily examines Collins's revisions of this ode, concluding "that Collins is striving toward a more generalized conception of action in the ode: in its ultimate form this action will replace individuals entirely and will . . . replace them with the abstract. . . ."[1]

Discussions have also focused upon the "lady" to whom the ode is addressed. After Joseph Warton first read or heard the ode in late May 1746 when he saw Collins at the races near Guildford, he wrote his brother Thomas: "It is addressed to a lady who was Ross's intimate acquaintance, and who by the way is Miss Bett Goddard."[2] Many years later, after telling William Hymers that he had recently examined Collins's first manuscript of the ode on the death of Colonel Charles Ross, Thomas Warton reported: "The lady to whom this Ode is addressed was Miss Elizabeth Goddard, who then lived at or near Harting, in Sussex."[3] These two references by the Wartons suggest that they knew Miss Goddard. Years later, when Thomas reviewed his brother's letter, he may have changed the informal *Bett* to the formal *Elizabeth* and stated with authority she lived in Harting, since Collins mentions that village in the last stanza of the ode. It seems doubtful that either knew more than the lady's name.[4] Sir Harris Nicholas, perhaps building upon Warton's statements, reported in 1830 (but without citing his source of information): "The lady is supposed to have been Miss Elizabeth Goddard, the intended bride of Colonel Ross, to whom he addressed his beautiful Ode on the death of that officer at the battle of Fontenoy, at which time she was on a visit to the family of the Earl of Tankerville, who then resided at Up-Park, near Chichester, a place that overlooks the little village of Harting, mentioned in the Ode."[5]

Discussions of Collins's development of the ode form, of his use of personification, and of the lady's identity fail to deal adequately with the poem's context and the poet's intention. As Howard D. Weinbrot has demonstrated in "William Collins and the Mid-Century Ode: Poetry, Patriotism, and the Influence of Context," this ode, as well as Collins's other four patriotic odes, must be read within the context of Britain's involvement in the War of the Austrian Succession (1740-48) on the continent and the consequent Jacobite rebellion in Scotland and northern England (1745-46). These were desperate years for the English, as Weinbrot has shown. Newspapers and magazines reported the recruitment of troops, the movement of regiments from Britain to Flanders, and the opposition to the war on the continent. Since Collins's uncle, Lt. Colonel Edmund Martin, was actively involved in battles in Scotland and Flanders (1746-48), Collins would also have a direct source of information available to him. Placed within the historical context, then, Collins's patriotic odes written after early May 1745 through the fall of 1746 show us that Collins was aware of issues and events and used his role as poet to comment on and thus commemorate battles and issues such as death in battle, "mercy," "liberty," and "peace."[6]

Nowhere does Collins reflect his knowledge of national affairs and the nation's distress over the loss of life at Fontenoy more clearly than in his selection of Charles Ross as the subject of his "Ode to a Lady on the Death of Colonel Ross in the Battle of Fontenoy." By choosing the Battle of Fontenoy as his subject and Charles Ross as his representative of those who fell in that battle, Collins seeks a wide audience and in particular the attention of prominent political figures who might offer him patronage.

Charles Ross, the second son of George, 13th Lord Ross, was born on 9 February 1721. On the death of his great-uncle, General Charles Ross, he succeeded to the estate of Balnagowan in 1732, because of an entail executed by General Ross in 1727.[7] The estate, located in Ross-Shire, Scotland, was worth £400 a year. On 1 December 1739, he received a commission of second lieutenant in Colonel Charles Douglas's Regiment of Marines. When this regiment received orders for the West Indies, Ross was "saved from the dishonour of not liking to go to the West Indies when it was his turn, by Sir R[obert]'s giving him a lieutenancy" in the foot guards. His promotion to lieutenant and his transfer to Colonel Blakeney's Company in the Third Regiment of Foot Guards commanded by the Earl of Dunmore was issued on 1 July 1740. In little over a year, on 20 October 1741, Ross was promoted to the rank of captain in the Third Regiment of Foot Guards, or the Scots Guards.[8]

While still a minor, Ross was elected a member of Parliament for the County of Ross in the general election of 1741. In this election he stood against the incumbent Hugh Rose, the brother-in-law of Duncan Forbes. One of the forty-three petitioners against "Undue Elections" following the 1741 election, Hugh Rose of Kilravock, challenged the election of Ross, his petition setting forth that some persons entitled to vote had been denied their rights, that others not entitled to vote had been permitted to vote, and that Charles Ross was permitted to vote although he was then under the age of twenty-one "and therefore, by an express Provision of an Act of Parliament, declared neither capable of electing, or of being elected, a Member of Parliament"; and charging further that even though the petitioner was duly elected by the barons and free holders, Ross was returned.[9] Hugh Rose's petition was referred to the Committee of Privileges and Elections, but before Rose's petition was acted upon, the House voted that the Committee hear no more petitions.[10]

After Charles Ross took his seat in the House of Commons, he aroused criticism in the County for his "constant way of voting with ye Opposition."[11] He also aroused the criticism of Horace Walpole, who noted that Ross, despite Robert Walpole's personal intervention to save him from being sent to the West Indies in 1740, "voted against us" in the division over the election of the chairman of the Committee of Privileges and Elections on 16 December 1741. The distress of the Court Party in losing this first crucial vote to the Opposition Party (and by only four votes) is reflected in Horace Walpole's letter to Horace Mann. That Walpole, and apparently his father also, had expected Ross's support and were therefore particularly angered by his vote is demonstrated by Walpole's statement "They cry 'Sir R miscalculated'; how should he calculate, when there are men like Ross, and fifty others he could name!"[12] In January 1742, the first Lord Egmont heard "that Lord Ross had been sent up from Scotland to influence his son Mr. Charles Ross, but Mr. Ross had a very good estate independent of his father."[13] According to the Duke of Argyll, young Ross and his older brother "were very violent in their opposition to the Lord Orford, even against their Father Ld Ross his directions & advice."[14]

Ross, described as "a very handsom young man,"[15] was in London in 1741 and throughout 1742. In May 1742, England began to send regiments to Flanders under Field Marshall the Earl of Stair to support Maria Theresa. At the beginning of May 1742, the Scots Guards was ordered to be in readiness for foreign service. On 17 May 1742, King George II reviewed the three battalions of the Guards that had been chosen for the expedition; and on 26 May, the men in the designated battalions embarked at Woolwich and landed at Ostend. For the rest of 1742, these troops saw no action. They went into winter quarters at Brussels, and about the middle of February 1743, they marched out of Brussels by way of Maestricht to the Rhine.[16] Ross, however, was not among those in his regiment sent to Flanders in 1742. In October 1742, Ross was listed in the Cockpit list of ministerial supporters. In the division of 10 December 1742 concerning the question of taking the Hanover troops into British pay, however, he was listed among the opposition absentees.[17] Not until the end of February or early March 1743 did he leave London. In a letter dated 25 February 1742/43, Ross writes: "I set out for Flanders very soon, what to do the Lord knows."[18]

Small wonder Ross questioned why he was being sent to Flanders. The English army had not stirred out of its quarters until it was finally reenforced by the Dutch and Austrian troops in 1743. Ross must have joined this "pragmatic army," as the joint forces were called, somewhere between Maestricht and Aschaffenburg on the Main, where the army ultimately set up camp. It was here on 27 June 1743 that George II led his troops to victory over the French army near the village of Dettingen. Ross took part in this battle.

By 6 December 1743, Charles Ross had returned to London from Flanders, for on that date he and another young army officer supported the opposition motion "that the Hanoverian Troops should not be continued in the service of Great Britain" after 25 December 1743. At this time, Argyll remembered, Ross "made the memorable speech in the House of Commons against the Hanover Troops & even gave evidence of the Facts he insisted upon."[19]

The apprehension and anxiety with which the Ministry anticipated this meeting of Parliament in December 1743 are reflected in letters written by Henry Fox to his friend Charles Hanbury Williams. On 6 October 1743 Fox wrote Williams:

> I send You two Letters (I fancy wrote by Ld Chesterfield) upon a very melancholy [subject] because I fear it will be found a very serious Subject, & of great Consequence. The ill Blood Hate Contempt & Anger which ev'ry day encreases against the Hanoverians in the Army, & which You may be sure will communicate like Wild fire here, already makes some of our best Friends talk of employing them again as impracticable, & yet without them, what that We wish can be practicable either in the Field or in the Closet? I wait with Impatience to see the D. of Marlborough, who is expected ev'ry day, & is returning not quite well of his hurt, but I believe in no danger

> of ev'n long confinement or restraint from it, a very great Favourite with the Army, & I am afraid in proportion *au contraire* elsewhere. The less You talk of these Dissentions the better, for our Ministers know not what Language to hold, nor what remedy to apply, but continue to know no more than I do, & not so much as any body who has a Correspondence with a sensible Subaltern.[20]

On 19 November 1743 Fox sent an urgent letter to Williams:

> The Parliament meets for certain on *Thursday the first of December* which will be a Day of the greatest Importance. I am sure You will not fail of being there when I tell You that I desire it absolutely necessary . . . . I shall say no more now, & don't expect another Letter from me, for, *Affaires on brouillers commes les notres* are for Discourse not for Letters.[21]

When debate opened on the floor of Parliament, Charles Ross argued that the English and Hanoverian troops could not act together. He, like many others who argued against the continuation of the hiring of Hanoverian troops to augment the English forces on the continent, charged that partiality was shown members of the Hanover forces. Ross's evidence consisted of "a story of a drunken quarrel between an English groom and a Hanoverian soldier on their march into winter quarters, after which the latter had been acquitted by his commanding officer before the party aggrieved had an opportunity of being heard, and procuring his evidence." When accused by Lord Charles Hay, a Colonel in the guards, that this speech reflected adversely on "the great person [George II] who commanded" at Dettingen, Ross asserted that he meant no such reflection and that "the instance he gave happened after that great person left the army."[22] Ross's evidence, Horace Walpole concluded, was "very triffling" and "did not at all add any weight to the arguments of the Opposition."[23] According to the first Lord Egmont, it "appeared to the House very frivolous."[24] Ross maintained his opposition to the hiring of 16,000 Hanoverian troops for the army in Flanders and voted against the government again on 10 January 1744.[25]

Sometime in early 1744, Ross rejoined his regiment in Flanders where the operations of the English troops were purely defensive. In what was perhaps a reconnaissance operation Ross saw action during the summer of 1744, for the following news item appeared in the *Daily Advertiser* on 20 August 1744: "A few Days ago Hon. Charles Ross, Esq; Member of Parliament for the Shire of Ross in Scotland, and Captain of Horse, was taken Prisoner near Lisle in Flanders, after being shot in the Arm and wounded, but was afterwards Exchang'd."[26] In giving Philip Yorke news of the army in Flanders on 11 August 1744, Thomas Birch relays information, or perhaps just gossip, about Charles Ross which not only clarifies the item in the *Daily Advertiser* but also conveys an attitude towards Ross no doubt shared by Yorke:

> Capt. Ross, who was so petulant in your House against the Hanoverians last Winter, & a Friend of his, who was a Volunteer, met with very ill Fortune in an idle Frolic of going to reconnoitre, or rather stare at the Fortifications of Lisle; for the latter was blown into Atoms by one of the great Guns, & Ross wounded by a Musket-Ball, & taken Prisoner. Their Friend Stanley prudently return'd home two Days ago, to avoid all ye Accidents of War.[27]

Charles Ross was with his regiment in Flanders when it took part in the action at Fontenoy in May 1745. After that terrible battle in which 7,000 to 10,000 allied soldiers were killed or wounded, Colonel John Munro reported to his uncle Duncan Forbes:

> Poor Charles Ross of Balnagown was shot with a musket ball through the belly, I believe early in the action; my Servants found him in a Ditch, and I sent him away to our head Quarters, where he died that night.[28]

When the news from the battle of Fontenoy reached London, the fate of Charles Ross was invariably noted: "We hear the following Officers were kill'd in the late Action in Flanders, viz. . . . Capt. Ross, Member of Parliament for the Shire of Ross, a very fine young Gentleman of a good Estate."[29] Referring to the battle of Fontenoy, Walpole wrote Mann on 11 May 1745 [O.S.]: "We don't allow it to be a victory on the French side; but that is, just as a woman is not called *Mrs* till she is married, though she may have had a dozen natural children. In short we remained upon the field of battle three hours—I fear, too many of us remain there still! . . . We never lost so many officers." He then enumerated officers who "are certainly killed." Among those he listed is "young Ross."[30]

Even months after Charles Ross's death, readers of London newspapers were occasionally reminded of his name and fate. On 5 September 1745, the *Daily Advertiser* reported: "Hugh Ross, Esq; of Kilravock, who represented the Shire of Ross in Scotland in the last Parliament, intends to offer himself again a Candidate for the same, in the room of the Hon. Charles Ross, Esq; kill'd at the Battle of Fontenoy." As late as 17 November 1746, about the time of the bi-election, the *Daily Advertiser* reported: "We hear that Hugh Ross, Esq; of Kilravock, and _____ Ross. Esq; are Candidates for the Shire of Ross in Scotland, in the room of Charles Ross, Esq; kill'd at the Battle of Fontenoy."[31]

Since Charles Ross left London for Flanders while Collins was still at Oxford, it is doubtful that Collins knew Ross personally. It is possible, however, that the two men met during one of Ross's visits to London in 1744. Even if they had not met, however, Collins knew people who must have known Ross. Indeed, Ross had achieved some notoriety in his short life: young, handsome, independent, landed, Ross cut a romantic figure. No doubt because of Ross's nationality, because of his opposition to Walpole and to the support of the Hanoverian troops, and because of his associations with Duncan Forbes, James Thomson's first patron, Thomson knew Charles Ross. Certainly Chesterfield, Lyttelton, and Pitt knew Charles Ross.

William Pitt, as young Ross, vehemently opposed the policy of paying the Hanoverian troops. During the long and warm debate over the motion to disband the Hanoverian troops on the evening of 6 December 1743, William Pitt rose to speak in favor of the motion and to support the two officers whose evidence had been refuted or belittled by Lord Charles Hay and Major-General John Campbell. "Our countrymen," Pitt is reported to have said, "must dread the renewal of these indignities they have been forced to swallow." After speaking of his Majesty's "military ministers," Pitt claimed that the "gentlemen of the army who have been bred to sentiments of honour" had been "very nice on these points." What they have presented is the "truth," but "they have suppressed part of it." It is "very wise in them to speak with caution," Pitt concluded, for "the hand of power is over them."[32] In the House of Commons, William Pitt continued to attack Carteret as "the Hanoverian troop-master . . . an execrable, a sole minister, who had renounced the British nation and seemed to have drunk the potion . . . which made men forget their country."[33] He was joined in his opposition to the hiring of Hanoverian troops by George Lyttelton.

In the House of Lords, Chesterfield had answered Carteret's defense of his policies, and with Edmund Waller, a member of the House of Commons who had moved the Opposition motion against the Hanoverian troops on 10 December 1742, had published anonymously *The Case of the Hanover Forces in the Pay of Great Britain* (December 1742). Chesterfield and Waller continued their attack outside the Houses of Parliament by publishing a sequel to their first pamphlet: *A Vindication of a late Pamphlet entitled, The Case of the Hanover Troops considered; with some further observations upon those troops* (January 1743). When Lord Horatio Walpole answered their pamphlets with the anonymous pamphlet *The Interest of Great Britain Steadily Pursued in Answer to a pamphlet, entitl'd, The Case of the Hanover forces impartially and freely examined. Part I* (March 1743), Chesterfield and Waller continued the debate by publishing *A Farther Vindication of The Case of the Hanover Troops: in which the uniform influence of the Hanover-rudder is clearly detected and expos'd: being a full answer, founded on facts and dates, to a pamphlet call'd, The Interest of Great Britain Steadily Pursu'd* (December 1743).[34]

In the early months of 1745, these politicians had been with the poet James Thomson frequently. Both Lyttelton and Pitt, according to Thomas Davies, "attended the rehearsal of Tancred and Sigismunda with great assiduity; they had a sincere value for the amiable author. Their instructions were heard by the players with great respect, and embraced with implicit confidence."[35] Thomson's biographer, Grant Douglas, claims that these men not only helped Thomson produce his play but also had "to prune and alter the text before it was suitable for the stage."[36] In a letter signed *Bellaria* published in the *Daily Post* on 26 April 1745, the claim was made that "A very remarkable new Lord of the Treasury [George Lyttelton] was proud of appearing its Foster Father, and attended at the public rehearsals; the first night of the performance [18 March 1745] this celebrated person and his friends in the Box with him (all very lately most flaming Patriots) were seen clapping their hands at the following remarkable speech: First of You All . . . To Quite Mistakes."[37] Less than a month later, the Battle of Fontenoy was fought.

Whether or not Collins knew Ross, Collins must have been aware of Ross's opposition to the government. Even if Collins had not met James Thomson as early as May 1745, he was familiar with his work and had sent a copy of *Tancred and Sigismunda* to Joseph Warton before 19 April 1745.[38] He would have known of the connection between Ross, Thomson's Scots friends, and the Patriots. By selecting Charles Ross, then, as the subject of his poem commemorating the Battle of Fontenoy, and thus honoring all those who fell in the battle, Collins not only sought an audience but also sought to gain the attention and approval of James Thomson and undoubtedly the patronage of Pitt, Lyttelton, and Chesterfield. By celebrating Charles Ross, William Collins also took a political position, or at least made a political statement of his own.

NOTES

1. *William Collins and Eighteenth-Century Poetry* (Minneapolis: Univ. of Minnesota Press, 1981), pp. 90-91.
2. John Wooll, *Biographical Memoirs of Joseph Warton* (London, 1806), pp. 14-15, note m.
3. *The Gleaner*, ed. Nathan Drake (London, 1811), IV, 477.
4. It should be noted, however, that one of the subscribers to Thomas Warton the Elder's *Poems on Several Occasions*, edited by Joseph Warton and published in March 1748, is a "Miss Goddard."
5. "Memoir" in *The Poetical Works of William Collins*, ed. Sir S. E. Brydges, The Aldine Edition (London, 1830), p. xxxviii. If an Elizabeth Goddard were visiting in Uppark, she probably was there with the Earl's eldest son Charles and his wife. In 1742 Charles had married Alicia Astley, the third daughter of Sir John Astley, 2nd Bt. of Patshull, Staffs. and Everley, Wiltshire. A longtime Tory and supporter of the Pretender, Astley was a member of the House of Commons from 1727 to 29 December 1771. Because he was in the House when Charles Ross was a member and because he consistently voted against the government, Astley must have known Ross (Romney Sedgwick, ed., *The History of Parliament: The House of Commons 1715-1754* [New York: Oxford Univ. Press, 1970], I, 424). Because the Earl of Tankerville was a member of the Knights of St. Andrew, or the Order of the Thistle, a distinctively Scottish order of knighthood, he must have had strong Scottish connections. His membership would indicate he knew many Scots. The Earl may well have known Charles Ross and his family. Thus, even if an Elizabeth Goddard were not visiting Uppark, that family would surely have known Ross and have been informed of his death. It would not have been necessary for the lady, whoever she may have been, to have lived in or near Harting, or even to have been visiting there, for Collins to have used the name of that village. Indeed, a chief resident of Harting, John Caryll of Ladyholt, the grandson of Alexander Pope's friend, must have known about Ross, more than likely knew Ross, and certainly knew of his death because of Caryll's ties to Parliament, to the Army, and to Scotland, and in particular to his family's connections in Ross-Shire.
6. *Context, Influence, and Mid-Eighteenth-Century Poetry* (Los Angeles: William Andrews Clark Memorial Library University of California, 1990), pp. 3-39. Weinbrot discusses in detail the battle of Fontenoy and shows what a profound impact that battle had in both England and France. It was a subject of discussion as well as a subject of poetry, particularly in France. Thus, Weinbrot places the ode in both its historical and its literary context.

7. George Crawford, continued by George Robertson, *General Description of the Shire of Renfrew* (Paisley, 1818), p. 519; "Private Remarks on the Ensigns; E. Lowdon Officers, 31 May 1745" by the Duke of Argyll, P.R.O. S.P. 41/16.
8. P.R.O. W.O. 25/133, f. 246; W.O. 25/90, f. 128; W.O. 25/134, f. 169; W.O. 25/20, f. 136. Horace Walpole to Horace Mann, 16 December 1741, in *Horace Walpole's Correspondence with Sir Horace Mann*, ed. W. S. Lewis, Warren Hunting Smith, and George L. Lam (New Haven: Yale Univ. Press, 1954), I, 245.
9. *Journals of the House of Commons, 25 June 1741-19 September 1745*, XXIV, 29-30. Among the McKenzie Papers in the British Library (Add. MS 39,190, ff. 30, 31-31v, 33, 35, 37) are letters concerning Ross's election: one letter from Robert Munro to William McKenzie, three letters from Charles Ross to Laird of Suddie, and one letter from Charles Ross to William McKenzie, all written in May 1741.
10. "Complaints against Undue Elections," *Gentleman's Magazine*, 12 (1742), 44; Sedgwick, I, 14; *Journals of the House of Commons*, XXIV, 129.
11. William MacGill, ed., *Old Ross-Shire and Scotland from Tain and Balnagown Documents* (Inverness, 1909), no. 268.
12. The first division of the new parliament foreshadowed the fall of Robert Walpole, for he lost on the important vote to choose a chairman of the Committee of Privileges and Elections. It was this committee which acted on most of the petitions against election returns. Walpole had used the committee to increase the majority of the ministerial party, a process which was called "Weeding the House" (Sedgwick, I, 14). Horace Walpole began his letter describing the vote with the admonition "Remember this day!" The Opposition's candidate, Dr. George Lee, defeated Robert Walpole's candidate, Giles Earle, by four votes and thus assured that the petitions filed by members of the opposition would be heard. The vote, according to Horace Walpole, was "two hundred and forty-two against two hundred thirty-eight—the greatest number I believe that ever lost a question" (Walpole's *Correspondence*, I, 243-44).
13. *HMC Manuscripts of the Earl of Egmont. Diary of the First Earl of Egmont [Viscount Percival]* (London, 1923), III, 243.
14. P.R.O. S.P. 41/16. Ross's father was a Commissioner of Trade, a position no doubt secured for him by the interest of Robert Walpole.
15. Letter from Edward Digby to John Ward, dated London, April 16, 1743, Bromley Davenport of Capesthorne Muniment, John Rylands University Library of Manchester. Digby, a member of Parliament from Warwickshire, indicates in this letter that he knew Charles Ross and that Ross was having his portrait painted.

16. Sir Frederick Maurice, *The History of the Scots Guards from the Creation of the Regiment to the Eve of the Great War* (London, 1934), I, 117-18.
17. Sedgwick, II, 393.
18. MacGill, ed., *Old Ross-Shire*, no. 723.
19. Walpole's *Correspondence*, II, 355 n. 1; P.R.O. S.P. 41/16.
20. The Lewis Walpole Library, Yale University, CHW Papers 48-10914, ff. 56-56v.
21. See n. 20, ff. 72v-73.
22. Philip Yorke's *Parliamentary Journal* in William Cobbett, *The Parliamentary History of England from the Earliest Time to the Year 1803* (London, 1812), XIII, 140-41.
23. Walpole's *Correspondence*, II, 355-56.
24. *HMC Egmont Diary*, II, 278.
25. Sedgwick, II, 393.
26. One cannot always trust the accuracy of accounts in London newspapers. Ross, for instance, was a captain in a regiment of foot and not in a regiment of horse. Nor is Collins accurate since Ross was a captain, not a colonel.
27. BL Add. MS 35,396, ff. 239v-240.
28. *Culloden Papers* (London, 1815), No. CCXLIII, p. 200.
29. *Daily Advertiser*, 10 May 1745. *Gentleman's Magazine* also reported among those killed: "Capt. Ross, a fine young gentleman, member for the shire of Ross" (15 [May 1745], 276).
30. Walpole's *Correspondence*, III, 42-43.
31. Both these accounts are inaccurate. In fact, Hugh Rose, the brother-in-law of Duncan Forbes, who had lost his seat to Charles Ross in 1741, stood against Sir Harry Munro of Foulis, Ross. Munro defeated Rose on 11 December 1746. Harry Munro, the son of Colonel Sir Robert Munro, married Anne, the daughter of Hugh Rose, on 13 January 1753 (Sedgwick, I, 392; II, 281).
32. *Yorke's Parliamentary Journal*, XIII, 141-42.
33. Quoted by Basil Williams, *The Whig Supremacy 1714-1760* (Oxford: Oxford Univ. Press, 1939), p. 234. In a letter to Williams dated 26 November 1743, Fox anticipates that "The Opposition, (for they divide, and as they give out divide 210), will object to approving a Treaty You have not seen, condemn what they suppose (and know) to be contain'd in it, abuse the Conduct at & after Dettingen, & renew all Pitt's Talk of Wars of Ambition & Acquisition, and unattainable Ends pursu'd, by mad Means, & propose to leave out or put in Words as they shall think best suit their purpose" (See n. 20, ff. 75-75v).

34. Publication dates have been established through the "Register of Books" in *Gentleman's Magazine*, 12 (December 1742), 664; 13 (January, March, December 1743), 56, 168, 672. There is an interesting pamphlet in Yale University's Beinecke Library concerning the Hanoverian Troop debate entitled *A True Dialogue between Thomas Jones, Trooper, lately return'd from Germany, and John Smith, Serjeant in the First Regiment of Foot-Guards. London: Printed for B. C. in Pater-noster-Row. M,DCC,XLIII.* At one point they discuss the debate in Parliament about the Hanoverian troops:

> **Troop.** Ay, with all my Heart; and I don't see how any Englishmen can vote for him.—By Jove, if I was in Parliament and had a Regiment, I would lose it sooner than vote for 'em, and so will ev'ry Officer in Parliament, I believe—at least if he would shew his Face in the Army afterwards. I am sure any *English* Officer that votes for 'em, will be hooted all along the Line, and call'd MYNHEER HOCH DEITCH.
> **Serj.** They say you had an E——h M—— in the Army that promoted all this.
> **Troop.** Ay, he was the Ringleader of it all; he talk'd *German*, dress'd like a *German*, and drank like a *German*; we call'd him the *Hanover* Hero. . . . (pp. 13-14)

35. *Dramatic Miscellanies* (London, 1785), I, 78-79.
36. *James Thomson* (London, 1951), p. 235.
37. Quoted by Arthur H. Scouten, ed., *The London Stage 1660-1800, Part 3: 1729-1747* (Carbondale: Southern Illinois Univ. Press, 1961), II, 1160. George Lyttelton had been appointed a Lord of Treasury on 25 December 1744 (*Gentleman's Magazine*, 14 [1744], 676). James Sambrook identifies Bellaria as Benjamin Victor and comments that the speech he refers to—lines spoken by Osmond, "I here renounce those Errors and Divisions/That have so long disturb'd our Peace, and seem'd,/ Fermenting still, to threaten new Commotions—/By time instructed, let us not disdain/To quit Mistakes"—"could almost be Lyttleton making his peace with Pelham." Sambrook further comments that the "reconciliation between former Whig opposition and post-Walpole Whig administration is the political context of *Tancred and Sigismunda*" (*James Thomson 1700-1748: A Life* [Oxford: Oxford Univ. Press, 1991], p. 239).
38. Joseph Warton wrote to his brother Thomas that Collins had sent him Akenside's Odes "with Tancred & Sigismunda" (BL Add. MS 42560, f. 3.). The letter is undated; the BL has dated it "Before Ap. 19, 1745."

SELF-PRAISE AND THE IRONIC PERSONAL PANEGYRIC OF PETER PINDAR

JEANNE GRIGGS

In the hands of John Wolcot, who wrote under the pseudonym "Peter Pindar" and was popular in both England and America during the 1780s and '90s, the genre of ironic personal panegyric, which consisted of blame-by-praise of a public figure based on panegyrical conventions, became less conventional and more comic than ever before. Rather than using the classical technique of self-praise, a technique by which speakers blame themselves with their own praise, traditionally a comic indication of the speaker's lack of modesty, Peter Pindar invented a new, satiric basis for judging, one based on the trustworthiness of his own pseudonymous persona.

Peter Pindar tried to assert his trustworthiness by arguing that he was an objective judge of his prominent satiric victims, rather than a hired flatterer. Like Charles Churchill, a satirist of the 1760s, Wolcot attempted to make money from his satires; P. M. Zall points out that "he looked upon the writing of satire as a trade . . . to be practised not only by polishing old skills but also by using them in whatever new combinations might appeal to the fancy of the public."[1] This seemingly mercenary attitude was considered admirable by other late eighteenth-century writers. Leo Braudy relates that Goldsmith and Johnson were "praising what was written for money (i.e., to please a public) and attacking what was written for fame (i.e., to please a patron)."[2] Because posts like the laureateship had been tainted for decades with accusations that the laureates produced only insincere flattery for pay, patronage and the writing of panegyric had fallen into deep disrepute.

Consequently, Peter Pindar's explanations of his disinterestedness are lengthy and involved; several of his satires are largely dedicated to this type of explanation.

Pindar established his reputation by writing ironic personal panegyrics on King George III. Although he took unprecedented liberties with the dignity of his royal victim, his carefully created persona is eternally full of protests as to the satirist's good-natured motives, and so the effect of much of his irony is as much comic as satiric—it merely amuses as often as it judges. Pindar's blame-by-praise of the king and other public figures was possible largely because of readers' prior acquaintance with the stories he told; his outrage was an amplification of the discontent of others. His topics were well suited for general readability, as "the majority of Wolcot's selections betray him as climbing on a bandwagon or as following traditional materials," according to Robert Vales.[3] Thus, Pindar's satire of public figures involved in government gives us an indication of what some of the typical attitudes towards government were in the 1780s and '90s.

Most notably in his satire of George III, Pindar returns frequently to several topics of ridicule, all of them major topics for contemporary criticism. Habitually employing the method of blame-by-praise in which the satirist praises vice or folly as though it is a virtue, he often praises the king for his economy or frugality, a trait which, in the 1780s, "was a constant theme for satire," according to John Brooke.[4] Pindar also praises the king for his over-attentiveness to domestic detail; Thomas Wright points out that "it appears that both the king and the queen were in the habit of attending to various minutiae of domestic economy which are more consistent with a low station in life than with the public dignity of the Crown."[5] In many satires, Pindar comically describes the king's stuttering manner of speaking, and then praises him for his "wit" in speaking this way. This was a broad ironic stroke, as John Brooke says that "the wits made fun of his habit of ending his remarks with 'what? what? what?'"[6] Another topic for Pindar's ironic praise of the king is his undignified behavior. Wright reports that

> King George was remarkable for slovenliness of manners, for his ungraceful and undignified carriage, for a love of entering into conversation with people of all ranks, and for the volubil-

> ity with which he poured upon them his naive and often pointless questions.[7]

In addition, as Wright says, the king's "constant practice of taking the air in unceremonious excursions"[8] led to circumstances in which his behavior could easily be ridiculed as less than dignified. His manner of governing also left him open to criticism; George III was very involved in the administration of his government. And, as J. D. Griffith Davies notes, "such active participation in the business of government was bound inevitably to strain to breaking point the constitutional maxim—'the King can do no wrong.'"[9] Rather than criticizing the king for the way in which he rules the country, however, Pindar ridicules him for his personal traits, for his physical rather than his kingly body.

Pindar's favorite topics for ridicule also appear in his satires of lesser public figures. Ever the literary opportunist, he recycled his ridicule of over-attentiveness to trivial detail or obsession with frugality in satires on both literary and political figures of his day.

By choosing to repeat stories currently popular with the general public, Pindar provided the general reader with the means to reconstruct his ironic praise as ironic mention, described by linguists Dan Sperber and Deirdre Wilson as a way of interpreting irony in which "the speaker mentions a proposition in such a way as to make clear that he rejects it as ludicrously false, inappropriate, or irrelevant."[10] The blame he implied, however, could be reconstructed as something less severe than is usual in satire, perhaps blame only from a certain perspective that the victim does not appear to have considered. Shoshana Felman argues, in her discussion of speech acts in the writing of J. L. Austin, that "the laughter provoked by a joke turns the reader into an accomplice."[11] And since accomplices side with their associates, they are, in fact, persuaded by their own laughter. The eighteenth-century rhetorician George Campbell asserts that "wit and humour . . . commonly concur in a tendency to provoke laughter, by exhibiting a curious and unexpected affinity."[12] This unexpected affinity is found in what Felman calls the "heroi-comic genre, in which the comic is based . . . on the incongruous mix of the sublime and the ridiculous."[13] A mix of the sublime and the ridiculous is evident in the writing of John Wolcot as "Peter Pindar," whose very pseudonym suggests such a mix.

In order to make his readers laugh, Pindar needed to establish a sense of perspective on how he thought his satiric victims should behave. Not satisfied with the probability that readers had already heard the stories he was retelling, Pindar filled some of his satires with his own self-praise, which he contrasted to his ironic praise of his victims.

Pindar's use of the conventions for ironic personal panegyric is essentially limited to ironic praise of the supposed "hero's" deeds. Pindar's ironic personal panegyrics do not compare the victim to an ideal of heroism, however. The classical ideals of virtue, heroism, and noble character are invoked only in order to establish the satirist's credentials. These credentials are strengthened by Pindar's self-praise, which is opposed to ironic praise of his announced victim. But Peter Pindar does not take himself too seriously; his self-praise is inevitably based on an obviously inflated claim of sublimity and genius. The joke of his pseudonym lies largely in the fact that his fame was a modern type, one that, as Braudy puts it, "took on a stance of isolated virtue to praise a human nature unbeholding to anything the aristocratic world had explicitly valued in terms of tradition and class alignments."[14] Pindar, a panegyrist of the old aristocratic world, would seem to be a singularly inappropriate literary model for a late eighteenth-century ironic panegyrist, except for what William Fitzgerald describes as "his insistence on his own presence, importance, and ability, reiterated at every turn,"[15] an insistence echoed by Peter Pindar. There is considerable irony just in the fact that these plain-style poems were being produced by the pseudonymous namesake of the model of sublime poetry.

In a typical passage of self-praise from a 1787 poem entitled *Ode Upon Ode*, Pindar's demeanor is at odds with his claims:

> Reader, when tired, I'm fond of taking breath:
> Now, as thou dost admire the true sublime,
> And consequently my immortal Rhyme,
> 'Tis clear thou never canst desire my death.[16]

Such ironic self-praise establishes the satirist's right to judge in the new, humble manner of the late eighteenth century—as he is capable of recognizing his own presumptions and faults, he is therefore also qualified to judge the presumptions and faults of others.

Pindar's self-praise in *Ode Upon Ode* is a device for explaining his ironic praise of George III. This satire is, as Grzegorz Sinko points out, a "parody of a panegyric"[17] by the official ode-writer for the king, the poet laureate, Thomas Warton.

The focus of the satire is on ironic praise for the actions of the king and queen, who had bestowed on two performers, Madame Mara and Mrs. Siddons, the favor of requesting their performances at a private audience. These actions were the subject of a minor court scandal at the time; the king and queen were accused of abusing their royal privilege. It was said that the performers were forced to stand in the presence of the king and queen for the length of their entire visit, without being invited to sit down and rest. Stanley Ayling notes that at this time "nobody sat in the intimacy of the king's presence, except as a very unusual mark of condescension."[18] But Robert Huish, a contemporary of Pindar's and a defender of George III, insists that Mrs. Siddons "was allowed three pauses, to go into the next room and refresh herself."[19] Pindar, at any rate, incorporated the scandal into his praise of the king and queen. Madame Mara, he says, "though faint . . . was not suffer'd to sit down;/Such was the goodness, grandeur of the Crown!" (207-208). Not only was she forced to remain standing, she also failed to receive any pay:

> For Mara's Chaise-hire and unrivall'd Note,
> Out of their wonderful benevolence,
> Their bounteous Majesties gave—not a groat. (220)

Mrs. Siddons's fate is virtually identical. After reading to the king and queen she came out

> "So rich!"—How rich?—"As rich as she went in."
> Such are the Stories twain. Why, grant the fact,
> Are Princes, pray like common folks to act? (257-60)

Having called the royal couple's notorious parsimony[20] a common trait, Pindar goes on to explain that "Even I, the Bard, expect no gifts from Kings,/Although I've said of them such handsome things" (265-66). The irony in this statement is double-edged; Pindar has not said "handsome" things about the king, but he has said things that might help the king behave more handsomely.

Pindar employs what at first might seem to be a standard satiric apologia for writing this satire of the king. In an imaginary dialogue with the laureate, Pindar has the laureate caution him to "say nothing in your Rhymes about a King:/If Praise, 'tis lies; if Blame, a dangerous thing" (1085-86). Rather than a standard apologia, in which the satirist claims to be so honest that he cannot help but expose hypocrisy and vice, no matter how dangerous it might be, Pindar's version is simply more indirect blame of the king, a man who can only be praised through "lies."

Pindar's disinterestedness in flattery and his interest in the truth is established by his declaration that

> The King (good man!) and I should never quarrel,
> Even though his Royal Wisdom gave the Laurel
> To Mister Tom-a-Stiles, or John-a-Nokes. (1235-37)

Pindar's reassurances that he is disinterested and has good judgment are ploys that assure an ironic reading of his praise of the king's and queen's behavior towards Madame Mara and Mrs. Siddons.

In *An Apologetic Postscript to Ode upon Ode*, also published in 1787, Pindar delivers a more traditional satiric apologia to help readers reconstruct his ironic praise for another action of the king's. "Thus I to Kings my Ode of Wisdom pen,/Because your Kings have souls like common Men" (91-92), Pindar claims. He praises his own bravery in being able "to say to Majesty, 'O King!/At such a time, in such a place,/ You did a very foolish thing'" (102-104). Following these passages of self-praise, Pindar relates a story about the king's visit to an old woman's cottage and his wonder at her apple dumplings, a story which shows the king's interest in the lives of common people, for which he was often ridiculed. The topic of apple dumplings, in addition, is provided for the pleasure of readers accustomed to criticism of the king for the plainness of his table.[21]

The story includes some of Pindar's usual ironic praise of the king's speech: "lo! the Monarch, in his usual way,/Like Lightning spoke: 'What's this? what's this? what? what?'" (119-20). The king's wisdom is also praised ironically; he is called "the Solomon of Britain" (139) when he reveals himself ignorant of the process by which an apple gets into a dumpling. Pindar implies that the dignity of the king is damaged by his lessons in dumpling-making, as "the Palace seemed the lodging of a Baker" (146).

A short section of ironic praise for the king's actions as he learns to make apple dumplings is surrounded by Pindar's offers to reform the king, offers that make the irony virtually foolproof. Pindar calls himself "Soul-physician to the King" (223), using a common satiric metaphor, the satirist as healer of society's ills.

In both the *Ode upon Ode* and the *Apologetic Postscript*, Pindar's ironic attacks on the king are made within the context of advice to the poet laureate, Thomas Warton. This ironic advice is featured more prominently in the *Instructions to a Celebrated Laureat*, which also appeared in 1787. Pindar criticizes and attempts to correct what he sees as the faults in Warton's previous praise of the king, eventually offering yet another story of the king's "heroic" deeds and adding some ironic praise for the king's "noble" character.

Pindar's ironic praise of George III is surrounded by direct attacks on him, attacks based on Warton's previous praise of the king, and which reflect some decline in reverence for kingship. Pindar asks Warton:

> How canst thou seriously declare
>   That George the Third
> With Cressy's Edward can compare,
>   Or Harry? (65-68)

It is inappropriate to compare George III with mighty rulers preceding his time, Pindar declares, dismissing the entire possibility for even ironic praise of George III on this basis. He reduces the king to a comic figure instead when he searches for good things to say about him:

> George is a clever King, I needs must own,
> And cuts a jolly figure on the Throne. (69-70)

Left with a jolly but unheroic figure to praise, Warton is depicted as unable to think of any flattering things to say about the king, so Pindar suggests a new topic:

> I'll tell thee what to say, O tuneful Tom.—

> Sing how a Monarch, when his Son was dying,
> His gracious eyes and ears was edifying
> By Abbey-company and Kettle-drum:
> Leaving that Son to Death and the Physician. (73-77)

Although offered as a topic of praise, this passage is clearly blame of George III's lack of action during the recent illness of his son, who had since recovered.

Pindar defends his own good humor in the course of this satire. When he has Warton point out that the "King has virtues that should make thee stare" (49), Pindar responds:

> Is it so? Then the sin's in me;
> 'Tis my vile optics that can't see:
> Then pray for them, when next thou say'st a prayer. (50-52)

While it is a show of good humor, Pindar's seeming humility is qualified by a suggestion that perhaps Warton sees more in the king than is there.

Pretending to demonstrate praise of the king for Warton's edification, Pindar includes a sample "birth-day ode" in the *Instructions*. The story Pindar tells would have been familiar to contemporary readers, as it is based on a true incident; Vales says that it is "what George III actually did on his birthday: the monarch had visited Whitebread's brewery."[22] The praise, of course, is ironic, and features Pindar's favorite topics. George III's unthinking avarice is apparent from the moment he proposes the visit to Queen Charlotte: "Clearly we must, must, must see Whitbread brew;/Rich as us, Charly; richer than a Jew" (121-22). One implication of this speech is that the identification of George III with a commoner could work on more than just the monetary level to which he calls Charlotte's attention.

While Pindar describes the king's actions at the brewery as "noble" (130), he depicts them as undignified; George III's concern with common people and common occupations is shown to be inappropriate to his office:

> And now his curious Majesty did stoop,
> To count the nails on every hoop;
> And, lo! no single thing came in his way,

That, full of deep research, he did not say,
"What's this? hae, hae? what's that? what's this? what's that?"
(217-21)

Concerned with minor details, George III can perform no deeds of heroism. He is too busy examining little things to see the big picture and choose an appropriate action for the circumstances. This lack of perspective is displayed when he feels the need to take notes on what he has been shown at the brewery. The notes are reproduced, for comic effect, and they include the following:

Mem.
To remember to forget to ask
Old Whitbread to my house one day.
Mem.
Not to forget to take of Beer the Cask,
The Brewer offer'd me, away.——

Pindar praises the "wisdom" of the author of such notes:

Now having pencil'd his Remarks so shrewd,
Sharp as the Point indeed of a new Pin;
His Majesty his watch most sagely view'd. (260-72)

This praise of George III's intellect is clearly inappropriate in the context of his actions, as he must dully remind himself to be as miserly and grasping as Pindar always portrays him.

Warton is given the job of expressing the doubt that should have been raised in the minds of readers: "Is this an action, Peter, this a deed,/To raise a Monarch to the Sky?" (488-89). Pindar protests that the act is "wise" and "glorious" (497) and says he hopes that other kings will "never do more foolish things" (500), which, while not quite a contradiction of his praise, is adequate explanation of the ironic intent.

Pindar continued to address Warton in order to praise the king ironically in *Brother Peter to Brother Tom* (1788), in which his explanations of the irony and defense of himself play an even larger role in the satire. In addition, he makes fun of contemporary panegyric, identifying it as empty flattery, and the office of the Laureate, insinuating that the current holder is willing to flatter anyone, for a price.

To show how pointless and silly it is to write panegyric in 1788, Pindar admonishes Warton to praise "some Royal Virtues, real or invented" (9) and to "forge Arms to prove a loftiness of birth" (14), as former panegyrists did. Panegyric, identified with the laureateship, now has only negative connotations, as Pindar points out by saying "since truth belongs not to the Laureat trade,/'Tis strange, 'tis passing strange, thou didst not flatter" (185-86). It is too late to reinvigorate the genre of panegyric, Pindar declares, asking Warton "no more to strike us with amaze,/Thy courtly tropes of Adulation blaze" (549-50) and urging him to let "no more ideal Virtues ride sublime/(Like Feathers) on the surge of Rhyme" (557-58).

Pindar theorizes that the king himself asked Warton to cease writing panegyric on him, as George III was wondering "what have I done that he should praise me?" (568). As is his habit, Pindar brings the king down to a human scale:

> I do not marvel that the King was wroth,
> Knowing himself no bigger than a Lath,
> To find himself a tall gigantic Oak:
> 'Twas too much of a Magic-lantern stroke. (591-94)

Not only does this humanize George III, it ridicules the idea that panegyric can be properly addressed to such a common sort of man.

In this context, the irony in Pindar's own praise of the king is virtually unmistakable. Addressing Warton, he points out that one subject for praise could be George III's recent declaration that no work, even entertainment, could take place on Sunday:

> Thomas, thy Lyrics might have praised the King
>   For making Sinners mind the Sabbath-day;
> Bidding the idle Sons of Pipe and String,
>   Instead of scraping Jigs, sing Psalms, and pray:
> Thus piously (against their inclination)
>   Dragooning souls unto Salvation. (41-46)

This mention of an actual action of the king is further exaggerated by a technique Pindar originated: the recommendation that his subject go further with the deed for which he is praising him. "Great act!" (125) Pindar calls it, saying "ere long 'twill be a sin and shame,/For Cats to warble out an amorous flame" (125-26) on Sunday, and adding that soon the king will decree that any animal who makes noise on a Sunday should be imprisoned for the crime: "to Bridewell shall be sent all cooing Pigeons" (131). Pindar's vaunted agreement with the king and his recommendations for going further with the decree make George III's decree seem arbitrary and ridiculous, an idiosyncratic and childish whim rather than a decision that should become law.

The actions of the king are ironically praised by a new comparison, to the actions of the Prince, his son. Ridiculing George III for what was widely perceived as a lack of scholarly interest, Pindar declares that "No man binds Books so well as George the Third" (363), whereas "the Prince upon such labor looks/Fastidious down, and only readeth Books" (367-68). To help ensure that this praise will be read ironically, Pindar follows it with praise by comparison on two favorite topics for praise of the king: his parsimony and his unkingly interest in livestock. While, Pindar claims, the king dreams "of pretty, Jewish, economic schemes" (376), the Prince would "rather give a Coat away than sell it" (378). Moreover, the king is great in "Pig-wisdom" (383), while the Prince "ne'er finger'd at a fair a Bullock's ribs,/Nor ever ogled a Pig's face" (387-88). These are the ways, the satirist concludes, that "a Father hath excell'd a Son!" (390). Since Pindar is customarily ironic about the "nobility" of such actions, the comparison of the king's actions to those of the Prince is actually praise for the Prince, in addition to the usual ridicule of his father.

Pindar's claim of good nature is bolstered by his humor. In addition to his explanations that he does not hate the king, that it is the satirist's duty to point out the king's faults, and that kings should set a good example to their subjects, all traditional defenses for satirists, he jokes about his public relationship with George III:

> The World may call me Liar, but sincerely
> I love him; for a Partner, love him dearly:
> While his great name is on the firm, I'm sure
> My credit with the Public is secure.
> Yes, beef shall grace my spit, and ale shall flow,
> As long as it continues George and Co.;
>   That is to say, in plainer metre,

"George and Peter." (213-20)

Rather than making a traditional claim to love the man but not the methods, Pindar deflects serious criticism by joking about how much money he has made ridiculing George III and how grateful he is that the king continues to act in a way that discerning citizens like Pindar consider ridiculous. Pindar's self-praise, his claim that he aims his satire at George III because "Bards must have Subjects that their genius suit" (631), puts the justification for his satire on an individual level. He makes a claim for his own genius, rather than claiming, as satirists had since classical times, that he criticizes for the good of his society.

Peter's self-praise and his defense of his individual judgment of the king are actually the subject of *Peter's Pension*, also published in 1788. Although he promises to turn down any offered pension, Pindar informs George III that he has misjudged by not offering his patronage to Pindar. "My Art would suit your merits to a T" (412), Pindar claims with subtle irony, intimating that his rarely revised journalistic productions are an appropriate medium for praise of a king like George III. This king is as heroic as this Pindar is sublime, the poet insinuates.

Pindar offers two stories of George III's deeds in order, as he assures the king, to "give the lie" (421) to stories "that dared attack you, and your fame devour" (422). The stories, of course, do not show the king acting in a noble or heroic manner. They contribute to Pindar's self-praise by showing that his loyalty cannot be bought; that he will tell the truth about George III as he sees it.

Pindar explains his attitudes towards the king most completely in *Expostulatory Odes to a Great Duke and A Little Lord* (1789). Claiming to reply to various criticisms of his satire on George III, Pindar employs a comic tone and scattered irony to explain his satiric purposes.

Pretending surprise that he can be accused of an "unloyal line" (34), Pindar pronounces: "Now be it known to all the Realms around,/I would not lose my Liege for twenty pound" (37-38), a comically small amount. Becoming more serious, however, he points out that

> In general, Authors are such coward things,
> They fear to speak their sentiments of Kings

Till those same Kings are dead. (41-43)

Pindar is braver than most; he argues that "if we wish a Monarch to reclaim,/In God's name let us speak before he's dead" (47-48), an argument that supports his own practice.

Pindar also defends himself against the charges that he never finds anything to praise about the king and that his satire is mean-spirited. For the first, he praises George III for one very noble deed: "An act is his my heart with rapture hails—/George gave the world the Prince of Wales" (197-98). The siring of a successor to the throne is the very least that is expected of a king, and the sting of such a compliment is increased by the knowledge that Pindar had unfavorably compared the king to his son in *Brother Peter to Brother Tom*. As for the sting of his satire, Pindar declares that "to mine, Charles Churchill's rage was downright rancour" (363) and that Churchill and Junius were both satirists "whose pleasure 'twas to stab" (373), whereas Pindar says that his "humour of a very different sort is./Their Satire's horrid Hair-cloth; mine is Silk" (375-76). While the urbane irony of Pindar's satire contrasts with the invective with which the names of Churchill and Junius were associated, Pindar exaggerates the harmlessness of his satire.

As always, Pindar repeats stories about the deeds of George III to justify his satire. He mentions a new scandal, a royal experiment in livestock-feeding that displays the king's parsimony. In this case, he says, the stories need no elaboration:

> If Kings will sell the Hares and Boars they kill,
> And Snipe and Partridge blood for mammon spill,
> Denying thus themselves a dainty dish;
> And go themselves to market with their Fish.
> If Kings, ambitious of a Blacksmith's fame,
> Not wondrously ambitious in their views,
> Instead of mending Empires, make Horse-shoes . . .
> If with their bullocks Kings delight to battle;
> On hard Horse-chesnuts make them dine and sup,
> Resolv'd to starve the nice-mouth'd cattle
> Until they eat the Chesnuts up . . .
> I must laugh out, and all the World must hear it. (656-81)

Peter's response to these deeds of George III is, he says, not as harsh as Churchill's or Junius's would be. He sees the humor in the king's actions and conveys that point of view to the readers, as if blame is now almost as inappropriate as praise. The old heroic standards of nobility are largely irrelevant to Pindar's comic presentation of the actions of George III.

Throughout his career, Pindar attempted to keep current his self-characterization as a disinterested and good-natured satirist. In 1795, in a preface to *The Royal Tour and Weymouth Amusements* (subtitled *A Solemn and Reprimanding Epistle to the Laureat*), he addressed the new poet laureate, Henry James Pye, saying:

> Let no man say I hate our Kings and Queens,
> Princes, and drawing-rooms, and levee-scenes;
>   Despise the bows and curtseys, whisper'd talk.
> I love the mummery from my very soul:
> Daily I spread its fame from pole to pole. (p. 321)

This laughing sort of defense of the satirist's good nature emphasizes the events to be described—and so the readers' reactions to them—over any claim of desire for reform of the king.

The practice of Peter Pindar shaped the tradition for ridicule of George III during the 1780s; Vincent Carretta argues that "the growing numbers of attacks, frequently more comic than satiric, on the king's mortal body indicate the royal body's enhanced security."[23] Rather than any decline in reverence for the institution of kingship, satire of the 1780s and '90s chronicles a rising awareness that an individual is sitting on the throne of England, a focus that deflects the threat of serious republican upheaval.

Some of Pindar's most amusing attacks on George III's actual body include mocking praise for his physical "endowments" and his "regal" speech. His ridicule of the king's prominent eyes and his habit of repeating words like "what, what, what," became well-known and other satirists attempted to imitate it.[24]

Some of Pindar's first attempts at mocking the eyes and speech of George III appear in *The Lousiad*, published in individual cantos beginning in 1785. Pindar describes the angry king when "just like two Billiard-balls his Eyes 'gan roll" (162). Rather than a stoic hero, he is a comic, bug-eyed figure. His character is revealed by the way he speaks; upon discovering the louse, for instance, the king sputters:

> "How, how? what, what? what's that, what's that?" he cries,
> With rapid accent, and with staring eyes:
> "Look there, look there; what's got into my house?
> A Louse, God bless us! Louse, louse, louse, louse, louse."
> (175-78)

This kind of kingly stuttering appears, to similar effect, in many of Pindar's satires. The repetition of "what, what, what" was, in fact, characteristic of George III; contemporary historians insist that it stemmed from the king's attempt to give common people an opportunity to respond to a royal question, rather than remain tongue-tied with reverence in his presence. Pindar's reproduction of this verbal tic, therefore, was another kind of stab at the king's concern for common people.

Pindar's satire on the speech of George III was imitated by the anonymous authors of the *Probationary Odes for the Laureatship* (1785). In Number 12, "Ode by Major John Scott, M. P. &c. &c.," the king's appearance is linked to his speech, as it is in *The Lousiad*:

> Lo! how thy white eyes roll!
> Thy whiter eye-brows flare!
> Honest soul!
> Thou'rt witty, as thou'rt fair! (22-25)[25]

The king's speech is also reproduced, a bit more simply than in Pindar's satire:

> Methinks I hear
> In accents clear,
> Great Brunswick's voice still vibrate on my ear—
> "What—what?—what?
> "Scott!—Scott!—Scott!
> "Hot!—hot!—hot!
> "What?—what?—what?"
> Oh! fancy quick! oh! judgment true! (53-60)

The contrast of the three repeated words (including the characteristic "what?") with the epithet "great" identifies the irony of the praise.

In the *Instructions to a Celebrated Laureat*, Pindar enjoys his joke on George III's speech at greater length. Creating a circumstance in which the king's ignorance has been revealed, Pindar makes him bluster:

> "True," said the cautious Monarch with a smile:
> "From malt, malt, malt: I meant malt all the while."—
> "Yes," with the sweetest bow rejoined the Brewer,
> "An't please your Majesty, you did, I'm sure."—
> "Yes," answered Majesty with quick reply,
> "I did, I did, I did, I, I, I, I." (286-91)

The king's speech seems to be outpacing the king's thoughts here; Pindar frequently characterizes him as a slow thinker but a fast talker. His speech sometimes demonstrates an apparent lack of thought:

> Now did his Majesty so gracious say
> To Mister Whitbread, in his flying way,
> "Whitbread, d'ye nick th' Excisement now and then?
> Hae, Whitbread, when d'ye think to leave off trade?
> Hae, what? Miss Whitbread's still a Maid, a Maid?" (336-40)

Far from "gracious," these questions portray an unthinking man who is blurting out whatever comes into his head. Pindar intensifies the insult by an exaggerated comparison of George III to a "waterspout" (364) or an "elephant" (367), sucking up answers and then spitting them all out on the queen. This inelegant comparison is a marked contrast to the avowed heroism of Pindar's subject.

Later in his career, Pindar merely alludes to George III's speech for comic effect. In *The Royal Tour and Weymouth Amusements*, Pitt visits George III at Weymouth, giving the king news of a defeat, and George manages to say only "Hae, hae? what, what? beat? what, beat agen?" (89). The rest of what the king says is not recounted, but it is ironically praised by the observation that when Pitt departed, he "left the Royal Wit" (106).

A twist on Pindar's mocking of the speech of George III comes in a late satire, the 1795 *A Royal Visit to Exeter*, which is written entirely in an attempt to recreate a heavy German accent. The inhabitants of Exeter, the speaker relates, "lang'd to zee the King;/To hear un zay zom *marv'lous* thing!" (10-11). The joke, so late in Pindar's succession of jokes on how badly the king speaks, is an obvious one.

Satire on the physical holder of what is supposed to be a dignified office also appears in Pindar's mocking praise of the president of the Royal Society, in *Sir Joseph Banks and the Emperor of Morocco* (1788). The pursuit of a butterfly, in this satire, literally reduces the president to pratfalls: "oft fell Sir Joseph on the slippery plain" (167). His head is ironically praised because it is "strong, like old Homer's Shields, in tough bull-hide,/And brass well temper'd" (250-51). When a "clan" of bees stings him in the head, Banks is compared to heroic Ajax:

> What was thought death and tortures by the clan,
>   Was only tickling the Great Man.
> Thus round big Ajax raged the Trojan host,
> Who might as well indeed have drubb'd a Post. (255-58)

The comparison to Ajax does not reflect well on Banks, accompanied as it is by the invidious epithet "great man" and the simultaneous comparison to a post.

Another topic for Pindar's attack of Banks was also a favorite topic for attack on the mortal body of George III; this is mock praise of an action that reveals the satiric victim's lack of dignity, or even childishness. Pindar shows that the deeds Banks performs are not like the deeds of classical heroes:

> I can't say that Sir Joseph Lions kills,
> Hugs Gnats, or the blood of Hydras spills;—
>   But then, most manfully he eats a Bat. (20-23)

While eating a bat is the kind of action that might be undertaken as a result of a boyhood dare, Pindar says Banks does it "manfully," calling attention to the boyishness of it. As if that is not enough to make the point, Pindar shows even more explicitly the childishness of Banks' actions. As Banks pursues the Emperor of Morocco butterfly, he is

> Darting his arms now here, now there, so wild,
> With all the eager raptures of a Child,

Who with broad anxious eye a bauble views,
And, capering legs and hands, the toy pursues. (183-86)

Rather than a serious scientist, Banks is reduced to a baby who wants a toy.

George III, rather than babyhood, is reduced only to childhood by Pindar's pen. In *The Royal Tour and Weymouth Amusements*, the king, leaving the garden of an inn, "knocks every apple that he spies, to ground;/Loads every Royal pocket; seeks his chaise;/Plumps in" (40-42), like a boorish schoolboy. Such behavior may be the privilege of the king, Pindar intimates, but it does not become him. Throughout this satire, George III is called "Cesar," in an attempt to make his often childish behavior contrast with the heroic standard suggested by the name. For instance, when he sees some fishermen, "Cesar" jumps into the sea with them:

He marvels at the fish with fins and scales;
He plunges at them, seizes heads and tails;
Enjoys the draught: he capers, laughs aloud,
And shows his captives to the gaping crowd. (184-89)

Like a child, "Cesar" plays with the fish, thinking them his "captives," although the fisherman have brought them ashore. The king's lack of dignity is explicitly compared to childishness:

Hark! Cesar and the little Children talk;
Together laugh, together too they walk:
The Mob around admire their pleasant things,
And marvel that children talk as well as kings. (297-300)

The punch line, of course, is implied by the last line of the second couplet.

Pindar is not the only satirist of his time to characterize George III's lack of dignity as childishness; one of the *Probationary Odes*, number 18, an "Irregular Ode by the Marquis of Graham," satirized the way ministers changed under George III:

Thus Ministers, are in,—are out,
Turn and turn about.— (103-104)

Describing changes in his government as if they are part of a child's game is another way of making George III seem childish.

Related to the topic of a childish lack of dignity is another favorite topic for ridicule of George III, ridicule of a preoccupation with what is characterized as trivial detail. In addition to the king, the Marquis of Buckingham and two writers, Boswell and James Bruce, are ridiculed for this kind of preoccupation.

The way that George III was most often criticized for what satirists saw as a preoccupation with trivial detail was through ridicule of the attention that he paid to commoners. In *A Panegyric on a Court* (1789), Thomas Gilbert ironically praises the king for this kind of stooping:

> With what a tender Heart and melting Eye
> He hears the Widows Groan, and Orphans Sigh! (5-6)[26]

That such praise is ironic is established later in the satire by more praise of some of the king's well-known "glaring virtues" (45).

Similarly, in *Probationary Ode* number 18, George is pictured as he

> Now shears his people, now his mutton;
> Now makes a peer, and now a button. (38-39)

The king does not have the wit to differentiate between such tasks, the satirist insinuates. He is even more ridiculously pictured

> peeping at the windows,
> Each star-light night,
> Looks with delight,
> And sees unseen,
> And tells the Queen,
> What each who passes out or in, does. (56-60)

Such satire creates a picture of a ruler who has no priorities.

Pindar often satirizes George III for his lack of priorities. In *The Royal Visit to Exeter*, the king goes to church, and, looking around him, says

> "Neat, neat; clean, very clean;
> "D'ye mop it, mop it, Measter Dean?

Mop, mop it every week*?" (142-44)

A footnote informs the reader that "*This observation really took place at Exeter, as well as at Salisbury, some years since" (p. 474). Not only is the king concerned with trivia in one instance, then, but in others. Pindar is invoking the satiric strategy of inexpressibility, by intimating that George III indulges in this kind of silliness anywhere he goes, and that this is just one example out of many.

The Marquis of Buckingham is satirized in a similar way in the second part of *Criticisms on the Rolliad* (1790), a long satire on various members of Parliament. Like his king, he is pictured as a man without priorities:

> How amiable, how interesting, is the condescension of the illustrious marquis, while he assists the old women in his neighbourhood in making bone-lace![27]

Like making apple dumplings, such a task should be below the purview of a nobleman; it is a task fit only for old women.

Pindar's satire of George III often creates an image of him as a man without priorities. In *The Lousiad*, all five cantos of which are concerned with the king's preoccupation with trivia, he is the same kind of figure as Banks, pursuing the butterfly:

> thou, so modest (wonderful to tell!),
> On lunar trifles art content to dwell,
> Flies, Grasshoppers, Grubs, Cobwebs, Cuckoo-spittle;
> In short, delighted with the world of little. (II, 73-76)

That the word "trifles" is used in this passage is a clue to the irony of the praise.

The irony of Pindar's praise of Boswell in *A Poetical and Congratulatory Epistle to James Boswell Esq., on His Journal of a Tour to the Hebrides with the Celebrated Doctor Johnson* (1786) is unmistakable, as he who satirizes George III for paying too much attention to trivia pretends to praise Boswell for his "pen's minutiae" (55). Calling him a "charming Haberdasher of Small Ware" (114), Pindar sarcastically gushes "how are we all with rapture touched, to see/Where, when, and at what hour, you swallowed Tea" (121-22). Explicitly comparing

Boswell to George III, Pindar says that the king admires his book because he himself knows such useful things as "which Gardener hath most cabbages and peas" (131). Pindar's sarcasm and his comparison of Boswell's interest in trivia to George III's help readers to easily identify his praise as ironic.

Even more easily identifiable ironic praise of someone who is preoccupied with trivia appears in the 1790 *Complimentary Epistle to James Bruce, Esq.* This satire is accompanied by a dedication in which Pindar compares Bruce's literary deeds to those of Boswell: "while Mr. Boswell entertains only with a breakfast on Spaldings . . . the sublimer Bruce treats us with a Dish of Lion" (p. 350), Pindar says. And even more impressive, "while Mr. Boswell amuses us only with his Drunken-bout, and consequently a simple Emetic scene, the soaring Bruce greets us with the more important history of a thundering Diarrhoea" (p. 350). Rather than "soaring," Pindar conveys the distinct impression that both Bruce's deeds and his literary efforts are generally low.

What satirists perceived as ignorance and a lack of respect for learning provided a related topic for satire on George III, who was frequently portrayed as a person too ignorant to tell whether a matter was trivial or not. Pindar portrays the king as unself-consciously ignorant and smug in the first canto of the *Lousiad.* George III is shown encouraging English boys to attend Gottingen, in Hanover, because he thinks it a better school than Cambridge (this is also a side accusation that the king retains some German allegiance): "So says the Ruler of us English fools/Who cannot judge like him of Wisdom's Schools" (147-48). By pretending to praise the judgment of George III, the satirist accuses him of thinking that one school is pretty much like another.

The ignorance of George III is even more exaggerated in "An Apology for Kings," one of the satires in Pindar's 1791 collection entitled *The Remonstrance.* In this satire, Pindar repeatedly claims that the stories he is telling illustrate the "ignorance" (85) of kings. One of these stories is about George III's ignorance of art. Upon being shown a statue of Sesostris, the king responds:

> "Sir Sostris, hey? Sir Sostris? 'pon my word!
> Knight or a Baronet, my Lord?
> One of my making? what, my Lord? my making?" (69-71)

When he is gently corrected, the king pretends that he knew who Sesostris was all along, saying "I need not that, I need not that, be told" (76). Shown another statue, of Hercules, the king says he admires it, but then reveals his preoccupation with trivia by exclaiming "the Queen, Queen's coming: wrap an apron round him" (82). Rather than a work of art, we assume, George III is able only to see the figure of a naked man. This story, the satirist explains, shows that "information to great folk is hell" (88) and "that saucy Statues should be lodged in sacks" (92). Pindar's portrait of George III is that of a man who does not know enough to know his own ignorance.

A variation on Pindar's ironic praise of those whom he describes as preoccupied with trivia or too ignorant to tell the difference between trivia and more important matters is his praise of George III for his interest in plants and livestock. Political commentators and satirists, both visual[28] and verbal, considered these interests trivial ones for a head of state. George III was so often popularly derided as "farmer George," that ironic praise on this topic was practically foolproof.

The king as farmer is one of the topics for ironic praise of him in number 18 of the *Probationary Odes*. Juxtaposition makes the satirist's criticism clear:

> Hence, too, when eas'd of Faction's dread,
> With joys surveys,
> The cattle graze,
> At half a crown a head— (62-65)

In other words, George III is more concerned with cattle than with the threat of faction. The triviality of such a concern is emphasized in another of the *Probationary Odes*, number 20, in which George III is ironically praised for not being as tyrannical as the Tudors, who lived in a threatening castle with a moat. Now,

> The level'd ditch no more its jaws discloses,
> But o'er its mouth, to feast our eyes and noses,
> Brunswick hath planted pinks and roses. (15-17)

The suggestion that "farmer George's" concern with appearance masks his insidious political concerns is a serious one. Later in this same satire, George is compared to Cincinnatus, a common comparison in satires of the 1780s and '90s:

> Oh! Europe's pride! Britannia's hope!
> To view his turnips and potatoes,
> Down his fair kitchen-garden's slope
> The victor monarch walks like Cincinnatus.
> See heavenly Muse! I vow to God
> 'Twas thus the laurel'd hero trod—
> Sweet rural joys! delights without compare!
> Pleasure shines in his eyes,
> While George with surprize,
> Sees his cabbages rise,
> And his 'sparagus wave in the air! (28-38)

Unlike Cincinnatus, of course, George III is portrayed as a dull-witted person whose intellect is barely capable of contriving a vegetable garden, much less keeping a nation in order.

Pindar exploited the "farmer George" topic whenever possible. In *The Royal Tour and Weymouth Amusements*, he portrays the king as unable to resist bargaining with farmers who are going by with swine, hay, and bullocks. The moment he spots them, the satirist exclaims

> See great Cesar run:
> He stops the Drover; bargain is begun.
> He feels their ribs and rumps; he shakes his head:
> "Poor Drover, poor; poor, very poor indeed."
> Cesar and Drover haggle, difference split:
> How much? a shilling: what a royal hit! (147-52)

The incongruity of seeing "great Cesar" run eagerly to bargain for livestock with one of his subjects is a clue to the irony of calling such a leader "great."

The king's love of a bargain is an often-satirized trait. Rather than necessary economies, satirists portrayed George III's many money-saving ventures as examples of his own personal niggardliness. The most famous of these ventures, of course, was the American tax on tea, a public issue, but satirists preferred to focus, instead, on what they presented as George's private miserliness.

Pindar's *Royal Tour and Weymouth Amusements* is full of satires on the king's miserly ways. He performs "acts that should bid his Poet's bosom flame,/And make his spendthrift Subjects blush with shame!" (73-74). The word "spendthrift" is deeply ironic, as it is contrasted with ironic praise for the king's order that his provisions should be sent by coach to Weymouth: "all, all from Windsor greets his frugal Grace,/ For Weymouth is a damn'd expensive place" (141-42). While at Weymouth, George even refuses charity to men who have served him. He is described talking to a sailor who lost a leg, as he tells the king, "hard fighting for my Country and my King" (241). When the sailor exclaims "oh had I money, Sir, to buy a leg!" (233), George's only response is to say "No money, hae? nor I. Go beg, go beg" (224) and again, "I've got, I've got no brass" (254). The repetition of phrases and colloquial diction further diminish the king's stature. He is shown ordering fisherman to bring fish to his lodging and enjoying a boat ride without paying these people anything. "What a bore that Kings their slaves should pay!" Pindar observes, adding ironically "sufficient is the honour of the day" (309-10). This kind of satire has a cumulative effect; Pindar has previously given examples of the king's unwillingness to pay for services in other satires, such as the third canto of *The Lousiad*, in which George gives only "a tumbler worth a groat" (459) to someone whose hospitality he has enjoyed. Pindar even concludes *The Royal Tour* by saying "Thus, thus I sing of Royalty, unpaid" (319), making it appear that the king's deeds are so noble he cannot resist recounting them, but getting across the insinuation that George III cannot bring himself to pay anyone, even those he could hire to flatter him.

Pindar's most sustained ironic praise of George III's parsimony appeared along with *The Royal Tour and Weymouth Amusements* in 1795, in the "Ode to the Mill." A subtitle informs the reader that this mill was "erected in Windsor-park, for grinding corn at a cheap rate for the poor." Pindar makes the irony of his praise obvious by reminding readers of his previous satire of the king for denying relief to the poor. Pretending to defend the king from accusations of niggardliness, Pindar brings up a new subject for criticism, as he asks "Worms, know ye not that Hanover is poor,/The favourite spot of our most gracious King" (39-40). Blaming Hanover for the actions of George III in England was a convenient way to insinuate that the king did not have England's best interests at heart. A sarcastic footnote to the poem reiterates its ridicule: "This most astonishing charity soon expired. The Children of Famine poured in too plentifully upon the Royal munificence; which very soon must have reduced Majesty to the same most pitiable situation" (362). The exaggeration of a king's being reduced to the status of a beggar displays the ridiculousness of George III's continual money-grasping.

The topic of niggardliness is connected to most of the other topics for ironic praise of George III, as an odd 1792 ironic attack on the king by praise of a Chinese ruler, the *Odes to Kien Long*, demonstrates. Addressing Kien Long, Pindar uses examples of George III's actions to point out that the Chinese ruler is concerned with loftier subjects:

> Nor dost thou cut down Palaces to Pens,
> Nor sendest unto market Cocks and Hens;
>   Nor to a butcher sellest Pork and Beef:
> Nor Wool nor Egg-merchant, O King, art thou;
> Nor dost thou watch the Girl who milks the Cow,
>   For fear the Girl might sip, and prove a thief;
> Nor settest Traps to save thy fowls and eggs,
> And catch thy loyal Subjects by the legs.
> Nor doest thou go a shopping, mighty King:
> I know that thou despisest such a thing;
>   Yes, to expose such meanness thou art loth.
> Thou scorn'st to pride thyself on buying cheap;
> And for some trifle a huge pother keep,
>   An ounce of blackguard, or a yard of cloth.
> Nor dost thou (which some people may deem strange)
> Send Pages with a halfpenny for change;
> Nor dost thou (which would be a crying sin)
> Cheat of his dues the Parson of Pe-kin. (5-22)

All of these kinds of non-heroic deeds are familiar from Pindar's other satires, and the repetition of his accusations here links criticism of the king for being overly concerned with trivia, particularly farming, to criticism of his habitual niggardliness.

Criticism of niggardliness also exposes the more serious accusation of tyranny in two of Pindar's satires on English noblemen, *A Commiserating Epistle to James Lowther, Earl of Lonsdale and Lowther* (1791) and *A Poetical Epistle to Benjamin Count Rumford* (1801). Pindar ironically praises the Earl of Lonsdale, a mine owner, for his unwillingness to pay damages to a man who brought an action against him for the collapse of his house, due to tunneling in Lonsdale's mines. To provide some context for his ironic praise, Pindar prefaces the satire with copies of some letters between Lonsdale and the people of Whitehaven, showing Lonsdale's petulance and determination to punish the man who brought the action by closing down the mines, an act that would put most of the town's inhabitants out of work. Lonsdale, the satirist ironically asserts, is in the right:

> though he most spiritedly suspends his Coal-works for a Time, to show the Power of his Vengeance; lo, he promiseth to open them again, on Condition he has full liberty to undermine any Houses that may impudently stand in the way of his Coal for the future. What an Act of Humanity! (p.3)

Calling such actions humane works with the evidence of cruelty provided in the letters to provide readers with the context necessary to reconstruct the irony.

Count Rumford was a similarly irresistible subject for satire on the topic of niggardliness. A member of the Royal Society, Rumford was a "practical philanthropist and expert on alms and beggery" who "grew forever associated with a liquid of his invention distributed as Rumford Soup to the poor of Western and Eastern Europe."[29] The soup was made from the cheapest possible ingredients, in order to feed a large number. Pindar pretends to praise Rumford for his "power inventive" (75) which enables him to get "dinners from a jakes" (76). A footnote explains that Rumford's "first and grand object" is "to complete the sum total of human happiness, by a prudent attention to economy" and that "he shows how, from the (recre-) (excre-) mentitious parts of human food and fuel, as well as from many hitherto shamefully neglected natural productions, to derive a wholesome, and pleasing, and nutritious diet" (134). After explaining in this way that Rumford actually forces the poor to eat excrement, Pindar goes on in the footnote to praise him for a feat never before achieved: "the Count has at last succeeded in the detection of that grand desideratum before and since the days of the great Martinus Scriblerus, viz. the extraction of sunbeams from cucumbers" (134). This reference, of course, identifies Pindar's satire with that of the Scriblerus circle and ridicules Rumford as the worst kind of imposter, since Swift had already pictured scientists who tried to extract sunbeams from cucumbers as ludicrously deluded.

Like George III's preoccupation with trivia, Rumford's is linked to his obsession with cheapness. His efforts to teach the poor to chew their food well so it will go farther are praiseworthy, Pindar pretends: "what gratitude, what thanks, to thee are due,/Instructing a great Empire how to chew!" (107-108). In the future, Pindar adds, the workhouses will hire "sharp Overseers" who "shall at the table stand" (111) and watch to see that the poor chew together. This prediction of the future puts Rumford's advice in perspective, showing its short-sightedness and stupidity.

Appearing late in Pindar's career, the *Epistle to Count Rumford* employed some of his favorite topics from the previous decade. Most of Pindar's best and most popular satires were published from 1785 to 1795; his reputation grew quickly and had spread far by the turn of the century. P. M. Zall (in the introduction to his selection of Pindar's poems) recounts the tale, popular, he says, in 1800, of the little colonial American girl reading a book who is asked by a British nobleman "what do you read, my dear?" Her reply is supposed to have been "Sir, the Bible and Peter Pindar."[30] Although his fame was as ephemeral as his topics, Pindar's jokes present opportunities for some very serious laughter, and his ironic technique was ingenious, adapted to the circumstances of the last decades of the eighteenth century.

Readers of Pindar's ironic panegyrics were able to reconstruct his ironic praise as blame because of his many explanations of his subjective point of view. One of the limitations of this strategy, however, was that it confined Pindar's satire to the personal.

Peter Pindar helped to create what Thomas Lockwood calls a "tendency for the mode of the satire to appear more directly and definitely as the manner of speaking of the satirist"[31] at the end of the century. John Wolcot was widely known as the creator of Peter Pindar, although this does not mean that the irony of his ironic panegyrics was "Romantic."[32] He attempted to use rhetorical irony—one definition of which is blame-by-praise, according to Norman Knox[33]—but found that his authority as a satirist was difficult to establish. His ability as a gossip and comedian, on the other hand, was unquestioned, and he played those roles to the hilt, establishing "Peter Pindar" as a kind of literary gadfly, harmless but irritating in his attacks on the greatest public figures of his day.

NOTES

1. P. M. Zall, "Peter Pindar 'Redivivus,'" *Notes and Queries*, 196 (19 July 1952), 320.
2. Leo Braudy, *The Frenzy of Renown: Fame and its History* (New York: Oxford Univ. Press, 1986), p. 390.
3. Robert L. Vales, *Peter Pindar (John Wolcot)* (New York: Twayne, 1973), p. 162.
4. John Brooke, *King George III* (New York: McGraw Hill, 1972), p. 205.
5. Thomas Wright, *Caricature History of the Georges* (London, 1868; repr. New York: Benjamin Blom, 1968), p. 460.
6. Brooke, p. 296.
7. Wright, p. 403.
8. Wright, p. 464.
9. J. D. Griffith Davies, *George the Third: A Record of a King's Reign* (London: Nicholson and Watson, 1936), p. 148.
10. Dan Sperber and Deirdre Wilson, "Irony and the Use-Mention Distinction," in *Radical Pragmatics*, ed. Peter Cole (New York: Academic Press, Inc., 1981), p. 308.
11. Shoshana Felman, *The Literary Speech Act*, tr. Catherine Porter (Ithaca: Cornell Univ. Press, 1983), p. 113.
12. George Campbell, "The Philosophy of Rhetoric," in *The Rhetoric of Blair, Campbell and Whately*, ed. Morris Golden and Edward P. J. Corbett (New York: Holt, Rinehart and Winston, 1968), p. 161.
13. Felman, p. 115.
14. Braudy, p. 387.
15. William Fitzgerald, *Agonistic Poetry: The Pindaric Mode in Pindar, Horace, Hölderlin, and the English Ode* (Berkeley and Los Angeles: Univ. of California Press, 1987), p. 19.
16. John Wolcot, *The Works of Peter Pindar, Esq., pseud.* (London: J. Walker, 1812), p. 436, lines 995-98. All subsequent references are to this edition and line numbers are indicated in the text.
17. Grzegorz Sinko, *John Wolcot and His School: A Chapter from the History of English Satire* (Wroclaw: Zaklad Narodowy, 1962), p. 47.
18. Stanley Ayling, *George the Third* (New York: Knopf, 1972), p. 190.
19. Robert Huish, *The Public and Private Life of George the Third* (London: Thomas Kelly, 1821), p. 454.
20. The king and queen were often ridiculed for their avaricious dispositions. See especially Ayling, p. 184, for comments on royal dining habits; Brooke, p. 205, for a discussion of the king's money struggles with parliament; Davies, p. 309, for a description of the king's frugality in opposition to his spendthrift children; and Wright, pp. 412-15, for a

discussion of the ways the king's and queen's frugality was satirized in contemporary prints.

21. Thomas Gilbert (*A Panegyric on a Court . . .*) is one of the many satirists to ridicule George III for his lack of interest in fashionable dishes at the royal table.
22. Vales, p. 45.
23. Vincent Carretta, *George III and the Satirists from Hogarth to Byron* (Athens: Univ. of Georgia Press, 1990), pp. 266-67.
24. The imitation of George III's speech in *Instructions to a Celebrated Laureat* is so successful that "it is mainly because of this poem that George's 'What's this? hae? hae? What's that? What's this? What's that?' became almost proverbial and often imitated," according to Grzegorz Sinko.
25. *Probationary Odes for the Laureatship*, with *Criticisms on the Rolliad and Political Miscellanies* (London, 1785), p. 22, lines 22-25. All references are to this edition, and line numbers are noted in the text.
26. Thomas Gilbert, *A Panegyric on a Court . . .* (London, 1789), p. 28, lines 5-6.
27. *Criticisms on the Rolliad* (London, 1785), p. 29.
28. Carretta describes what he says is a representative example of a 1780s print featuring a comic representation of George III as "farmer George" on p. 271.
29. Sinko, p. 69.
30. P. M. Zall, *Peter Pindar's Poems* (Columbia, S.C.: Univ. of South Carolina Press, 1972), p. 1.
31. Thomas Lockwood, *Post-Augustan Satire* (Seattle: Univ. of Washington Press, 1979), p. 150.
32. Ian Watt defines "romantic irony" as a kind of irony that "always involves the writer himself, and his attitude both to the world and to his creation; and it is subjective in another sense, because it usually involves the writer in an internal counterpoint between his feeling and his reason" in "The Ironic Tradition in Augustan Prose from Swift to Johnson," in *Stuart and Georgian Moments*, ed. Earl Miner (Berkeley and Los Angeles: University of California Press, 1972), pp. 180-81. In "Techniques of Irony in Light of the Romantic Theory," *Rice University Studies*, 57 (1971), Ernst Behler says that the Romantics "conceived of [irony] as a device permitting the author to rise above the confines of his work, to appear within it, and to exhibit deliberately the fictional character of his creation" (p. 7). Lilian Furst's argument, in *Fictions of Romantic Irony* (Cambridge, Mass.: Harvard Univ. Press, 1984), is that by the end of the eighteenth century, a distinction could be made between a satirist, who

judges others, and an ironist, who sees faults in others but is implicated in them too. And in *English Romantic Irony* (Cambridge, Mass.: Harvard Univ. Press, 1980), Anne K. Mellor argues that irony makes the reader consider multiple possibilities. She calls the ironist a skeptic and discusses the "background of political revolution and post-Enlightenment distrust of the capacity of human reason to ascertain the laws of nature, or, indeed, any absolute truths concerning the ways of the world" (p. 4) from which Romantic irony developed.

33. Norman Knox, *The Word Irony and its Context, 1500-1755* (Durham: Duke Univ. Press, 1961), p. 16.

LAURENCE STERNE'S TREATMENT OF A NEW MODE OF DISCOURSE: NONVERBAL COMMUNICATION IN *TRISTRAM SHANDY*

ERIC J. FRISCHHERTZ

In English society, more than in other Western European countries, court behavior filtered into the middle classes.[1] The English primogeniture system was different from that of other Western countries; members of aristocratic families were actually forced to enter a lower social class. James Laver points out in *The Age of Illusion* that "the chief difference, perhaps, between England and the rest of Europe was that a title was inherited only by the eldest son. . . . The sons of dukes and marquises were allowed to call themselves Lord . . . but *their* sons were merely Mr. So-and-so, the peer family's name."[2] Therefore, in England the "trickle-down" of court behavior occurred even more rapidly than in other European countries because of the great amount of interaction between the nobility, the untitled nobility, and the bourgeoisie; court behavior quickly found its way into middle-class society. The social historian Norbert Elias says that "as early as the eighteenth-century, there was no longer any considerable difference of manners between the leading bourgeois groups and the courtly aristocracy."[3] The incorporation of the bourgeoisie into the ceremony of court society, and thereby into the realm of nonverbal communication, completely revolutionized discourse in European society.

Diplomacy became the preferred method of social interaction. In court, the lower ranking courtier had constantly to make himself aware of the most minute actions of his superiors. Also, for the bourgeoisie, the new "art of observing people"[4] became equally important because interaction between aristocracy and the bourgeoisie was so frequent that treating social relations as a diplomatic proceeding became a standard practice in society. Elias points out that

> Both the courtly bourgeoisie and the courtly aristocracy spoke the same language, read the same books, and had, with particular gradations, the same manners. And when the social and economic disproportionalities burst the institutional framework of the *ancien regime*, when the bourgeoisie became a nation, much of what had originally been the specific and distinctive social character of the courtly aristocracy and then also the courtly bourgeois groups, became, in an ever-widening movement . . . the national character.[5]

With the onset of this new social understanding, the affective emotionalism of the Middle Ages came to an end. The bourgeoisie became ardent observers of body language just as the courtiers had, and this concern develops throughout the seventeenth century and culminates in the eighteenth century. In both rhetorical areas and intellectual movements, nonverbal communication and gesture became central topics of social concern.

The rising concern in the seventeenth century with classical rhetoric —especially with the physical aspects of rhetoric: delivery, gesture, and body language—indicates that courtly behavior influenced the other classes of society. For example, in 1605,

> in his *Advancement of Learning*, Francis Bacon greatly criticizes Aristotle for his omission of the gestures of the body and astutely observes the dispositions and inclinations of the mind: "For Aristotle hath very ingeniously and diligently handled the factures of the body, but not the gestures of the body, which are no less comprehensible by art, and of greater use and advantage. For the lineaments of the body do disclose the disposition and inclination of the mind in general, but the motions of the countenance and parts do not only so, but do further disclose the present humour and state of the mind and will."[6]

But while Bacon criticizes Aristotle for ignoring gesture as an integral aspect of rhetoric, he does not offer further insight into the problem. It is John Bulwer who begins to systematize guidelines for gesture in rhetoric. In 1644 Bulwer published his *Chirologia . . . Chironomia*, a work completely concerned with gesture. In fact, as Wilbur Samuel Howell indicates, "'Chironomia' is a word out of Quintilian meaning 'the law of gesture,' and Bulwer characterizes this law as 'The Art of Manual Rhetoric.'"[7] Howell continues, "As for the 'Chirologia,' Bulwer interprets that as the natural language of the hand and body—that is, the meaning that writers have fixed upon such gestures as wringing the hands, shaking hands . . . and so on."[8] Bulwer's book is a collection of illustrations and rules which discuss gesture in rhetoric, and Bulwer declares his theory about gesture in the opening lines of *Chirologia*:

> In all the declarative conceits of gesture whereby the body, instructed by nature, can emphatically vent and communicate a thought, and in the propriety of its utterance express the silent agitations of the mind, the hand, that busy instrument, is most talkative, whose language is as easily perceived and understood as if a man had another mouth or fountain of discourse in his hand.[9]

And in the opening of "Chironomia" he says, "in regard to the rhetorical properties of the hand, man may well be called *Chirosophus*, [that is], *Manu sapiens*, hand wise."[10] Bulwer presents ideas which are somewhat revolutionary to the time. Before the seventeenth century, and even during most of that century, rhetoricians were not concerned with gesture as a rhetorical mode. David McNeill says that "*Chirologia* and *Chironomia* were the first [books] on gesture published in English."[11] James Cleary, the editor of *Chirologia . . . Chironomia* (1974), describes the radical nature of Bulwer's theory:

> Bulwer is convinced that the language of the hand has a meaning and power with which man must reckon. With a sense of near awe, Bulwer exclaims that the language of the hand escaped the curse at the confusion of Babel and that it was sanctified and made a holy language of our Savior's hands.[12]

Bulwer's theory also extends to the idea that gesture is a language "which is common to all men."[13] He includes numerous rules for the novice rhetorician including the respective ideas which body motions convey. For example, he says, "TO WRING THE HANDS is a natural expression of *excessive grief* . . . TO THROW UP THE HANDS TO HEAVEN is an expression of *amazement* . . . TO SMITE SUDDENLY ON THE LEFT HAND WITH THE RIGHT is a declaration of some *mistake*,"[14] etc. Bulwer's text is revolutionary because in it he attempts to systematize gesture for the purposes of rhetoric, and after the publication of Bulwer's work, studies and theories about gesture and body language began to flourish—especially in the eighteenth century.

In the eighteenth century, rhetoricians began to see new horizons for rhetoric. Numerous schools of thought arose throughout the century, but perhaps the most famous and influential of these is the elocutionary movement which spanned the century. Traditionally, classical rhetoric consists of five parts: invention, arrangement, expression, memory, and delivery (or action). The elocutionary movement exalted delivery and action to the point that gesture and body language overshadowed all other aspects of rhetoric. In essence, the elocutionary movement subordinated what the speaker says to how he says it.

Examples of the principles of the elocutionists are most evident in the writings of the leading figures of the movement, and one of the most important elocutionists is John Henley. Henley's theories focus on the rhetorical aspects of gesture to the point that he forsook the content of a speech in favor of a strong delivery, and Henley also initiated the idea that just as there is a proper way to speak, so is there a proper way to gesture. Along with these revolutionary ideas, Henley elevated rhetoric, but especially delivery, to an aesthetic level. He says, "'the true masters of speaking and action' are the masters of music and painting, inasmuch as the principles of both are 'taken from philosophy . . . and in particular from mathematics.'"[15] Henley's theories were not necessarily those of all other elocutionists, but they are emblematic of the direction of the elocution movement; and the radical nature of his writings clearly illustrates the degree to which gesture and body language and the comprehension of these new modes of communication had affected British society.

Probably the most influential, and certainly the best known of all the elocutionists, is Thomas Sheridan. Howell says that "to this day the diminishing bands of elocutionists still consider him the most important figure among their predecessors, and many of them would go so far as to call him not only the originator but the greatest apostle of this school of rhetoric."[16] The most probable reason for Sheridan's distinction is that, to a great extent, he broke with the theory of his fellow elocutionists. While he still saw delivery as the most essential aspect of rhetoric, he took a different approach to justify his argument. Sheridan imagined his rhetorical theory to be a complement to John Locke's *Essay Concerning Human Understanding*.[17] In his *Course of Lectures on Elocution* Sheridan says that

> it is impossible we can think with precision, till we first examine whether we have precise ideas annexed to such terms: and it is equally impossible to communicate our thoughts to others with exactness, unless we are first agreed in the exact meaning of our words.[18]

Sheridan attributed the above discovery (that each signifier must have an exact, agreed upon signification) to Locke. From this excerpt, it is obvious that Sheridan was more concerned with language than the other elocutionists (Henley, for example). Sheridan was aware of the tenets of the elocutionary movement and of the importance of delivery, but he felt that the first step to making the elocutionists' cause universal was to systematize the understanding and the use of language.

Despite his Lockean concerns with verbal language, though, Sheridan did not neglect gesture as one of the most important aspect of rhetoric. The seventh lecture in Sheridan's *Course of Lectures on Elocution* is devoted entirely to gesture, and in the opening of this chapter he discusses the relationship of body language to communication as a whole. He reexamines the function and the idea of language in terms of human understanding of gesture. He posits that communication is not exclusively verbal, but that verbal and nonverbal communication are inseparable aspects of human dialogue. He says:

> Hitherto, language has been considered, as addressed to the mind thro' the ear, by means of words, and tones. But Nature . . . has also made it visible to the eye as well as audible to the

> ear. So that the deprivation of either sense, should not wholly prevent the exercise of man's nobler faculties.[19]

So while Sheridan was concerned with gesture and the function of gesture in communication to a great degree, his overall concern was with communication. Sheridan's desire was to make communication possible for every person who spoke the English language. He wanted to abolish regional accents, idioms, and expressions which he believed impeded discourse between Englishmen from areas of the country other than London. This is the difference between Sheridan and Henley: Henley was concerned with the individual, with exalting the rhetorical skills of one person over another; Sheridan, on the other hand, is concerned with language on a universal scale, with implementing the study of language as an educational discipline. Howell says Sheridan "envisioned a school of oratory . . . for the study of correct English speech, and he thought that students would flock to it."[20]

Along with Sheridan's concern for a systematic approach to the study of English was his similar desire for England to establish schools for gesture in rhetoric. He believed that such schools were essential to the study of the language because speech and gesture are not separate entities in Sheridan's mind. Anticipating twentieth-century research which proves "There is no separate 'gesture language' alongside of 'spoken language,'"[21] Sheridan realized that verbal and nonverbal communication are coeval aspects of language which must be learned together to achieve a full richness and quality in communication. He says of a system of rules:

> Were there masters to teach this, in the same manner as other arts are taught, such a system of rules, would not only be useful but necessary. And indeed, without such a system of rules, to qualify persons for the office of instructing pupils methodically in the art, we can never hope to see proper masters arise among us.[22]

Of course these masters never arose, yet despite the failure of the elocutionary movement, its great influence on rhetoric and the theories of language and communication in the eighteenth century indicate the extent to which the behavior of court society permeated all levels of English society. The flourishing of literature and theories about gesture and the power of body language were the direct results of a changed social construction in the method of interaction and communication. The rise of court society initiated this change into British culture, and theorists and scholars such as John Bulwer and Thomas Sheridan are examples of the dissemination of the new interactive behavior in British culture.

In fact, it is the overwhelming social concern with gesture and reading body language that preoccupies Laurence Sterne in *The Life and Opinions of Tristram Shandy, Gent.* By the 1760s, when Sterne published the first two volumes of *Tristram Shandy*, the attachment to gesture as a mode of communication was so strong in English society that Sterne decided to satirize it in his new book. Sterne understands the effectiveness of gesture in communication, and he knows how to create vivid imagery in the text through narrative description of gesture and action. But in many cases in *Tristram Shandy* Sterne presents over-examination at its most ridiculous. He takes the Renaissance concern with closely observing everyone else at court down to the most minute detail and shows the reader how ridiculous it can become. Tristram narrates physical observation unto absurdity—often presenting his characters in a very unflattering light.

Laurence Sterne uses nonverbal communication in *Tristram Shandy* in ways which, upon cursory examination, seem contradictory. In one instance Sterne employs intricate gestical descriptions to develop the characters and the cultural scenario of the novel. Sterne illustrates that nonverbal communication is, in some situations, more expressive than verbal communication. Walter's frequent frustration with his inability to express himself to his brother Toby is a fine example of gestical communication superseding verbal communication. Sterne's minutely detailed descriptions of Walter's gestures and actions as he attempts to regain his composure are much more expressive than any speech Walter could have given. In this scene nonverbal communication is more expressive than verbal communication because, by communicating with the reader nonverbally, Sterne invites the reader to formulate his own opinions about the personalities of his characters based on their physical actions because, contrary to Bulwer's belief, gestures are not universally translated in the same fashion. Gestical language is the antithesis of Locke's hope for verbal language (that every signifier will have a precisely defined signified) because there is no standard, no dictionary, of nonverbal communication which gives people a common frame of reference.

On the other hand, Sterne uses gestical communication to satirize the elocutionist theories about it. Gesture is not, in every case, a form of communication. Sometimes, as when Slop falls from his horse, a person's gestures are simply the result of clumsiness or negligence. Sterne points out in this satire that if gestures are taken too seriously, if every person is watched with a camera-like eye, misinterpretations will occur as will interpretations which are the result of unintentional signifiers. Sterne rejects the elocutionists' notion that gesture is a language superior to verbal language. Sterne sees gesture as a supplement to verbal communication. Like Sheridan, Sterne places the same importance on verbal and nonverbal communication.

In *Tristram Shandy* Sterne expresses his sensitivity to the changes in rhetoric and especially to the tenets of the elocutionary movement by displaying an overwhelming concern with body language and nonverbal communication in describing the actions and language of his characters. The elocutionists' focus on gesture in rhetoric is a new approach to communication, and Sterne considers the ramifications of this theory from the outset of his book. Sterne published *Tristram Shandy* from 1760 to 1767: a period which is both the height of the elocutionary movement and the apex of the influence of courtly etiquette on English society. Gesture as an aspect of communication is prevalent now not only among the gentry, but among the bourgeoisie as well. Sterne presents his concern with gestures and action in the epigraph to *Tristram Shandy*: translated by James A. Work it is as follows: "It is not actions, but opinions concerning actions, which disturb men,"[23] which indicates that Sterne is concerned with kinetic rather than static energy, i.e., gesture and body language. And this concern leads to Sterne's examination the efficacy of body language in character and plot development as well as in the creation of a cultural context, but despite his use of body language as a literary device, Sterne rejects the vogue elocutionist theories concerning gesture. And last, through a satire of the principal philosopher of *Tristram Shandy*, Walter Shandy, Sterne considers the theories of language and communication of John Locke which are the basis and foundation of eighteenth-century linguistics.

Sterne immediately thrusts the reader into a scenario of physical rather than verbal action. From the outset, he focuses on the physical actions of his characters. Sterne's presentation focuses on his characters' movements, inviting the reader to draw assumptions about the personalities of the characters he describes. Sterne reiterates the theme of the epigraph later in the book. He says, "Attitudes are nothing, madam,——'tis the transition from one attitude to another ——like the preparation and resolution of the discord into harmony, which is all."[24] With this statement Sterne indicates to the reader the necessity of noticing a series of actions and basing one's opinion on this rather than on the rigid posture of a character at any given moment. The reader must consider the actions of the character even more than what the character may be saying in order to form an accurate opinion about his personality. An example of this concept is Sterne's description of Walter's recurring anguish over his inability to make his brother Toby understand what he is saying. Rather than attempting to describe Walter's frustration through a description of his emotions, Sterne communicates Walter's state of mind through a detailed description of his gestures and actions from which the reader is invited to draw an opinion:

> my father played the same jig over again with his toe upon the floor——pushed the chamber-pot still a little farther within the valence—gave a hem——raised himself up upon his elbow and was just beginning to address himself to my uncle *Toby* ——when recollecting the unsuccessfulness of his first effort in that attitude,——he got up upon his legs, and in making a third turn across the room, he stopped short before my uncle *Toby*. (p. 200)

In this description, Walter's actions indicate nothing in themselves. It is only in the reader's mind, through his *opinions* about Walter's actions, that the function of Sterne's narration comes into effect. *Tristram Shandy* is a break from the didactic tradition of previous eighteenth-century novels such as *Robinson Crusoe* and *Pamela*. Sterne challenges the reader to interpret the actions of his characters. He does not attempt to enter the minds of his characters; he allows the reader to come to an understanding of his characters through an interpretation of their actions. In the above example, Sterne illustrates Walter's effort to regain his composure before making another attempt on Toby's reason simply through a description of his body language, and it is not any single action in the description—neither the toe jig, nor the push of the chamber pot, nor the pacing around the room—which communicates Walter's frustration. Sterne reveals Walter's anxiety through a combined description of Walter's physical actions.

Sterne's use of physical description stems from his concern with rhetoric. He understands the power of nonverbal communication and of describing gesture in a literary text. John Traugott, in *Tristram Shandy's World*, asserts that Sterne "was a rhetorician and not a 'novelist.'"[25] To support this theory further, Sterne says in *Tristram Shandy*:

> Writing, when properly managed, (as you may be sure I think mine is) is but a different name for conversation . . . no author, who understands the just boundaries of decorum and good breeding, would presume to think all: The truest respect which you can pay to the reader's understanding, is to halve this matter amicably, and leave him something to imagine. (p. 77)

So Sterne's writing is an interactive process, a communication between the author and the reader—a conversation. Oratory or rhetoric is the formal interactive process of communication between the speaker and his audience. During an oration there is the physical presence of the audience to remind the speaker of this interaction, but in writing, the author must imagine his audience. Sterne is a master of this art in that he describes in brilliant detail not only the words of his characters but their physical actions—giving the reader a sense of immediacy and physical presence to the characters through the medium of the text.

Sterne's sensitivity to gesture in rhetoric enables him to reproduce a functional communication exchange with a reader who physically participates in only half of the conversation. Of course part of this is due to the technique of having Tristram address both the reader and his "listeners" on frequent occasions throughout the book, but the most significant innovation which Sterne attempts is his experimentation with body language and gesture as a form of literary communication. For Sterne realizes, as recent studies by the sociologist Ray Birdwhistell have proven, that "To focus exclusively upon the *words* humans interchange is to eliminate much of the communication process."[26]

Traugott points out that in *Tristram Shandy* there are two types of eloquence. He says:

> At Bobby's bier . . . Trim's artless clap on the heart with mere gesture is contrasted with Walter's rhetorical harangue. *Both* are impressive . . . Walter's rhetoric is, precisely as is Trim's gestical eloquence, an emotional experience and communicable, though nondeterminate, expression.[27]

While Sterne certainly does not downplay the expressive qualities of verbal communication, he does understand, as does any good orator, that physical actions can be just as communicative as verbal ones. Both Trim's body language and Walter's verbal language precisely illustrate their emotions, and Sterne's juxtaposition of the actions of these two characters indicates his desire to incorporate a functional union between the two modes of communication. Sterne presents a theory of language that is strikingly similar to the doctrines that Thomas Sheridan also advocates. Both see the need for a balanced union between verbal and gestical language. Sterne is not a disciple of Sheridan; he has his own theories regarding language, and he incorporates these theories into his text, not into an oratorical practice. Eugene Hnatko agrees with this assertion when he says that one

> of the qualities we expect of conversation, the reliance upon a non-verbal context for speech, consisting of and indicated by gesture, insinuatives, and attendant nuances of voice inflexion, is approximated or suggested wittily and brilliantly by a number of devices without violating the requirements of more formal written prose.[28]

Sterne is not an orator (like Sheridan); he applies his ideas about language and communication in a purely written format. He breaks with the rhetoricians of the eighteenth century in that he not only agrees that body language is necessary in communication, but he takes this a step further by describing these actions in writing rather than applying them in oratory.

Fernando Poyatos discusses the different functions of gesture and nonverbal communication in literature in his *New Perspectives in Nonverbal Communication*. He points out, for example, that

> writers . . . try to convey to their readers the perception of blushing and goose flesh, of perspiration and tears, of soft skin and body scents . . . the degree in which bodily phenomena are granted true literary and intellectual values depends entirely on the artistry of the narrator, and this quality plus the activities themselves deserve serious attention.[29]

Body language and physical description deserve "serious attention" because of the enormous thematic importance that such descriptions can have on a text. Nonverbal communication can be used to satirize or caricature or simply to present a more striking image than is possible through verbal communication. The first of three points which Poyatos asserts that are relevant to *Tristram Shandy* is that nonverbal communication in literature can distort realism: "This is the literary . . . rendering of physico-psychological reality, meant to ridicule, to offer a caricature of reality, to gratuitously exaggerate it or truly show what the eyes cannot see."[30] For example, in *Tristram Shandy*, Sterne communicates the circumstances of Tristram's accidental circumcision by a window through a gesture provided by Corporal Trim: "Trim, by the help of his forefinger, laid flat upon the table, and the edge of his hand striking a-cross it at right angles, made a shift to tell his story so, that priests and virgins might have listened to it" (p. 266). Sterne undercuts the seriousness of the fact that a windowpane has fallen on a child's penis and creates a humorous situation instead through Trim's comic nonverbal communication.

The second aspect of the function of nonverbal communication in literature, according to Poyatos, is the capability of the author to help define the historical and social setting of the novel by describing the actions of the characters in terms of the cultural norms of the setting: "the nonverbal behaviors of the characters communicate to the reader . . . the space setting, or cultural context."[31] In *Tristram Shandy*, a good example of Sterne's use of this technique is in his introduction of the relationship of Toby and Trim. He describes Toby's reliance upon Trim in even the most insignificant aspects of life:

> [Toby] was sitting with his maps, &c. about him,——being somewhat of the smallest [table] for that infinity of great and small instruments of knowledge which usually lay crouded upon it;——he had the accident, in reaching over his tobacco-box, to

> throw down his compasses, and in stooping to take the compasses up, with his sleeve he threw down his case of instruments and snuffers . . . 'Twas to no purpose for a man, lame as my uncle *Toby* was, to think of redressing all these evils by himself,——he rung his bell for his man *Trim*. (pp. 66-67)

This comical scene illustrates the social status of both Toby and Trim—one the gentleman, the other the servant. And while Toby and Trim have an obvious friendship (illustrated throughout the book), Trim never infringes upon this privilege, but always remains stalwart in his station as servant. Trim is literally of lower rank than Toby—he is a corporal, and Toby is colonel. The above quotation indicates the function of both Toby and Trim in relation to each other. Nothing is said between the two men to indicate any social standing; Sterne explains their relationship through the *actions* of Toby. While Sterne refers to Trim as Toby's "man," indicating that he is a servant, Sterne also indicates how much Toby relies on Trim—to the extent that he requires Trim for even the most trivial occurrences.

The last point that Poyatos makes which is relevant to *Tristram Shandy* is his assertion that nonverbal and gestical descriptions are integral to the development of character: "when the author himself wants the character to seem real and unique we should expect him to maintain that unique personality from the beginning to the end of the story in a progressive, logical way."[32] Sterne accomplishes this in part through a consistent use of nonverbal description to describe his characters' personalities. For instance, he describes a discussion between Walter and Toby concerning Walter's theory that infants should be delivered feet-first because of the dangers related to head-first delivery: "Are these dangers, quoth my uncle *Toby*, laying his hand upon my father's knee, and looking up seriously into his face for an answer . . . greater now o' days?" Unsatisfied with Walter's response and refusing to be drawn into an argument with Walter,

> My uncle *Toby* instantly with drew his hand from off my father's knee, reclined his body gently in his chair, raised his head till he could just see the cornish of the room, and then directing the buccinatory muscles along his cheeks, and the

> orbicular muscles around his lips to their duty——he whistled *Lillabullero*. (pp. 116-17)

Whistling "Lillabullero" is Toby's stock response to any attempt by Walter to drag him into an argument, and he has an opportunity to whistle a great many times throughout the book. Toby's nonverbal act illustrates something about both Walter and Toby: Walter is willing to instigate a fight with his brother in order to vindicate his own philosophy, and Toby, who detests Walter's philosophizing, refuses to allow his brother to draw him into a conflict. This whistling is significant because it illustrates the power of nonverbal communication. Toby's whistling indicates to Walter that Toby refuses to speak, and Walter must break off the argument because his foe will not participate. Toby, who continually plays at war games with Trim, is at heart a pacifist, and each of these nuances of character is revealed through the simple act of Toby's whistling.

While Sterne uses nonverbal communication and body language throughout *Tristram Shandy* to develop various aspects of plot and character, the best example of nonverbal description through which Sterne defines characters is in the relationship of Toby and Trim. Helene Moglen says that "Trim and Toby's relationship offers the only example of communication on an explicitly verbal as well as mute, intuitive level."[33] On one occasion Trim and Toby excite each other into a frenzy through their gestures. One animates the other until both their passions are aroused. The two are asserting that their names have nothing to do with their character as men:

> Bless your honour! cried *Trim*, advancing three steps as he spoke, does a man think of his christian name when he goes upon the attack?——Or when he stands in the trench, *Trim*? cried my uncle *Toby*, looking firm——Or when he enters a breach? said *Trim*, pushing in between two chairs——Or forces the lines? cried my uncle, rising up, and pushing his crutch like a pike——Or facing a platoon, cried *Trim*, presenting his stick like a firelock——Or when he marches up the glacis, cried my uncle *Toby*, looking warm and setting his foot upon his stool. (p. 214)

By mimicking the actions they are talking about, both Trim and Toby are drawn into the excitement of the battlefield which they love so much. Simply mentioning the military maneuvers is not enough; they physically act out the scenario in order to decide determinately that their names have nothing to do with their personalities.

That Toby and Trim exhibit an unusual reliance on a nonverbal context for communication is evinced by the fact that Trim performs actions for Toby without being asked. The cue is nonverbal. Sterne says, for example, "My uncle *Toby* laid down his knife and fork, and thrust his plate from before him, as the landlord gave him the account; and *Trim*, without being ordered, took away without saying one word, and in a few minutes after brought him his pipe and tobacco" (p. 293). And on another occasion, when Toby is having trouble delineating the "perfections" of the Widow Wadman, Sterne again indicates the nonverbal relationship of Toby and Trim: "Prithee, *Trim*! said he, taking up his pipe again,——bring me a pen and ink: *Trim* brought paper also" (p. 453). In this instance Toby provides a verbal cue for Trim, but, as in any relationship which sometimes exists on the nonverbal level, some things, which are unstated by the speaker, the listener understands. Such is the case between Trim and Toby. They know each other so intimately that their communication extends beyond speech. Like a skilled orator, Trim is adept at picking up the nonverbal messages of his superiors, and he reacts to these accordingly.

While Sterne is a rhetorician, he is not a follower of the predominant rhetorical movement of the mid-eighteenth century, i.e., he is not an elocutionist. Sterne recognizes nonverbal communication as an important literary device and uses this device prodigiously in *Tristram Shandy*, but he does not place so much importance on gesture that it overshadows verbal communication. The meticulous focus on body language in eighteenth-century England is a result of the behavior of courtiers in the Renaissance. The examination of characters such as the Shandys and their acquaintances in the same minute detail with which courtiers observe each other is ridiculous. For example, Sterne describes Doctor Slop as he falls off of his horse:

> He cross'd himself ✝ . . . He had better have kept hold of the pummel.——He had so;——nay, as it happen'd, he had better have done nothing at all;——for in crossing himself he let go

> his whip,——and in attempting to save his whip betwixt his knee and his saddle's skirt, as it slipp'd, he lost his stirrup,——in losing which, he lost his seat;——and in the multitude of all these losses . . . the unfortunate doctor lost his presence of mind . . . tumbling off it [his pony] diagonally, something in the manner of a pack of wool, and without any other consequence from the fall, save that of being left . . . with the broadest part of him sunk about twelve inches deep in the mire. (pp. 75-76)

Sterne wrote a letter explaining this passage and his ideas on satire to a friend who criticized Slop's fall. He says, "I will reconsider Slops fall & my too Minute Account of it—but in general I am perswaded that the happiness of the Cervantic humour arises from this very thing—of describing silly and trifling Events, with the Circumstantial Pomp of great Ones."[34] The humor of Slop falling off his horse stems from his insignificance. The reader expects the author to devote such painstaking attention to the physical actions and gestures of only the most important public figures. With his description of Slop's fall, Sterne undercuts the immense importance that such figures as John Henley place on gesture as a mode of communication. People perform many actions and gestures which have no special relevance to their communication, and, as is the case with Dr. Slop's fall, sometimes gestures have no relation to communication at all.

Sterne continues his satire of the elocutionists with a comic look at rhetoricians, and specifically the elocutionists' theories of gesture. Sterne accomplishes this with the character Trim. The satire is aimed at rhetoricians who place more importance on how they say something than what it is they are saying. Helene Moglen says "Trim is an orator";[35] he has a knowledge of the postures and gesture which the elocutionists recommend in rhetoric, yet Sterne never indicates where Trim learned the art of rhetoric. Moglen says that "his knowledge of the gesture which can most effectively be employed, the posture which can most eloquently be assumed, is intuitive,"[36] and it is this intuition which Sterne examines in minute detail presenting Trim, like Dr. Slop, as a ridiculous figure.

The best example of Sterne's satirizing Trim the orator is in his description of Trim's posture and gestures as he reads a sermon:

> before the Corporal begins, I must first give you a description of his attitude . . . He stood before them with his body swayed, and bent forwards just so far as to make an angle of 85 degrees and a half upon the plain of the horizon;——which sound orators, to whom I address this, know very well, to be the true persuasive angle of incidence . . . He stood . . . with his body sway'd . . . his right leg firm under him, sustaining seven-eighths of his whole weight,——the foot of his left leg . . . advanced a little,——not laterally, nor forwards, but in a line betwixt them. (pp. 85-87)

Sterne continues the description in such painstaking detail in order to point out that it is ludicrous to treat oratory as a physical science which can be mastered simply by contorting the body in accord with the "rules" of gesture and posture. Also of importance to this scene is the fact that Sterne commissions Hogarth to illustrate it (p. 85), an illustration in which Walter and Toby are smoking—oblivious to Trim, and in which Dr. Slop is asleep. Trim certainly does not have a captive audience. Simply because a speaker adheres to the accepted rhetorical conventions does not mean that he will be an effective communicator.

Sterne also considers John Bulwer's assertion that the hand "speaks all languages, and as an universal character of reason, is generally understood and known by all nations"[37] because, as John F. Sena says, in

> the seventeenth and eighteenth-centuries . . . the interest in and demand for establishing a universal language became pervasive [and] . . . numerous theoretical arguments were advanced for the ability of gestures to function as a universal language—by Michel Le Faucheur and Thomas Sheridan as well as by Bacon and Bulwer.[38]

When he wrote *Tristram Shandy*, Sterne had obviously considered and rejected the idea of gesture as a universal language, because on more than one occasion in the book, nonverbal communication breaks down. For example, the Widow Wadman's lascivious scheme to acquire Toby is undermined when Toby misses the nonverbal invitation in her attempt to flirt with him. She says:

> I am half distracted, captain *Shandy* . . . holding up her cambrick handkerchief to her left eye . . . a mote——or sand ——or something . . . has got into this eye of mine——do look into it . . . In saying which, Mrs. *Wadman* edged herself close in beside my uncle *Toby*, and squeezing herself down upon the corner of his bench, she gave him an opportunity of doing it without rising up. (p. 407)

And Toby responds in all sincerity by saying, "I protest, Madam . . . I can see nothing whatever in your eye" (p. 408), foiling Wadman's advances with his misinterpretation of her actions.

Evident from Sterne's satire of Trim as an orator and because of his refusal to allow nonverbal communication to continue unimpeded in *Tristram Shandy* is the fact that Sterne has no faith in the rhetorical theories of the elocutionists, nor does he have any faith in the prospect of a universal gestical language; and twentieth-century socio-linguistic research has vindicated Sterne's opinions. Ray Birdwhistell says in 1970, "Insofar as I have been able to determine, just as there are no universal words, no sound complexes which carry the same meaning the world over, there are no body motions, facial expressions, or gestures which provoke *identical* responses the world over."[39] This is not to say that Sterne is opposed to gesture as a mode of communication. He is simply opposed to the idea of gesture as a more important method of communication than speech. Gesture, body language, and physical descriptions play an integral part in *Tristram Shandy* in the development of plot, setting, character, and satire.

Sterne continues his examination of language by addressing John Locke's theory of language proposed in the *Essay Concerning Human Understanding*. Like Bulwer and Sheridan, Locke sees the possibility of "the perfection of language."[40] Unlike Bulwer, though, he does not believe that this is a process which he can effect singlehandedly. He says, "I am not as vain to think, that any one person can pretend to attempt the perfect *Reforming* the *Languages* of the world, no not so much as that of his own Country, without rendering himself ridiculous."[41] Despite this disclaimer, Locke creates a set of rules for people to follow which he believes will lead to the perfection of language:

> A Man should take care *to use no word without a signification*, no Name without an *Idea* for which he makes it stand . . . *Secondly* . . . those ideas he annexes them [words] to, if they be *simple* must be clear and distinct; if *complex* must be *determinate* . . . *Thirdly* . . . [people] must also take care to *apply their Words*, as near as may be, *to such* Ideas *as common use has annexed them to* . . . *Fourthly* . . . it is sometimes necessary for the ascertaining the signification of Words, to *declare their Meaning*.[42]

Locke differs from Bulwer and the elocutionists in more than his modesty about the perfection of language. He does not assert the possibility for the creation of a perfect universal language, but he does believe that the various languages can be perfected through the application of his rules.

Sterne responds to Locke's theory of language through a feigned allegiance to Locke in the character of Walter Shandy, the philosopher of the tale, but as Traugott says:

> Whereas Locke would resolutely analyze all ideas and exactly determine the significations of words in order to reconcile necessarily isolated minds . . . Sterne's purpose is to demonstrate and describe the constant frustration of such analysis, the impossibility of determining meaning apart from a context of human situations . . . [by] opposition to Locke's rationalism, and by explicit ridicule, too, Sterne maintains the case for wit, rather than rational analysis, in determining human motives.[43]

Sterne's theory of communication differs from Locke's in that Sterne does not see the possibility of a language which is so perfect that it is capable of expressing every human emotion and thought in even the most taxing of situations. Sterne illustrates this point by placing Walter in scenarios of great physical and emotional stress. Sterne then presents his opinion of an appropriate reaction—the complete failure of language to express Walter's sentiments. Traugott says, "both he [Sterne] and Locke had a hypothesis of human communication, and Sterne chose to exhibit his by putting the logic of Locke's to the test of description in dramatic situations."[44]

One characteristic which presents Walter as the Lockean of *Tristram Shandy* is his near mystic faith in words—especially names. Walter's concern with Tristram's name is the locus of this personality trait. After Tristram's nose is crushed by forceps during delivery, Walter decides to counteract this ill turn of events concerning Tristram's fate by giving him a regal name which will overshadow the defects of his nose. "He shall be christened *Trismegistus*" (p. 202), Walter says. After this plan backfires and Tristram is christened with his present name, of which Walter "had the lowest and most contemptible opinion of it of any thing in the world——thinking it could possibly produce nothing in *rerum natura*, but what was extremely mean and pitiful" (p. 39), Toby and Trim, as the voice of reason (oddly enough), discuss Walter's theory of names and point out the foolishness of such a superstition:

> For my own part, *Trim*, though I can see little or no difference betwixt my nephew's being called *Tristram* or *Trismegistus*—— yet as the thing sits so near my brother's heart, *Trim*,——I would freely have given a hundred pounds rather than it should have happened . . . [Trim replies] 'Tis all a fancy, an' please your honour——I fought just as well . . . when the regiment called me *Trim*, as when they called me *James Butler*. (p. 213-14)

That all words have a definite signification which can be applied even to human personalities is ridiculous, and even Toby and Trim know that a person's actions are not determined on the basis of the signification of words. Melvyn New says, regarding "Walter's efforts to explain the universe to 'the world,'" that "failure and frustration meet . . . every attempt to impose, by theory and by word, a systematic or mechanical order on the chaotic state of [his] . . . own experience."[45] Consider Walter's ridiculous attempt to order and direct Tristram's life —to attempt to control Tristram's fate—through language in the form of the *Tristrapoedia*, which is always chronologically behind Tristram's age.

The final way in which Sterne satirizes Locke's theory of language through Walter is by placing him in situations in which Walter's language completely fails him, and in which he can communicate through gesture and bodily action alone. For example, Sterne describes Walter's reaction when he discovers that Dr. Slop has crushed Tristram's nose:

> The palm of his right hand, as he fell upon the bed, receiving his forehead, and covering the greatest part of both eyes, gently sunk down with his head (his elbow giving way backwards) till his nose touched the quilt;——his left arm hung insensible over the side of the bed, his knuckles reclining upon the handle of the chamber pot . . . his right leg (his left being drawn up towards his body) hung half over the side of the bed, the edge of it pressing upon his shin-bone.——He felt it not. A fix'd, inflexible sorrow took possession of every line of his face.——He sigh'd once,——heaved his breast often,——but utter'd not a word. (p. 157)

Verbal language fails Walter on this occasion. There are no words to express the extent of his grief and anguish, but Sterne is able to communicate Walter's emotions through gesture and physical description, leaving the reader to empathize with Walter, not on a rational, but on an emotional level.

A final example of Walter in a situation where language proves insufficient is Walter's attempt to explain to Toby the significance of noses to human personality according to the theory of Hafen Slawkenbergius. Walter is discussing "the ingenuity these learned men have all shewn in their solutions of noses." "Can noses be dissolved?" Toby replies. Sterne describes Walter's reaction:

> ——My father thrust back his chair,——rose up,——put on his hat,——took four long strides to the door,——jerked it open, ——thrust his head half way out,——shut the door again . . . returned to the table,——pluck'd my mother's thread-paper out of *Slawkenbergius's* book,——went hastily to his bureau, ——walk'd slowly back, twisting my mother's thread-paper about his thumb,——unbutton'd his waistcoat,——threw my mother's thread-paper into the fire,——bit her sattin pin-cushion in two, [and] fill'd his mouth with bran. (p. 173)

Through intricate descriptions of gesture such as the excerpt above, Sterne undercuts Locke's idea that verbal language, when properly used in a rational manner, is sufficient for human expression. Locke does not take into account situations where reason can fail a person leaving that person with no ability to express himself. Sterne shows that while Locke's verbal theory does not function in reality, gesture can provide an outlet of expression to the person for whom reason has failed. Sterne's final assessment of Locke's *Essay Concerning Human Understanding* is that "It is a history-book, Sir . . . of what passes in a man's own mind; and if you will say so much of the book, and no more, believe me, you will cut no contemptible figure in a metaphysical circle" (p. 61).

But while Sterne uses nonverbal communication frequently in *Tristram Shandy*, the stance that he advocates in the end is moderation. He does not assert gesture as a new language with the potential to surpass speech in importance, nor does he present verbal communication as the best and most useful form of discourse. Language, verbal and nonverbal, is necessary for communication. In some situations one form of language is better than another, but neither verbal nor nonverbal communication so outstrips the other in importance that one should be discarded for the other. In *Tristram Shandy*, though, Laurence Sterne pushes nonverbal communication to its limit, using it for description, character development, cultural context and satire. Because of the multiple ways that he uses gestical language, Sterne creates a new tradition of nonverbal communication in literature.

NOTES

1. For more information on the development of nonverbal communication as a result of the centralization of power in Europe and the development of court society, see two works of Norbert Elias: *The Court Society* (Oxford: Basil Blackwell, 1983), and *The Civilizing Process: The History of Manners* (New York: Urizen, 1978).
2. James Laver, *The Age of Illusion: Manners and Morals, 1750-1848* (London: Weidenfield and Nicolson, 1972), p. 2.
3. Norbert Elias, *The History of Manners*, p. 36.
4. Norbert Elias, *The Court Society*, p. 104.
5. Elias, *Court Society*, p. 36.
6. John Bulwer, *Chirologia: or the Natural Language of the Hand and Chironomia: or the Art of Manual Rhetoric*, ed. James W. Cleary (Carbondale: Southern Illinois Univ. Press, 1974), p. xiii.
7. Wilbur Samuel Howell, *Logic and Rhetoric in England, 1500-1700* (Princeton: Princeton Univ. Press, 1956), pp. 340-41.
8. Howell, *Logic 1500-1700*, p. 341.
9. Bulwer, p. 15.
10. Bulwer, p. 155.
11. David McNeill, *Hand and Mind: What Gestures Reveal About Thought* (Chicago: Univ. of Chicago Press, 1992), p. 3.
12. Bulwer, p. xiv-xv.
13. Bulwer, p. xv.
14. Bulwer, pp. 32-33, 35.
15. Samuel Wilbur Howell, *Eighteenth Century Logic and Rhetoric* (Princeton: Princeton Univ. Press, 1971), p. 200.
16. Howell, *Eighteenth Century*, p. 214.
17. Howell, *Eighteenth Century*, p. 235.
18. Sheridan, *A Course of Lectures on Elocution: Together with Two Dissertations on Language; and Some Other Tracts Relative to those Subjects* (London, 1762), p. vi.
19. Sheridan, p. 113.
20. Howell, *Eighteenth Century*, p. 220.
21. McNeill, p. 1.
22. Sheridan, p. 100.
23. Laurence Sterne, *The Life and Opinions of Tristram Shandy, Gent.*, ed. James Aiken Work (New York: Odyssey, 1940), p. 1.
24. Laurence Sterne, *The Life and Opinions of Tristram Shandy, Gent.*, ed. Howard Anderson (New York: Norton, 1980), p. 200. All subsequent references to *Tristram Shandy* are to this edition.
25. John Traugott, *Tristram Shandy's World: Sterne's Philosophical Rhetoric* (Los Angeles: Univ. of California Press, 1954), p. xiii.

26. Ray L. Birdwhistell, *Kinesics and Context: Essays on Body Motion Communication* (Philadelphia: Univ. of Pennsylvania Press, 1970), p. 50.
27. Traugott, p. 61.
28. Eugene Hnatko, "Sterne's Conversational Style," in *The Winged Skull: Papers from the Laurence Sterne Bicentenary Conference at the University of York* (Ohio: Kent State Univ. Press, 1971), p. 234.
29. Fernando Poyatos, *New Perspectives in Nonverbal Communication: Studies in Cultural Anthropology, Social Psychology, Linguistics, Literature and Semiotics* (Oxford: Pergamon, 1983), p. 279.
30. Poyatos, p. 303.
31. Poyatos, p. 308.
32. Poyatos, p. 310.
33. Helene Moglen, *The Philosophical Irony of Laurence Sterne* (Gainesville: Univ. of Florida Press, 1975), p. 90.
34. *The Letters of Laurence Sterne*, ed. Lewis Perry Curtis (Oxford: Clarendon Press, 1935), p. 77.
35. Moglen, p. 90.
36. Moglen, p. 91.
37. Bulwer, p. 16.
38. John F. Sena, "The Language of Gesture in *Gulliver's Travels*," *Papers on Language and Literature*, 19 (1983), 146.
39. Birdwhistell, p. 34.
40. John Locke, *An Essay Concerning Human Understanding*, ed. Peter H. Nidditch (Oxford: Clarendon Press, 1975), p. 402.
41. Locke, p. 509.
42. Locke, pp. 512-15.
43. Traugott, p. xv.
44. Traugott, p. 4.
45. Melvyn New, *Laurence Sterne as Satirist: A Reading of Tristram Shandy* (Gainesville: Univ. of Florida Press, 1969), p. 103.

THE CELEBRATED ACADEMY AT PALGRAVE: A DOCUMENTARY HISTORY OF ANNA LETITIA BARBAULD'S SCHOOL

WILLIAM McCARTHY

From 25 July 1774 to June 1785, the Rev. Rochemont Barbauld and his wife, Anna Letitia, newly renowned for her poems and essays, kept a school for boys at the village of Palgrave in Suffolk. The school opened with eight pupils; in its last term under the Barbaulds, it served forty-one. Although its main "catchment area" was East Anglia (Norfolk and Suffolk), it drew pupils from as far away as New York and the West Indies; and although the Barbaulds were Protestant Dissenters, it drew sons of aristocratic Establishment families as well as sons of middle-class Dissent. It gave its pupils an education more diverse and more progressive (judged by late eighteenth-century liberal standards) than they could have obtained at all but a few of its counterparts, the local and national grammar schools; and a number of its pupils went on to become eminent in various callings in the nineteenth century. The school also elicited from Anna Barbauld her most original, most influential, and ultimately most famous books, *Lessons for Children* (four volumes, 1778-79) and *Hymns in Prose for Children* (1781). Although it remained open after the Barbaulds' departure, in the public mind the school was always associated with them: when, in 1808, Rochemont Barbauld died, the *European Magazine* remembered him as the man "who formerly kept the celebrated academy at Palgrave."[1]

Palgrave School marks a turning point in Anna Barbauld's biography. Its main literary products, *Lessons* and *Hymns*, both by themselves and through the enormous influence they exerted over other writers for children such as Maria Edgeworth, Sarah Trimmer, Hannah More, and Priscilla Wakefield, also transformed infant pedagogy in England and America—a story which is finally coming to be appreciated as it should be, after decades during which Barbauld was treated as fair game for ridicule in the rumor mills miscalled "histories" of children's literature. That is not, however, the story I wish to tell here.[2] My subject is Palgrave School itself. To estimate its cultural impact or its place in Anna Letitia Barbauld's life and work, one needs to answer the questions basic to understanding any school: Who came to it, and from where? What "physical plant" did it have? What was taught at it, and with what intentions? What careers did its graduates follow? These are the questions I address in this essay.

Hitherto, with one important exception, commentators on Barbauld have had little with which to answer them, apart from *Lessons* and *Hymns* themselves, a few of Barbauld's letters, and the spare account given by Lucy Aikin, Barbauld's niece and first biographer. The exception is C. W. Brodribb's 1935 essay, "Mrs. Barbauld's School," which described several school documents then in his possession (he was Barbauld's great-great-great nephew); they perished five years later in the bombing of London.[3] Recently two other documents have come to light: a printed playbill for a school performance of William Mason's *Caractacus* in 1782 (reproduced below), and a "List of Mr Barbauld's Pupils 25th March 1785."[4] These are supplemented by a few new letters from Rochemont and Anna Letitia to tradesmen, and from Anna Letitia to the parents of pupils or prospective pupils—in effect, the school's business correspondence. I print these letters in Appendix I. From all these documents, new and old, and from scattered references elsewhere, a good deal can be gleaned: the names of ninety-five pupils (listed alphabetically in Appendix III), the fees they paid (bills are printed in Appendix II), and the titles of books used in the school (Appendix IV).

Although *Lessons* and *Hymns* are the greatest fruits of Barbauld's pedagogy, they will figure in this essay only as evidence of the Palgrave curriculum. They are not, in fact, despite the destruction of Brodribb's family holdings, the only survivors among Barbauld's teaching texts. At least some of her contributions to John Aikin's *Evenings at Home* (1792-96) and some of the pieces gathered by Lucy Aikin in a posthumous volume called *A Legacy for Young Ladies* (1826) are probably relics of Palgrave, and others which are not at least suggest what her Palgrave teaching must have been like. Her essay "On Education" (1798) embodies her reflections on teaching; and a few teaching poems appear in the edition of her poems by me and Elizabeth Kraft.

Simply to marshal all this information is to present a fuller idea of Palgrave School than has yet appeared in print. However, the practices of a single school cannot really be appreciated except by comparison with those of contemporary schools; absent such comparisons, it is rash to assert that a school is distinctive for its curriculum or pedagogy. Therefore, besides marshalling what I know about Palgrave, I also suggest comparisons between it and several other schools—the Dissenting academies, three "great" schools, Eton, Rugby, and Christ's Hospital as recalled by Leigh Hunt, and three "progressive" schools described by school historian W. A. C. Stewart.[5] In these contexts Palgrave will appear to have been moderately "left-wing." Like the Dissenting academies (from which it descended), Palgrave treated its pupils better and taught them more than the Establishment schools did theirs; and its results suggest that it produced a different kind of adult from the boys who emerged from Eton or Westminster.

The Economical Part:
Place, House, Household, Fees

Palgrave is today, as it was in 1774 when the Barbaulds arrived, a small village located a mile south of Diss, the market town for its region. Palgrave is in Suffolk, Diss in Norfolk (the river Waveney, at this place a mere stream, separates them), and both lie two miles west of Scole, the coach station on the north-south road between Ipswich, the county town of Suffolk, and Norwich, the county town of Norfolk and cultural capital of East Anglia. Moreover, Palgrave lies at the head of the road that goes southwesterly to Bury St. Edmunds. Rochemont had accepted the post of minister to the Dissenting congregation at Diss, late in 1773, on condition that the leading congregants sponsor a school by means of which he could augment his income; and this location, twenty-two miles from each of the two county towns, must have looked likely to draw pupils from both.[6]

The house the Barbaulds took—or which, probably, was taken for them—had belonged to a well-known local antiquary, "Honest Tom" Martin. A daguerrotype of the building, dating perhaps to a few years before its demolition in 1860, shows a two-gabled brick structure in early Georgian style; on the long side of the front wing (the side facing the Bury road) are thirteen sash windows, three up and three down on each side of the entrance with one above it. The room that became the schoolroom was situated in the back of the rear (or south) wing; in its long casement window the pupils are said to have scratched their names.[7] Above it would have been the pupils' sleeping-quarters. The first eight pupils seem to have shared four rooms; at least, William Taylor and Frank Sayers shared a room, with one bed for both of them. In the front wing the Barbaulds and the "family" would have lived; besides servants, "family" may have included those pupils called "parlor boarders" (of whom more below). The house held twenty-two persons.[8]

In his acceptance letter Rochemont had stipulated that at least six pupils must be guaranteed "to come and live with me" for a year. He also required a writing master "who would regularly attend my boarders; for this part of the business I could not take upon me." He did not expect to have an usher—that is, an assistant teacher—"till the school were in a thriving condition"; and "as to a house-keeper, I shall gladly . . . bring one along with me." He means Anna Letitia, who will "take upon her the economical part."[9] Anna Letitia's duties in "the economical part" included keeping the school and family accounts (her account books perished in 1940), making beds, scolding maids, and overseeing the school's provisions—the ordinary tasks of a schoolmaster's wife, at least in a small school, and especially in a school just getting started.[10] Her household staff may well have begun as no more than two maids and a boy-servant, and it was probably never larger than four or five. In September 1775, when the school was indeed "in a thriving condition," it had an usher, and in 1783 a "French" usher; but of course the ushers were Rochemont's assistants, not hers.[11]

The expenses of opening a school, even a very small one, were not trivial—a point labored by Thomas James, headmaster of Rugby,

"Barbauld House," the Palgrave School building. Daguerrotype (ca. 1850s?), reproduced by courtesy of Mr. Simon Martyn.

writing in 1796 to would-be schoolmaster Samuel Butler: "For what is a hundred pounds a quarter (though it be undeniably certain) to enable you to provide house-rent, pay your rates, highway rates, church rates . . . and to buy all the provisions for at least two maids and a boy-servant. . . . Four pupils, three servants, man and wife . . . are nine times £25, or £225—but a pupil (with one glass of wine a day) is £30—exclusive of your own and wife's clothes, etc. . . . Then where is your furniture, of beds especially, to come from?"[12] James's prices reflect the wartime inflation of the 1790s, but, mutatis mutandis, his point remains good. Rochemont Barbauld stipulated that his six guaranteed pupils must each pay at least twenty guineas (£21) tuition, plus two guineas (£2-2-0) "entrance" fee; these would have amounted to £138 for the school's first year, which seems a close shave. In the event, however, Rochemont got more than he asked for: he started with eight pupils, who paid £25 each as well as the entrance fee. These better terms appear to have been average for their time.[13]

As the school grew, it took an additional category of pupils at a higher rate. These were "parlour-boarders"—known to fame as Anna Letitia's "little parlour pupils" (so she called them), "those little pupils I have under my own care," a class she was induced to take by the success of *Lessons for Children* and for whom she wrote *Hymns in Prose*. Actually, her "little parlour pupils" and the "parlour boarders" were not the same group. At Daventry Academy under her second cousin, the Rev. Thomas Belsham—whose tenure was contemporary with that of the Barbaulds at Palgrave—parlor-boarders were students received "upon advanced terms," who breakfasted and supped in the parlor rather than in common with the other students, and who were lodged in the best rooms. At Christ's Hospital, Leigh Hunt's school, a "boarder" in the 1790s was a pupil whose parents "were too rich to let [him] go on the foundation." At Palgrave, Anna Letitia's "little parlour pupils" paid the advanced sum of forty guineas, and the ones whom I can identify appear to have had rich parents; yet only two of the nine pupils listed in 1785 as paying forty guineas were "little," i.e., under age ten. Probably Anna Letitia's "own peculiar pupils" (as Lucy Aikin also called them) were drawn from the larger category of forty-guinea-paying parlor students. Or rather, added to it; for in reckoning the number of pupils to determine whether or not the school had vacancies, Rochemont did not count those of his wife.[14]

Yet, counted or not, they must have occupied house room. In 1785 there were nine parlor boarders, plus an additional thirty-two pupils on the common terms. In January 1779 Anna Letitia boasted to her brother that their house was "full from top to bottom" (*Works*, II, 22). Can it have held forty-one pupils (some of whom enjoyed private rooms), the Barbaulds, and their servants? Presumably some of the pupils—those who lived nearby—were day students only; and perhaps others, like some Eton pupils, boarded at "dames' houses" in the village.[15] The crowded house was, of course, success, and the numbers show that Palgrave School did very well. If in its first year its eight pupils brought in just under £217 (£25 plus two guineas per pupil), in its last under the Barbaulds it was earning at the rate of £1147 (thirty-one pupils at £25 each, one at thirty-five guineas, and eight at forty guineas). If its backers expected a profit from their investment in it, they probably were not disappointed.[16]

Mr. B's Menagerie: The Pupils

How many boys passed through Palgrave School during the Barbaulds' tenure there? In his advertisement for the school (Appendix I, item 1), Rochemont stipulated that he would take no more than thirty at a time. At different dates we catch glimpses of different numbers. In July 1774, the school opened with eight; in July 1775, it had sixteen "boarders" ("boarders" implies that there were additional day pupils); early in 1776 Anna Letitia expects to have twenty-seven pupils before vacation, and two more at midsummer; on 26 May 1780, thirty-two pupils were present; on 4 November 1783 and 29 April 1784, thirty-six copybooks were ordered for the school (Appendix I, items 14 and 17); the 1785 list of pupils names forty-one boys.[17] For reasons given in the headnote to Appendix III, I posit a total population of 131 pupils over the eleven years. This makes Palgrave a small school by comparison with Eton at the same period (250 pupils in 1775, but normally 300 to 500 at any one time [*ECR*, p. xx]), or even its smaller rival, Harrow (which in 1780 had only 139 pupils and in 1791 sank to 119), let alone Christ's Hospital in the 1790s (600, veritably a factory [Hunt, p. 49]).[18]

Ninety-five Palgrave boys are known by name, and many of those are known by parentage also (see Appendix III). From this information, much can be inferred (albeit often conjecturally) about the demography of the school. In age, the boys ranged from three-and-a-half (Denman) or even three (William Gell) to eighteen or even nineteen (both Bulwers, Basil William Douglas, perhaps Phipps). The ages of fifty-nine boys at their earliest recorded presence at Palgrave (not necessarily the same as their ages upon arrival there) can be determined: excluding C. R. Aikin, Anna Letitia's nephew and adopted son, who arrived at age two, four were age three to six; twelve were age seven to nine; twenty-two were age ten to twelve; fifteen were age thirteen to fifteen; and five were age sixteen or over. Their median age was just over 11.5.[19] These ages are probably typical for grammar schools of the time: at Eton, boys arrived at any age between four and sixteen, and many stayed till eighteen or nineteen; and at Harrow the ages of pupils on entry ranged from seven to fifteen, though very few stayed on after seventeen (*ECR*, pp. xxxiv-xxxv; *HSR*, p. xv). The fame of *Lessons for Children* and *Hymns in Prose* should not mislead us, as it apparently did Samuel Johnson, into imagining that Palgrave was chiefly an infants' school.[20]

The difficulties of travel ordinarily restricted grammar schools to those boys living within convenient distance of them. Probably most Palgrave boys came from East Anglia: of sixty-two boys whose origins are known, twenty-three came from Norfolk, six from Suffolk, and two from Essex.[21] But Palgrave was surprisingly cosmopolitan for a small village school; in geographical range, its catchment area resembled Eton's (*ECR*, pp. xx-xxi). Boys came from London (at least five, and perhaps nine), Berkshire (one), Buckinghamshire (one), Derbyshire (two or three), Lancashire (five), Surrey (four), and Warwickshire (one). Four, the Douglases, came from Scotland, perhaps two from Ireland, and one from Holland. There were also boys from Jamaica (the Glanvilles), perhaps from other places in the West Indies (?Browne, Elliott, and Malbrough), and even from North America (Foy, from New York).

Before attempting demographic generalities about the types of students the school attracted, I must take note of the personal connections which would have led some to go there. Personal connections account, in the first place, for all the Aikins (sons of Anna

Letitia's brother John, one of whom she adopted), Priestley (son of Joseph, her old friend from his years as a tutor at Warrington Academy), Enfield (son of William, also a friend from his years as tutor and rector at Warrington), the Glanvilles (probably sons of a Warrington Academy graduate friendly with Anna Letitia and her women friends at Warrington), and Hunter (grandnephew of her and the school's bookseller, Joseph Johnson). More remotely, they may also account for the Douglas boys, whose presence would otherwise be hard to explain: what would induce a Scottish earl to send his eldest son to an unknown school in East Anglia? Anna Letitia's Aikin ancestors lived in and near the town of Kirkcudbright in Galloway, seat of the Douglases, earls of Selkirk; one Aikin had a business dealing with a sister of the first earl of Selkirk. Moreover, Anna Letitia's brother John had been apprenticed in 1761-64 to Dr. Maxwell Garthshore, who came from Kirkcudbright himself; and thus, long after their Kirkcudbright ancestor had quit Scotland, they may have retained some contact with their country of origin.[22] On Rochemont's side, we are told that his court connections "brought them several sons of noblemen and gentlemen of fortune."[23] To the influence of Rochemont's father, the Rev. Theophilus Lewis Barbauld, rector of St. Vedast's Church in the City, who had served the royal family in Germany, we can probably attribute Corbett (grandson of Lord Bute, to whom Theophilus Lewis had appealed for patronage in 1763, with success), LaFite (son of a French woman of letters known to him in 1777, who went on to become a governess at the English court), and perhaps Lord Aughrim. Then, once pupils came to the school, their own family connections may have brought others. At least, it is noteworthy how many of them were related by marriage: Bunny, C. Marsh, Scott, and Martineaus; Dysons and William Taylor.

That the Barbaulds' affiliation with Warrington Academy had been expected to bring pupils is clear from the advertisement announcing Palgrave's opening in 1774: "PALGRAVE SEMINARY . . . will be opened . . . by the Rev. Mr. ROCHMONT BARBAULT, from WARRINGTON . . ." (Appendix I, item 1). In 1774 Warrington Academy was flourishing and respected among Dissenters; it also had specific connections with Norwich, going back to the Rev. John Taylor (1694-1761), who left Norwich in 1757 to become one of the first Warrington tutors. His grandson, Edward Rigby, brother of Anna

Letitia's friends Sarah and Lissy Rigby, graduated from Warrington in 1764 and migrated back to Norwich to practice surgery. Thomas Martineau, likewise, was a link to Warrington: His older brother, Philip Meadows Martineau, entered the Academy in 1765. And Martineau's uncle, Philip Meadows of Diss, was one of Palgrave School's original sponsors (Appendix I, item 1).[24]

Thus Palgrave School was envisioned, in part at least, as serving the same clientele that Warrington Academy served. It did serve that clientele: families of solid middle-class citizens such as Roger Kerrison, banker, alderman, Justice of the Peace for Norfolk, Treasurer to the Corporation of Guardians for the Poor, and Mayor of Norwich, many (but by no means all) of whom were Dissenters, and a small but noticeable number of West Indian planter families. But Palgrave served another clientele also. Of forty-eight pupils whose class origin is known or inferrable, just over half were sons of middle-class professional or commercial families; the rest were connected to titled families and landed gentry. By "professional" I mean medical, legal, clerical, or literary; the fathers of eleven pupils held these occupations. The fathers of fifteen pupils engaged in commerce, including the occupations of banker, insurance agent, wool-stapler, and manufacturer of cloth for export. Eleven pupils were related to landed gentry, including four sons of West Indian planters, and eleven to aristocratic families. This mingled class composition of its student body shows that the school catered neither to "strictly select families" (as one commentator has it) nor simply to the new mercantile orders. It catered, rather, to both.[25]

In this, Palgrave does not appear to have differed from "great schools" such as Eton. Eton, in fact, may actually have been more egalitarian in drawing its pupils from diverse social classes: Etonians were sons of titled and landed families (including plenty of West Indian planters) of course, but some were sons of apothecaries, auctioneers, bakers, bankers, booksellers, brewers, and even (once) a bricklayer (*ECR*, pp. xx-xxi, 607). It is a fallacy of school history-writing to suppose that the mark of a politically "progressive" school is its taking pupils from the middle and lower classes of society. No school in England, according at least to Leigh Hunt, was more egalitarian in its student body than Christ's Hospital ("more boys are to be found in it, who issue from a greater variety of ranks, than in any school in the kingdom" [Hunt, p. 50]); yet in its politics, Christ's Hospital was, again according to Hunt, "orthodox and high Tory" (p. 71). Palgrave, like the great schools, addressed a clientele primarily Establishment, whose sons were to be prepared to play their economic, political and ideological roles in society. It differed from the great schools not in taking different pupils, but in how it treated and what it taught its pupils.

Love towards the Rising Generation: The Politics of Discipline

Why did Establishment families, people often of property and influence, choose to send their sons to a school managed by Dissenters, people legally marginal, in preference to Eton, Harrow or Rugby? Lucy Aikin is surely correct in asserting that Palgrave School owed its success in large part to Anna Letitia's literary fame ("Memoir," p. xxiv). One indication of this is that Dorothy Gell, mother of two Palgrave boys and wife of a leading Derbyshire landowner, owned copies of every book Anna Letitia had published; still in her teens when her sons were born, Gell belonged to the half-generation of young women who were reading "Miss Aikin's" poems with enthusiasm.[26] *Lessons for Children* and *Hymns in Prose* would have confirmed public confidence in the school by demonstrating, in addition, Barbauld's genius for teaching.

But there was another reason as well. In 1790 Barbauld herself explained the motives of Establishment parents by reference to the very marginality of Dissent, which (she said) preserved it from the worst temptations of power and kept it relatively honest. Establishment parents, she suggested, feared the influence of their own schools: "If our rising seminaries should excel in wholesome discipline and regularity, if they should be schools of morality, and yours, unhappily, should be corrupted into schools of immorality, you will entrust us with the education of your youth, when the parent . . . wishes to preserve the blooming and ingenuous child from . . . early licentiousness."[27] Her interpretation was shared by Gilbert Wakefield, who, arriving at Warrington Academy as a tutor in 1779, found that "at least" one-third of its students were Church-of-England, sent there by parents who preferred a Dissenting academy to "the restrictions on opinion, and the relaxation of discipline in the *Universities*."[28]

That these views were accurate hardly needs laboring. The state of Establishment schools in England in the late eighteenth century is a standing topic of deploration by school historians. The regimen at those "great schools" to which Palgrave parents were afraid to send their sons consisted of punishment routinely by flogging (a practice well suited to the tastes of sadistic masters like Bowyer at Christ's Hospital), savaging of younger boys by older (when, for example, Tommy Denman went to Eton, older boys branded him with a hot iron for refusing to entertain them with a speech), drinking and gambling by the boys, and the reflex of all this, occasional violent uprisings of pupils against masters.[29] Even reform could wear an ugly face, as it did in the surveillance methods of Samuel Butler, master of Shrewsbury School in 1821: he demanded that his assistant spy on the pupils, informing him of "their habits, tempers, and dispositions, their haunts and propensities, that by an unseen yet watchful control I may prevent evil habits." And Thomas James, reformist master of Rugby in the 1790s, counselled whipping as a punishment for absence from school beyond the time allowed (Butler, I, xxvii, 41).

Against these practices it is a pleasure to set Anna Letitia's thoughts on discipline, from a luminous 1798 essay "On Education." In a passage which probably responds to the progressive theories of the Edgeworths on children's lying but which by implication indicts the regimes of the traditional schools, she demands that the autonomy of the child be respected even when it leads the child into subversive relationship with the adult world:

> There is nothing which has so little share in education as direct precept. To be convinced of this, we need only reflect that there is no one point we labour more to establish with children, than that of their speaking truth; and there is not any in which we succeed worse. And why? Because children readily see we have an interest in it. Their speaking truth is used by us as an engine of government—"Tell me, my dear child, when you have broken any thing, and I will not be angry with you." "Thank you for nothing," says the child; "if I prevent you from finding it out, I am sure you will not be angry." . . .

> Your child is the best judge how much occasion he has to tell a lie: he may have as great occasion for it, as you have to conceal a bad piece of news from a sick friend, or to hide your vexation from an unwelcome visitor. That authority which extends its claims over every action, and even every thought [recall Samuel Butler's insistence on knowing "their habits, tempers, and dispositions, their haunts and propensities"], which insists upon an answer to every interrogation, however indiscreet or oppressive to the feelings, will, in young or old, produce falsehood; or . . . it will produce a habit of dissimulation, which is still worse. The child, the slave, or the subject, who, on proper occasions may not say, "I do not choose to tell," will certainly, by the circumstances in which you place him, be driven to have recourse to deceit, even should he not be countenanced by your example.[30]

This should be taken not as indifference to truth-telling (Barbauld cared about truthfulness and aimed to inculcate respect for it), but as evidence of her profoundly unillusioned understanding of the politics of parent-child (or teacher-pupil) relations. In the light of that understanding, she defends the rights of children to personal privacy and their own agendas. Her willingness to entertain the child's agenda, and even what Dr. Butler would probably have called his "evil habits," is also manifest in a poem, "A School Eclogue," which mentions Palgrave and was probably written there, for the amusement of the boys. In a parody of Virgilian eclogue, Harry and Edward engage in a contest of rhyme. Soon it degenerates into an argument between them, in which each reproaches the other with his delinquencies. Edward has robbed an orchard, hiding the berries in his pocket, whence their juice oozed out and stained his trousers. Harry has played truant, sneaking off to the local fair—a whipping offense at Rugby under Thomas James. Moreover, both boys are guilty of adolescent sexuality (no doubt the real "evil habit" for which headmasters like Butler watched). To represent it acceptably, Barbauld has both boys declare themselves "lovers" of a girl named Anna, a classmate (although in fact no girls are known to have attended Palgrave School), for whom each has been moved by passion to commit some crime: writing her name in candle smoke on the ceiling over his bed, saving her from disgrace at lesson-time by conveying the answers to her. All of this is rendered from the assumed perspective of the boys, not of the master. Although they construe their actions as delinquent, they do so only because they know that the master will construe them that way—they are afraid lest they feel "the prowess of his arm," i.e., lest he whip them (*Poems*, Poem 79). The poem suggests how much Anna Letitia sympathized with the transgressive boyish energy which adults would repress. As she wrote to her brother in 1775, she could not help feeling a "partiality . . . for the little urchins" (*Works*, II, 6), a partiality fully shared by her husband.[31]

"The prowess of his arm" and a reference elsewhere in the poem to the boys' shoulders feeling "th' unpleasant goad" raise the question whether the Barbaulds in fact used corporal punishment at Palgrave. Probably they did not. Part of the poem's entertainment value to the Palgrave boys may well have been its representing Harry and Edward as liable—like almost all other schoolboys—to bodily punishment, from which they, its readers, would have been blissfully exempt. On the subject of discipline, as probably on most others, the Barbaulds may have felt their way gradually into new practices. Anna Letitia disapproved of flogging in 1803, and I cannot imagine that she could have countenanced corporal punishment as long as she held the views on children's rights which she published in 1798. The forms of punishment she and her husband are much more likely to have used are "fines and jobations," which the schoolboys in her poem also dread; both were used at Warrington Academy.[32] It is tempting to wonder, also, whether they instituted punishment by peers in the form of "juvenile trials" for offenses committed at the school. These were used by the radically innovative Rev. David Williams at his school in London in 1773, and Anna Letitia's brother John Aikin was charmed by the idea: he composed a fictional "juvenile trial" in response to a published collection of them, and inserted it in *Evenings at Home*.[33] As inducements to learning, the Barbaulds offered prizes for which students could contend at end-of-term examinations (see Appendix I, headnote and item 7).

They managed the school as a polity whose members, the pupils, were citizens. A document printed by Brodribb shows that Palgrave pupils had political rights and obligations. In Anna Letitia's hand, the document sets forth the "Privileges and Duties of the Captain of Palgrave School" (Brodribb, p. 733, undated). His privileges are to sit where he likes at dinner and supper, to have a place by the fire in the school or hall, a right to "the sitting-room above stairs," and a right to a double share of "any presents of fruit, etc.," given to the school. His duties, however, are strenuous: "He shall arbitrate in all disputes amongst the boys, reserving however a right of appeal to Mr. Barbauld"; he "shall hold himself obliged to set a good example in the school, both in morals and diligence, to protect the younger ones from oppression, to preserve order; and as far as is in his power to prevent anything from being done which may throw dishonour on the society. It will be the Captain's business to deliver the sense of the school when any favours are to be asked of Mr. Barbauld or any other person; or when any thing is to be said which relates to the whole body." His term of office is one session, and he is "to be chosen by a majority of votes, Mr. Barbauld always reserving to himself a negative on the candidate." The captain, upon being elected, formally pledges himself to behave "in such a manner as may best promote the reputation, order, happiness and improvement of the society to which I belong"; should he be guilty of "falsehood, breach of trust, oppression, habitual indolence, or any other capital fault, he will be immediately degraded from his dignity." Clearly the Palgrave captaincy is a version of the monitorial system, later standard in English public schools, in which "head boys" are "links" between the master and "the rank and file of the school."[34] The election of the captain anticipates the electoral system of school governance at Hazelwood School, 1819 (Stewart, p. 59); in this, the Barbaulds may have been pioneers. Their captain was clearly conceived as the chief magistrate of a model republic; the Palgrave school polity was designed to exercise the boys in something like "the Rights of Man and the Citizen." Also in citizenly responsibility: the captain's few privileges were outweighed by his many magistral obligations, and the boys, when they cast their votes, had to declare that they did so "unbiassed by any motives but a serious and deliberate regard to the merits of the candidates."

Ideas about polity always presuppose ideas about what can and should be expected from human beings. A regime of punitive authority implies a belief that human beings are "naturally" refractory and antipathetic to learning; they must be coerced. That ordinary schoolmasters acted on this Hobbesian faith well into the nineteenth century is evidenced by, for example, a letter to the *Monthly Magazine* in 1819, from an opponent of corporal punishment: "The system [he complains] . . . which is generally pursued in the government of day-schools, is neither more nor less than downright despotism: a succession of arbitrary, capricious, and often unreasonable commands, are imposed upon the pupil, and punctual performance expected; whilst the little trembler has no other motive to stimulate him to the discharge of his toilsome duty than unmingled fear of corporal punishment!"[35] The Barbaulds believed, with the Edgeworths and other late eighteenth-century educational innovators, that human beings are docible creatures who respond to friendly treatment. Anna Letitia's idea of a small boy's "nature" appears in a mock-Linnean "Description of a Curious Animal lately found in the Wilds of Derbyshire" which may be seen "at the Rev. Mr. B.'s menagerie, with many other young animals equally curious." "This little creature," she writes, resembles a monkey in its agility and "tricks." It

> makes a chattering noise, climbs, leaps and runs, and has a spring in its muscles equal to an antelope. . . . Will often shake its hair over the whole face, which gives it a look of peculiar wildness. Is very good-natured and playful, caressing to its keeper and every one who takes notice of it. Is however easily put in a passion, and when angry makes a threatening noise, but is soon put to flight by the least show of resistance. If seized, kicks with its hind legs: is however tolerably docile, considering how lately it has been caught. Feeds on fruits, roots, or flesh; will eat cakes or nuts out of the hand.[36]

Unsentimentally, the description resembles Lemuel Gulliver's account of the first Yahoo he meets. However, this "Yahoo" is charming even in its savage state and has potentials that Swift's does not. Although it needs at times to be curbed, its curiosity, playfulness, and caressing temper are native to it, and are the means by which it will be civilized. Anna Letitia buys into neither the myth of natural depravity nor the myth of natural innocence, and this ethical realism seems reflected in the political moderation of the school rules. The rules assume, for example, that there is a danger of younger boys being oppressed by older (she is not utopian in her ideas of human conduct), but they aim to avert the danger (or to remedy the offense) by fostering the older boys' capacities for fairness and social responsibility. Knowing, however, that those capacities don't always work, she reserves a right of appeal to a superior magistrate, Mr. Barbauld himself. Politically, the most obvious analogy is the idea of "mixed" government, of "checks and balances" within a system which, however, is managed by the governed themselves.[37]

Intelligent Citizens: Curriculum and Pedagogy

In educating its pupils for citizenship, Palgrave School carried on a Protestant Dissenting agenda which can be traced back at least to Milton's *Of Education* (1644). Milton hoped that his revisionary plan of study would produce "brave men, and worthy patriots, dear to God, and famous to all ages." Rochemont Barbauld believed that to educate children was itself a patriotic duty: "the true patriot . . . will gladly undertake a task whereby he may so essentially contribute to the welfare of his country. A prominent feature of his character is love towards the rising generation." A leading exponent of this traditional Dissenting ideal was Joseph Priestley, tutor at Warrington Academy from 1761 to 1767 and Anna Letitia's close friend. Priestley designed a course in "civil history and civil policy" to educate Warrington students for "active life." By "active life" he meant "all those stations in which a man's conduct will considerably affect the liberty and the property of his countrymen, and the riches, the strength, and the security of his country."[38] Priestley's ambition was to form "the able statesman, and the intelligent and useful citizen" (p. 12); he envisioned an elite of landed gentry and the upper ranks of the legal, military, and mercantile professions (p. 11), educated to an informed sense of their responsibility to the nation. Priestley justified his curriculum, as Milton had his, by appeal to its utility: "the studies of youth should tend to fit them for the business of manhood" (p. 9).

The Priestley curriculum corresponds closely, allowing for the difference between an academy and a grammar school, with what was taught at Palgrave. His principal subjects, civil history and "civil policy" (what later came to be called political science), were intended for young men; he discouraged their study before age sixteen (p. 13). As prerequisites to them, he names French (the language of international affairs), accounting, "practical mathematics," algebra and geometry (pp. 15-16). Latin and Greek, those foundations of the traditional grammar-school curriculum, are comparatively demoted in Priestley's scheme: "not absolutely necessary, but . . . very desirable" (p. 15). In defense of their demotion Priestley urges that Britons today need to write "with propriety in [their] native English" (p. 17); this amounts to recommending a course in English composition.

Like Priestley's, the Palgrave curriculum was designed for "Youth intended for any of the Professions, or Trade" (Appendix I, item 1). Rochemont Barbauld undertook to teach "Latin, French, English and Accompts, and, if any desired it, Greek, Algebra, Geometry and Geography." The list of books in Appendix IV shows that the school did teach Latin, French, English literature and composition, Greek, and arithmetic. History and geography, which were Anna Letitia's province, will be noticed shortly. The youngest Palgrave boys were taught to read and write, and all received fundamental religious instruction. Science, although not mentioned by Priestley, had a place at Palgrave, as it did at all the Dissenting academies. Besides studies, the boys exercised themselves in trap-ball and swimming, and perhaps kept gardens.[39]

Although Priestley's formulation of the curriculum may have provided the immediate incentive for Palgrave, Anna Letitia could have derived such ideas equally from her own family: her mother's father, the Rev. John Jennings (1687-1723), master of an academy at Kibworth, had taught essentially the same curriculum, and his pupil, the influential Philip Doddridge, had further disseminated it. Moreover, Anna Letitia's father, the Rev. John Aikin (1713-80), a Doddridge pupil (and thus a grand-pupil of Jennings), had been both master of a grammar school at Kibworth and one of the tutors at Warrington Academy. In the latter capacity, he emulated his teacher's teacher both in his curricular choices (he was a particularly good teacher of English literature) and in his encouragement of open enquiry by students into the subject-matter of his lectures—a practice suggesting a politics of relative equality between master and pupils.[40] Anna Letitia's ideas about curriculum would, in short, have been overdetermined before she came to Palgrave. Rochemont, too, although not from a Dissenting family and not a personal friend of Priestley, was a pupil of John Aikin at Warrington and a zealous convert to Dissent. Their curricular choices were perhaps not so much "choices" as a carrying-on of the curriculum dear to their social group, the intellectually conscious section of the liberal middle classes. Priestley's formulation of it was only its latest update.[41]

If the Palgrave curriculum carried on traditional Dissenting practice, it was decidedly innovative compared with the conventional grammar-school program. For a detailed account of that program we are indebted to the Rev. Thomas James of Rugby, who in 1798 showered advice and counsel on his former pupil, Samuel Butler, when Butler was about to become master of Shrewsbury. James listed lessons for each day of the week and recommended texts for them.[42] Comparing his total of sixty-eight texts with the Palgrave list in Appendix IV (fifty-one identified texts, but not complete) brings into high relief the prime curricular difference between Palgrave and Rugby: at Rugby, fifty-three percent of the texts used were "Classical," i.e., in Latin or Greek, or, if in English, concerned with Classical subjects; at Palgrave, the corresponding figure was only twenty-five percent. Conversely, at Palgrave almost seventy-five percent of the texts used were in English or French; at Rugby, less than half—and most of those were relegated to the school's leisure library. (James allowed just the half holiday to French [Butler, I, 28].) Leigh Hunt's account of Christ's Hospital is instructive, too. In school, Hunt had to choke down huge helpings of Ovid, Virgil, Terence, Homer, Cicero, Demosthenes, and Greek plays (p. 66); he says he hated them, and much preferred to read—out of school, of course—the English poets Spenser, Collins, Gray, and Shakespeare (pp. 68-70, 71-73). Hunt is here constructing that well-known figment of Romanticism, the Ephebe Poet as Unhappy Schoolboy. In view of the bad press that Palgrave School has suffered from historians influenced by Hunt's friend Charles Lamb (who coined the oft-repeated curse upon the "Barbauld Crew, those *Blights & Blasts*" of infant happiness), it is worth remarking an irony: Had Hunt gone to Palgrave instead of Christ's Hospital, he could have enjoyed his favorite poets in the normal course of his studies. Compared to those of the Establishment schools, Palgrave's was a "modern" curriculum—even (if Hunt's preferences are taken as the measure) a "Romantic" one.[43]

Yet, its modernism notwithstanding, Palgrave appears to have offered quite respectable Latin and Greek. That they were utilitarian rather than elegant is witnessed by William Taylor: they were not taught "with that nice attention to quantities and accents which constitutes so prominent a feature in the routine grammar-school exercises of the present day." Still, the facts that Charles Aikin could have recited an ode of Anacreon at three and a half and that Thomas Denman had begun Latin by age four and a half argue for serious attention to the subjects. Barbauld was not the kind of innovator who contemptuously throws aside the past. She appreciated the importance of the Classics in European culture, not only as points of reference for subsequent literature but as a moral force: "Books are a kind of perpetual censors on men and manners; they judge without partiality, and reprove without fear or affection."[44] Indeed, she wished the Classics to be more accessible to readers, and thought a short vernacular introduction to them for those without Greek or Latin would be useful (p. 73). But, always utilitarian in her educational judgments, she regarded learning languages for their own sake as an error in education. It was with very dry irony that she remarked, in an 1803 review of a conservative book on schooling, that "if the great object of the first eighteen or twenty years of life be to make a youth thorough master of two dead languages, the slow and steady method of our great schools, cannot be much improved upon."[45]

Anna Letitia's part in the teaching is much easier to identify than Rochemont's, and although I assume they discussed their pedagogy and curriculum and agreed on them, I will henceforth speak only of her work. She taught not only reading and religion to the youngest pupils, but geography and history, English composition and oratory, and natural science. Besides *Lessons for Children* and *Hymns in Prose for Children*, a few of her teaching texts or probable teaching texts survive: poems on the flora, fauna and human customs of other lands; two pieces which look like natural science lectures, "On Plants" and "Earth" (both published in *A Legacy for Young Ladies*); the lecture "On the Classics," mentioned above, which treats Classical texts as, inter alia, sources of history; and a fragment of a lecture on Islamic history. She wrote unknown numbers of "prologues, epilogues, and interludes" for plays performed at the school (Aikin, "Memoir," p. xxvii), and verses for boys to recite at examinations. A few other texts, such as a long essay "On the Uses of History" (also in *Legacy*), although apparently not written for Palgrave, probably testify to her teaching there.[46]

One of the glories of Palgrave was her teaching of history and geography. These were standard subjects in Dissenting academies, which tended to treat them politically, leading up to theory of government (Priestley's course in civil history and policy was more traditional than innovative in this respect), and which also aimed to comprehend the whole world: at Kibworth, John Jennings taught histories of Europe, Asia, Africa, and America, and at Daventry, students were introduced to the "different manners and customs of the different nations of the Earth."[47] (At Rugby, by contrast, Dr. James considered it impossible to teach more than Greek, Roman and English history; his geography, accordingly,was that of ancient Greece and Italy [Butler, I, 19, 29].) Eighteenth-century interest in geography tends to be construed today as an ideological reflex of imperialism, and while it can be seen in retrospect to have served the interests of Empire, serving those interests was not its only effect; it also enlarged European ideas of human culture. Barbauld was well aware, to be sure, of the intimate relation of maps to commerce and war (*Legacy*, p. 142), but her teaching of geography and history seems to have been informed by a spirit of pluralism and a sense of wonder. The study of history she justifies on the traditional ground that it shows the entire range of "human nature" (*Legacy*, pp. 121-22). She rejects the practice of bowdlerizing history for young readers because doing so deprives them of knowledge (p. 134); accordingly, she criticizes collections like *The Flowers of History*, which Thomas James used at Rugby (Butler, I, 41). She decries the habit of cultivating contempt for other nations: "There is no use, but a great deal of harm . . . in an arrogant assumption of superiority, and the clownish sneer of ignorance at every thing in laws, government or manners which is not fashioned after our partial ideas and familiar usages" (*Legacy*, p. 130). She kept herself well informed on historical and geographical research, following the work of Sir William Jones on Indian languages and literature (p. 84) and the publication in Europe of Islamic classics—and, it would appear, incorporating this information into her lectures.[48]

In her lectures she emphasized the materials and concepts of the subject, explaining from the ground up things that most teachers simply take as given. There is, thus, something of a "deconstructive" turn in her lecture "On the Classics," where she notes the embarrassments in the "real sources of History": deficient documents and "jarring accounts." To know the real sources of history "will sometimes unclothe and exhibit in its original bareness what the art of the moderns has drest up and rounded into form. It will show the unsightly chasms and breaks which the modern compiler passes over with a light foot; and perhaps make us sceptical with regard to many particulars of which we formerly thought we had authentic information. But it is always good to know the real measure of our knowledge" (*Legacy*, pp. 79-80). Before depending on chronology, which she treats as a fundamental concept of history, she explains what it is and notices different systems of chronology used by various peoples (*Legacy*, pp. 148-56). Indeed, "history" itself she constructs from the ground up, by relating it to the life circumstances of her pupils or readers—to the sights and signs of the English countryside: villages, fields, pathways, bridges, bits of old pottery dug up in gardens, all of which, in a thinking person, stimulate the curiosity and imagination which are the sources of historical inquiry (*Legacy*, pp. 118-21). Likewise she constructs geography, in a passage in *Lessons for Children*, by going on a walk with Charles from the front door of the school all the way to land's end. During the walk she notes that the village steeple behind them grows smaller and smaller (i.e., the concept of horizon) and how the ground changes around them (i.e., topography), and when they come to the ocean she tells him that this big thing is represented by the little space on the map at home (i.e., symbol-systems).[49]

She advocated active engagement in learning. Read history with a map, she urged ("On Female Studies," *Legacy*, p. 50); she adored maps herself, was thoroughly conversant with them, and—ahead of her time —called for historical maps which would represent a place as it was at a given historical moment (*Legacy*, pp. 139-40), a need met today by *Shepherd's Historical Atlas* but apparently unmet by the maps she knew. She also urged making charts, an idea inspired by Priestley's *Chart of Biography* (*Legacy*, p. 162). And she advised, for a mnemonic aid, using poetry (*Legacy*, p. 163; she quotes Johnson's *Vanity of Human Wishes* as an example of the power of verse to fix geographical names in the memory.) Her own poems on geography, presumably designed to be memorized by her pupils, associate plants and animals, and less often people, emblematically with locales:

> Oe'r Afric's sand the tawny Lion stalks,
> On Phasis' banks the graceful Pheasant walks,

> The lonely Eagle builds on Kilda's shore,
> Germanias forests feed the tusky Boar. . . . (*Poems*, Poem 75)

The lists of plants, animals and peoples are given, in several of the poems, in answer to a teacherly question:

> "Where Canada spreads forth her deserts hoar,
> Chilled by the polar frosts of Labrador,
> Where mighty lakes their azure wastes expand,
> And swell their wat'ry empire o'er the land;
> What tribes or wing the air or tread the plain,
> What herbage springs, what nations hold their reign?"
> (Poem 164)

Quite possibly the poems were meant for recitation, the pupils answering in chorus to the question, posed by her. In an age when the lecture and the grammar drill were the normal methods of transmitting knowledge in schools, this "interrogative system"—and in verse!—must have seemed a refreshingly "activist" innovation.[50]

The emblematism of Barbauld's geography poems leaves them open today to the charge of stereotyping—in fact, of dehistoricizing—the places and peoples they name (e.g., "the painted Indian" in "Canada"), but the impression they were designed to make, I would argue, is something more akin to appreciation of the world's natural and cultural variety. It may have been these and similar teaching poems that Lucy Aikin had in mind when she wrote of Barbauld's geography teaching that "she relieved the dryness of a study seldom rendered interesting to children, by so many lively strokes of description, and such luminous and attractive views of the connexion of this branch of knowledge with the revolutions of empires, with national manners, and with the natural history of animals, that these impressive lectures were always remembered by her auditors less among their tasks than their pleasures" ("Memoir," pp. xxvi-xxvii).

"Pleasures" is the key word. Nowhere in Barbauld's pedagogic writings does one come across the idea that learning—other than learning two dead languages with finical exactness—must be boring or painful. Rather, she befriended the idea that children could be, in the phrase of the time, "played into learning."[51] This was one of the traits of her teaching that her pupil Thomas Denman most gratefully

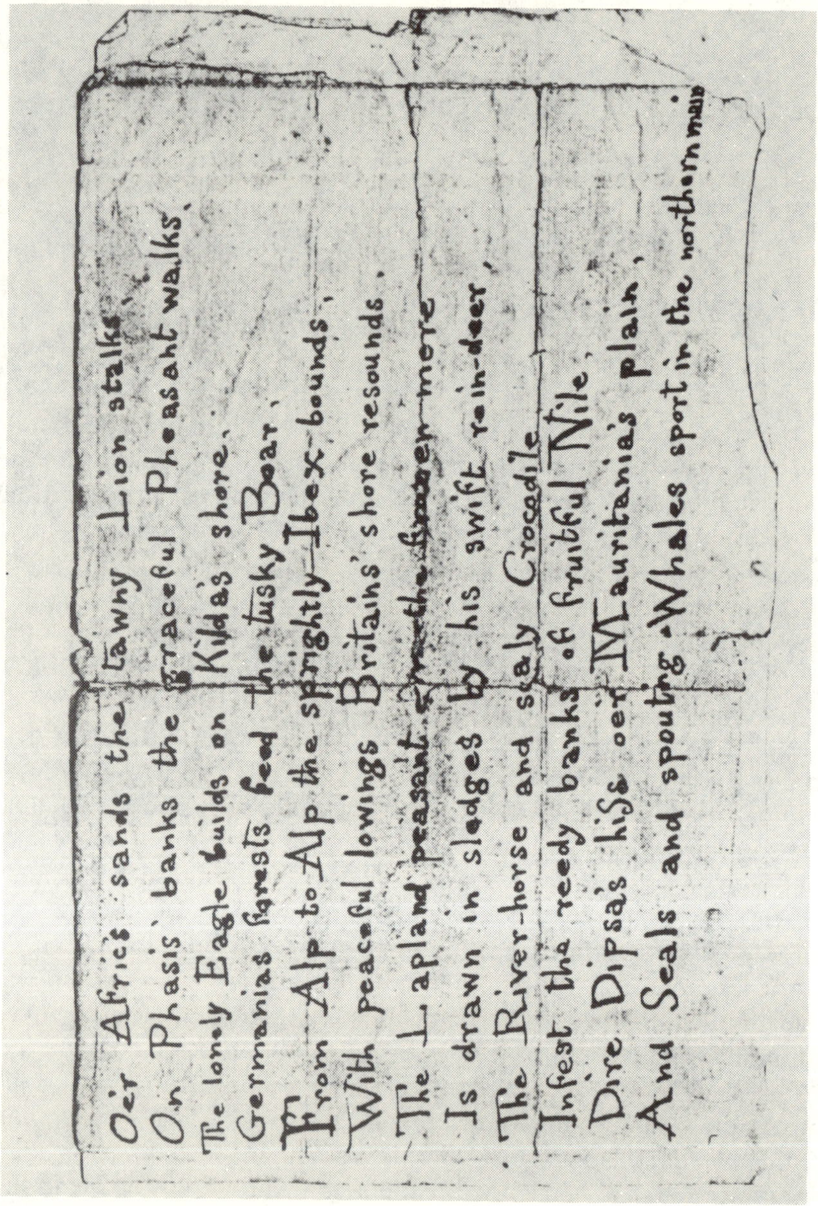

O'er Afric's sands the tawny Lion stalks,
On Phasis' banks the graceful Pheasant walks,
The lonely Eagle builds on Kilda's shore,
Germania's forests feed the tusky Boar,
From Alp to Alp the sprightly Ibex bounds,
With peaceful lowings Britain's shore resounds,
The Lapland peasant o'er the frozen mere
Is drawn in sledges by his swift rein-deer,
The River-horse and scaly Crocodile
Infest the reedy banks of fruitful Nile,
Dire Dipsas hiss o'er Mauritania's plain,
And Seals and spouting Whales sport in the northern main.

MS, "Animals, and their Countries," in ALB's print hand. Reproduced by courtesy of Mr. Conant Brodribb.

remembered: he told Julia Smith that Barbauld "br[oke] down the old by rote plans" and made lessons "more interesting."[52] Her practice here stands in sharp contrast to that at Christ's Hospital, at least as recalled by Leigh Hunt: the masters there made no effort at all, it seemed to Hunt, to get the boys to *like* their studies (Hunt, p. 67).

It was probably entirely by playing—by demonstrations, by physical activity engaging the child—that Barbauld taught natural science. In *Lessons for Children* we encounter a surprising number of science lessons, including experiments: in one, Charles has put a caterpillar into a box and is amazed to find, when he looks for it later, that the caterpillar is gone and a butterfly has replaced it; in another, Mamma indulges—no, encourages—him in the wicked pleasure of rubbing the cat's fur the wrong way in a dark corner on a cold night, to see it spark.[53] Elementary astronomy is taught several times in *Lessons*: the positions of the sun at different times of day, for example, and the phases of the moon (III, 7-10, 21-24). Geology and metallurgy make appearances (things that are "dug out of the ground" [III, 67-75], different metals and their industrial conversion into tools [III, 53-66]), as do botany (the structure of trees and flowers [III, 75-81]) and the Linnean idea of classification (the defining characteristics of animals [IV, 72-75]). The parodic Linnean description of the "curious animal" found in Derbyshire (discussed above) would also have served a pedagogic purpose, probably as a playful model for the boys to imitate.

Although she taught science by means of play, her reasons for teaching it were most serious. "The great laws of the universe, the nature and properties of those objects which surround us, it is unpardonable not to know: it is more unpardonable to know, and not to feel the mind struck with lively gratitude. Under this head are comprehended natural history, astronomy, botany, experimental philosophy, chemistry, physics" ("On Female Studies," *Legacy*, p. 51). In keeping with Dissenting tradition, Barbauld regarded the physical sciences as ways of knowing God: hence, learning "the great laws of the universe" ought to lead to "lively gratitude" for God's design.[54] And also for God's bounty to humans. A short piece titled "Earth" (*Legacy*, pp. 112-16) is an encomium on the earth, intended to arouse affection and admiration for the planet ("Who can reckon up the benefits supplied to us by this parent Earth,—ever serviceable, ever indulgent!"): it resembles what today would be an exercise in ecological consciousness-raising. Barbauld had been early impressed by a sermon of Joseph Priestley's, "On Habitual Devotion" (1767, published 1782), in which Priestley urged the importance of habitually connecting with God everything one experiences, and it was important to her personally to cultivate what she called "the devotional taste," a disposition to emotions of reverence.[55] To encourage Palgrave pupils to associate God-feelings with their daily experience is the intent of her most famous book, *Hymns in Prose for Children* (1781). She composed it, a hymn at a time, for the pupils to recite in their regular worship. Its doctrine is generically Christian (Jesus is the redeemer of humans from death), but its emphasis is on the God who made nature. There is at least as much botany in *Hymns* as in *Lessons* ("The thistle is armed with sharp prickles; the mallow is soft and woolly. The hop layeth hold with her tendrils, and claspeth the tall pole . . ."), and the botany is made instrumental to marvelling at the divine plan ("How can a small seed contain a plant? How doth every plant know its season to put forth? They are marshalled in order: each one knoweth his place . . .").[56]

Hymns was written for her youngest pupils, and was thus at once a manual of devotion and a reading book—a frequent combination in eighteenth-century children's literature, where early readers were likely to be catechisms. Barbauld herself had learned to read from a "wise book," not otherwise identified but probably a catechism; Joseph Priestley is said to have been able to recite all 107 questions of the Assembly's Catechism by the time he was four, and Gilbert Wakefield could say his catechism and read the gospels at three years old.[57] Leigh Hunt describes "a book used by the learners in reading" at Christ's Hospital, *Dialogues between a Missionary and an Indian*. "It was a poor performance, full of inconclusive arguments and other commonplaces" —and, as Hunt brilliantly shows, the master used it to torment students (p. 62). Religious instruction at the "great schools" may be suspected to have been mediocre at best: Hunt despised the preachers who bored him and his schoolmates on Sundays (pp. 56-57); Thomas James's advice on the conduct of prayers smacks of soulless routine (Butler, I, 35). But child readers of *Hymns in Prose* loved them dearly and never forgot them: one, Sir John Bowring (not a Palgrave pupil), dedicating his *Matins and Vespers* to Barbauld in 1824, declared that she seemed to him a saint, and remembered that her words had delighted him even to tears.[58]

Hymns was written for Barbauld's pupils, *Lessons* for just one pupil, Charles Aikin, her adopted son. (I assume that, once it was published, she used it in the school.) The original form of *Lessons* was presumably what her great niece described years afterwards as the lessons Barbauld had used in teaching her to read: "little sentences and stories" in print hand.[59] She and Rochemont adopted Charles in part because they wished to "follow [their] own plans" in education (*Works*, II, 10) and try things they did not (at least, yet) feel licensed to try with other people's children. Evidently one of those plans was to begin the teaching of reading at age two, for before Charles was two and a half the first volume of *Lessons* was ready to publish. That she took reading instruction so seriously is probably another defining difference between her pedagogy and that of the "great schools." At Rugby, Thomas James did not even consider the subject part of his charge: he delegated reading, along with spelling, to the writing master (i.e., the man who taught penmanship, a mere accessory to a school [Butler, I, 36]). Barbauld herself had learned to read competently by age two—an achievement which greatly impressed her mother, but which was perhaps not quite so prodigious as it would be today (cf. Priestley and Wakefield, above).[60] Her own reading pedagogy was based on the belief that children could learn much more rapidly than conventional schooling seems to have assumed; and because she taught them on that expectation, she was able to obtain the effect. A poem she wrote for Thomas Denman to recite at the Christmas examinations in 1782, when he was not yet four years old, testifies to the kind of verbal achievement she could foster. It also testifies to her sympathetic understanding of her pupils, in this case of the small boy's eagerness to be a big boy:

> Nay, nay, I'll not be held, let me come by,
> The boys all spout, so pray why should not I?
> I won't be called a baby any more,
> Next February I'm completely four. (*Poems*, Poem 74)

The piece goes on to mingle popular tales (the Babes in the Wood, Hop o' My Thumb: Charles Lamb would have been surprised) with current events (Admirals Rodney, Howe, and deGrasse are all mentioned), and it ends, amazingly, by incorporating a line from *Macbeth*:

> Smile on me now, and in another year
> I'll strut and fret my part with any here.

The poem is twenty lines long, and it is impressive testimony to her teaching expectations that she could ask a child not yet four to recite so complex a text.[61]

She could do it because she provided her pupils an abundant verbal culture, with frequent opportunities to become verbally adept. One of the documents described by Brodribb, a "weekly chronicle" of the school, must have served that purpose, to judge from the example he describes (p. 734). It contained verses (such as, perhaps, "A Schoolboy Eclogue," described earlier, and perhaps other surviving poems), "a set of bouts-rimés to be filled in by the next week," and prose "essays, letters, etc." "Letters" sounds like a letters-to-the-editor section, and it perhaps gave the boys a forum. There were also word games, of which Barbauld herself was fond: in one, a list of words or phrases is given, each representing the name of a boy (e.g., "a man behind a cave"), and the boys are to solve these puzzles by guessing the name ("Denman"). There would also have been riddles and "logogriphs," a kind of puzzle in which the clues consist of anagrams of the word to be sought. The "chronicle" probably included fables by Barbauld as well, some of which eventually found their way into *Evenings at Home* and *A Legacy for Young Ladies*: "The young Mouse," "The Wasp and Bee," "The Goose and Horse," "The Flying Fish," "The Phenix and Dove" in *Evenings*, "The Pine and the Olive" and "The River and the Brook" in *Legacy*. How many issues of the weekly chronicle Barbauld produced during her eleven years of teaching I do not know; probably it evolved in the course of experimentation or simply of having fun. But if it really was a weekly, it must have been the principal channel of her own verbal energies during those years, as well as a major stimulus to those of the boys.[62]

One subject on which we have specific testimony about her teaching is composition. The testimony comes from William Taylor, who in his memoir of his school friend and fellow-writer, Frank Sayers, remembered Barbauld's practice:

> On Wednesdays and Saturdays the boys were called in separate classes to her apartment: she read a fable, a short story, or a moral essay, to them aloud, and then sent them back into the schoolroom to write it out on the slates in their own words. Each exercise was separately overlooked by her; the faults of grammar were obliterated, the vulgarisms were chastised [i.e., stylistically purified], the idle epithets were cancelled, and a distinct reason was always assigned for every correction; so that the arts of enditing and of criticising were in some degree learnt together. (Taylor, *CW*, p. xii)

Also the arts of aural comprehension and memory. The practice of assigning paraphrastic themes was customary: at Christ's Hospital, composition exercise consisted of abridging essays from *The Spectator*.[63] If her method of correcting her pupils' writing now sounds jejune, we must notice that the very fact that Taylor troubles himself to describe it in 1823 suggests that it must have been novel to him, even fifty years afterwards. For one thing, it did not include corporal punishment for committing solecisms; for another, it appealed to the child's intelligence by giving reasons. Taylor himself regarded her instruction in writing as "the most valuable part of the discipline" at Palgrave School, and Sayers agreed: "Among the instructions bestowed at Palgrave, . . . he most valued the lessons of English composition superintended by Mrs. Barbauld."[64] As will appear below, her writing instruction may also have been the most influential part of the Palgrave curriculum.

Her pedagogy aimed to foster in her pupils skills relevant to such citizenly acts as public writing and speaking. These skills were further exercised in the school's annual dramatics exhibitions, which occurred usually in May, and in the examinations which were held in December (see Appendix I, headnote). At the examinations the pupils recited pieces from English literature, as well as pieces she wrote for the purpose (such as the poem recited by Denman, above). Here she was still teaching reading, but "reading" meant, at this higher level, "the art of declamation." Lucy Aikin asserts that "the neglected though delightful arts of good reading and graceful speaking were nowhere taught with more assiduity and success" than at Palgrave ("Memoir," p. xxvii). This tends to trivialize those arts; they were more than "delightful," they were ethically charged. The idea of the citizen orator was deeply embedded in the Dissenting mind; Barbauld herself was to practice in that role in her political pamphlets of the 1790s. Teaching the citizenly act of oratory ranked among the strongest traditions in Dissenting education. Here is another difference from the "great schools": Dr. James, at least, hated "rehears[ing]" speeches and counselled his protegé Butler against it (Butler, I, 28).[65]

The English literature on which oratory was practiced was itself a subject dear to Dissenting tradition.[66] Barbauld's friend William Enfield, tutor and rector at Warrington, compiled an anthology of pieces from English literature, *The Speaker* (1774), for recitation; Barbauld herself was to compile a similar anthology for women, *The Female Speaker* (1811); both argue in prefaces for the importance of studying one's native literature in youth. At Palgrave, therefore, the school year closed with dramatic performances and recitations by the pupils, on stage and in costume.[67] On the evening of 26 May 1780, for example, term ended with a prodigious collection of recitations. The playbill consisted of a complete performance of Rowe's *Tamerlane*, with interludes, during which were recited or performed "part of [Edward] Young's paraphrase of Job," "a song by Marloe," "a Fable of [Robert] Dodsley," Collins's "Ode to Fear," a scene from Samuel Foote's farce *The Commissary*, Milton's "L'Allegro," Pope's Third Pastoral, "Harpagon's Soliloquy" from *L'Avare* by Moliere (whether in French or English is not stated), a scene from Steele's *The Tender Husband*, Virgil's Fourth Pastoral, and a scene from *Henry IV, Part 1*.[68]

The plays and readings chosen for performance would be chosen as contributing to build other citizenly virtues besides oratorial skill. Barbauld no doubt acted on the same principle as the French educationist Mme. de Genlis: "It is hoped that young people may find lessons in them, both entertaining and instructive. Besides, in playing these pieces, in learning them by heart, several advantages may be found; such as, engraving excellent principles upon their minds, exercising their memories, forming their pronounciation, and giving them a graceful pleasing manner."[69] "Engraving excellent principles upon their minds" was Palgrave School's ultimate intention. It is evident at every level: in *Lessons for Children*, for instance, lessons like the famous description of the hare killed by hunters and the story of the boy who tortured a robin inculcate kindness to animals. Fables such as "The River and the Brook," in which a mighty river expresses contempt for its tributaries but learns their importance during a drought, are designed to rebuke class pride and encourage respect for others cavalierly deemed inferior. Principles and attitudes expressly political are developed by Barbauld's own "theatre of education," the two playlets, "Alfred, a Drama," and "Canute's Reproof to his Courtiers," both of which explore skeptically the uses and limits of regal power, the relations between kings and their subjects and kings and their advisors.[70] Typical of the way in which Barbauld's liberal political outlook entered into her teaching texts is a sentence in one of her letters "On the Uses of History" (*Legacy*, p. 119, not written for Palgrave). Tracing the discipline of history to sources in human traits such as curiosity, she imagines the questions a person would ask if he found himself in a strange place; and she is careful to note that if he were "a man of spirit, he would feel delighted on hearing the relation of the opposition made by an honest yeoman to an overbearing rich man on the subject of an accustomed pathway or right of common." At the period when enclosure was destroying precisely such accustomed rights as the right of common, this is a sentence pregnant with the ethos of political resistance.[71]

A good example of dramatic production inculcating an ideology of patriotic and critical citizenship is the school's performance, on 27 May 1782, of William Mason's *Caractacus*. The issues in this play are all public, having to do with the fate of a nation (Druidic Britain during the Roman conquest) and assigning responsibility for its fate to leading individuals—issues appropriate for boys being educated to stewardship of the public good. The play is intensely patriotic, deploying a language of heroic self-sacrifice similar to Barbauld's own in her poem "Corsica" (1769): "when we fight to save our country,/We fight the cause of Heav'n. The man that falls,/Falls hallow'd; falls a victim for the Gods. . . ." The object of this patriotism is a Britain conceived as a locus of "the native rights man claims from man" (p. 84) and hence worthy of being defended by self-immolating valor. These sentiments are congruent with Whig sentiments of the 1760s. (*Caractacus* was first published in 1759 and revised for the stage in 1776.) A performance in May 1782, when the nation was faced with the fact that a great national effort, the war to keep America, had failed, would carry a double meaning. In April, while the play was rehearsing, Anna Letitia wrote to William Taylor (then a former pupil), expressing disgust with public affairs: "I think an Englishman abroad must by this time be afraid of looking into a newspaper for fear of being made to blush for some new disgrace or bite his lips at some new loss."[72] Performed in this context, the play could appear to reaffirm British goodness and valor in the face of disheartening reality. In the play Britain is defeated, but not shamefully: "unconquer'd, undismay'd,/The British soul revives" (p. 81). In this performance, real losses were being ideologically recouped.

At the same time, the performance implied a *critical* patriotism. "Patriotism," indeed, was the special claim of liberals who during the 1760s and '70s opposed monarchical "encroachments" and "corruption" by George III and his ministers, and who believed that the war to keep America should never have been fought. They made common cause with the Americans, regarding them as defenders of the rights of all Englishmen against incipient tyranny; it was in part by American arguments against "taxation without representation" that they were moved, in 1780, to initiate the fifty-year campaign to reform Parliament. Theirs was the political stance of Joseph Priestley, the Warrington Academy tutors, and the Barbaulds, as it was, more or less, of Dissent in general.[73]

From performing *Caractacus* in May 1782, then, Palgrave pupils might have been expected to carry away not just patriotic feeling but also oppositionist ideas. One was that the authority of kings ought to be limited by submission to superior wisdom: in the play Caractacus, although a king, is subordinate to the Druids and Bards—that is, to the intelligentsia—whose understanding of events is better than his, and by whom, therefore, he ought to be guided. His rash refusal to be guided ruins Britain's last chance to repel Roman conquest. An analogy with George III, whose rashness brought on another kind of ruin, could be expected to present itself.[74] Another idea was that violence is inefficacious to resolve disputes: Caractacus has been defeated in battle before the play starts, and his personal bravery is nullified in the action of the play by Roman fraud. The failure of violence to solve anything in America would be likely to be thought of.[75] A third idea was that liberty is subverted when citizens yield to temptations of personal gain, as Vellinus does when he agrees to betray Caractacus to the Romans and as all Britons do when they accept Roman bribes. (Mason's line, "gold, vile gold,/Has now a charm for Britons" [p. 58] aptly expresses 1760s liberal anxieties about "corruption.") Perhaps the chief lesson

By the Young Gentlemen of Palgrave-School.

On MONDAY, May the 27th 1782, will be preſented
A TRAGEDY, called

CARACTACUS,

Written by MASON.

Caractacus, Maſter LOWNDES,
Elidurus, Maſter J. DOUGLAS,
Vellinus, Maſter OLIVE,
Arviragus, Maſter MOUNSEY,
Aulus Didius, Maſter BUNNY,
Druid, Maſter FIRTH,
Evelina, Maſter A. DOUGLAS,
Bards, Maſter Thompſon, Maſter S. Baddely, Maſter W. Marſh, and Maſter Coldham.

A PROLOGUE, written by the Rev. S. WESTBY,
Will be ſpoken by Maſter MOUNSEY.

To which will be added a *FARCE,* called

TASTE,

Puff, Maſter S. MARSH,
Carmine, Maſter J. DOUGLAS,
Bruſh, Maſter DYSON,
Novice, Maſter SCOTT,
Lord Duke, Maſter UPTON,
Alderman Pentweazel, Maſter WRIGHT,
Caleb, Maſter BROWN,
Boy, Maſter C. LE BLANC,
Lady Pentweazel, Maſter MOUNSEY.

To begin preciſely at Five o'Clock.——*No admittance behind the Curtain.*

BURY: Printed by RACKHAM, Bookſeller, Stationer, and Bookbinder.

Playbill for performances of *Caractacus* and *Taste*, 27 May 1782. Norfolk Record Office MC257/63/1/3, 684 × 4. Reproduced by permission.

Barbauld wished her pupils to take away from their experience of *Caractacus* was a will to be guided, in those shameful times, by the ideals of the play: to do better, when they should come to be in charge of events, than the leading men of the 1770s had done, and to do it by heeding the intelligentsia and studying to preserve public and private liberty.

Palgrave School was not alone in trying to instill civic spirit in its pupils. Linda Colley has argued that a "patrician patriotism" was inculcated by the public schools and universities. Their methods, however, appear to have differed so greatly from those of Palgrave as to constitute, in effect, different aims. At Eton, as Colley notes, Dr. Barnard (headmaster and provost 1754-81) devised the tradition of "having the most distinguished sixth formers—which in practice meant young men of rank intent on political careers—donate portraits to the school. This supplied a fillip to their own ambition, as well as an incentive to younger Etonians who spent their schooldays under the painted gaze of arrogantly superior alumni. . . ." Only one Palgrave boy, Bunbury, is known to have had his portrait painted. Barnard also "reorganised that peculiar Etonian ritual, the Montem," into a ceremony at which senior scholars were awarded military titles and "wore junior versions of British military uniforms." Although the elected head boy at Palgrave was indeed called a "captain," I cannot imagine the Barbaulds encouraging their boys to play soldier offstage. On the contrary: When Anna Letitia informs her brother that his son George has gone hunting with the usher, she takes pains to assure him that George has not "let off a gun." I have just argued that the school's production of *Caractacus* would have carried antimilitarist meaning. So does the poem she wrote for Denman to recite: its mentions of Admirals Rodney, Howe, and deGrasse are not in the least respectful. So does another poem probably composed for Palgrave, "Written on a Marble": It ridicules military "heroes" as "overgrown schoolboys/Who scuffle for empires and toys." So, emphatically, does her dialogue, "Things by their Right Names," which may have been written at Palgrave. In it, Charles asks Father to tell a story about "a bloody murder." Father obliges by telling a story which, as it unfolds, to the child's disappointment turns out to be a story of a battle. When reproached with committing a category error, Father replies, "I do not know of any *murders* half so bloody."[76] We might distinguish Eton patriotism from Palgrave civic spirit by describing the first as "custodial" and the second as "critical." Etonians were bred to dominate, Palgrave boys to oppose domination. No doubt this formula is invidiously simple; but, as will appear below, the after-careers of Palgrave boys and Eton boys did tend to diverge along just such lines.

Mother of My Mind: Outcomes

It is for Anna Letitia's achievements in teaching them that Palgrave pupils seem most to have remembered the school. I would emphasize that her pupils appear to have loved her teaching, and there is evidence enough that her pedagogy was imaginative and entertaining to discredit Charles Lamb's bromide. One of her best-documented pupils, Thomas Denman, regretted in later years that Barbauld's pedagogic "genius" (his word, reported by Julia Smith) was inadequately known to the public. And William Taylor is said by his biographer always to have regarded Barbauld as "the Mother of his mind."[77]

The Barbaulds' relations with their pupils did not necessarily end when the pupils—or the Barbaulds—left Palgrave. I have mentioned that Anna Letitia wrote to William Taylor in Germany in 1782; in fact, she continued to watch his development as a writer and a person for the next thirty years and more, sometimes praising and sometimes chiding him, just as when he had been her pupil. She promoted his work, reading his translation of Bürger's ballad *Lenore* (not yet published) to a literary salon in Edinburgh in 1794—an occasion always remembered for its impact on the young Walter Scott, who, although not present at it, credited to her reading his desire to become a poet. She surely played some part in Taylor's being engaged by her brother to provide commentary on foreign literature for the newly-founded *Monthly Magazine* in 1796, and in his being taken on to write for the *Annual Review* (1803-1809) and the *Athenaeum* (1807-1809), both Aikin family enterprises. When Taylor began to publish a serial spoof on Priestley in the *Norwich Iris* (1802), Anna Letitia expostulated with him; like a good pupil, he withdrew it. In 1812 she sent him a copy of her own poem *Eighteen Hundred and Eleven*, urging him to undertake some work more important than reviewing; in 1815 she praised his

English Synonymes Discriminated.[78] Taylor was not the only pupil she continued thus to nurture. It is likely that she helped others publish their first books. Frank Sayers's *Dramatic Sketches of the Ancient Northern Mythology* (1790) was published by Joseph Johnson, Barbauld's and the school's bookseller; William Gell's *Topography of Troy and its Vicinity* (1804), Thomas Douglas's *Observations on the present State of the Highlands of Scotland* (1805), and Isaac Weld's *Illustrations of the Scenery of Killarney* (1807, his second book) were published by Longman, in 1803-09 publisher of the *Annual Review* and *The Athenaeum*. It is probably not coincidental that books by Palgrave "old boys" were taken by the Aikins' "house publisher." She also kept in personal touch with former pupils: Rochemont Barbauld's diary for June 1802 registers visits by or to Lord Selkirk (i.e., Thomas Douglas), Weld (twice), John Scott Martineau, William Taylor (twice), several Pages, and Denman, and on visits to Norwich she always saw Thomas Martineau and his family. Anna Letitia's long-lasting relations with her former pupils "emulate[d]"—to borrow the words of education theorist Madeleine Grumet—"the continuous and extended relation of a mother and her maturing child."[79] When in 1777 she adopted Charles Rochemont Aikin as her son, she made him her pupil. But the converse happened as well: taking other people's boys as her pupils, she "adopted" them as her sons.

What, finally, were the effects of Palgrave teaching upon the pupils: was Barbauld's "intended learning outcome" achieved in them? Assessing the effects of teaching is hard enough today, with all the apparatus of follow-up interviews, surveys, and exit examinations; estimates of its influence two hundred years ago, on a student population for half of whose members we have no biographies, must inevitably be speculative. That said, I proceed to numbers. Comparison of Palgrave outcomes, known or probable (50 in total), with those of pupils at Eton (355 known or probable) and Westminster (306 ditto) during the same period yields some striking results. (See Table of Outcomes in Appendix III.) From Palgrave the three great professions, the Church, the Military, and the Law, claimed, respectively, twelve, five, and nine pupils each; or, expressed as percentages of the total of outcomes, 24, 10, and 18. For Westminster, the corresponding percentages are 32.4, 19.9, and 12.7; for Eton, they are 31, 23.9, and 9.3. That is, only into Law did Palgrave send proportionately more pupils than Eton or Westminster; to the Church and the Military it sent proportionately fewer—to the Military far fewer, a point to which I shall return.

Given Palgrave's emphasis on "active" citizenship, one might expect numbers of its graduates to have assumed places in government. In fact, ten Palgrave pupils (20% of outcomes) went into national or local government, a percentage similar to Westminster's (23.8%) but smaller than Eton's (26.2%). That Palgrave's contribution to domestic government could match (proportionately) that of a much larger, older, and better-funded school like Westminster suggests, perhaps, the success of the Barbaulds' emphasis on citizenship; that its contribution was smaller than Eton's suggests a comparative political marginality. More ambiguous was Palgrave's contribution to the work of empire. In my Table, "Empire" is defined as colonial administration and diplomacy; into this work Palgrave sent just three persons—an innocuous number, yet, expressed as a percentage of outcomes (6%), exactly comparable to the contributions of Westminster (7.8%) and Eton (6.2%). One of the Palgrave three alone, moreover, did or organized the work of multitudes: Thomas Douglas, earl of Selkirk, arranged for (and at times, personally led) the founding of colonies in Canada, inducing large numbers of Scots displaced by land enclosure to risk themselves in these ventures and displace the native Americans in turn. On the other hand, in that age when the British Empire was actively consolidating, the career of "Empire" might reasonably be defined to include entrance into the Military also; and Palgrave's contribution to Empire thus defined, because it sent so many fewer boys into the Military, was markedly less (16%) than Westminster's (27.8%) and Eton's (30.1%).[80]

Ten Palgrave pupils achieved entries in the *Dictionary of National Biography*—twenty percent of outcomes, twice as many (proportionately) as Etonians managed to do (9.9%). The fields in which they thus distinguished themselves (or, more accurately, were distinguished by the compilers of the *DNB*) also signal a difference between Palgrave and Eton: the newly-developing sciences of geology, chemistry and topography (Arthur and Charles Aikin, Weld), philology (Crabb and Taylor), classical archaeology (William Gell), literature (Sayers, Taylor), law and government (Denman, Charles Marsh), colonial administration (Thomas Douglas, Marsh again). With the exceptions of Denman, Douglas, and Marsh, Palgrave graduates distinguished themselves in "cultural" rather than directly political fields—unlike Etonians, who entered the *DNB* overwhelmingly for work in Government and the Military. That literature, in the broadest sense of the word—the production of published matter—figures largely among the activities of these and other Palgrave graduates is testimony to the success of Barbauld's verbally rich pedagogy. Arthur Aikin went on to edit a major literary review and to write on geology; his brother Charles published on vaccination and chemistry. Crabb and William Taylor both became Germanists and published books on English synonymy (Crabb's remained a standard work well into the present century). William Gell published a much-admired survey of the topography of Troy. Sayers published verse dramas based on Norse mythology. William Taylor (again), besides translating contemporary German writers, wrote copiously for the reviews; he was noted for the individuality of his style. In all, eleven Palgrave pupils (22% of outcomes) went on to publish writing of some kind, a number not only proportionately but even absolutely extraordinary: Westminster managed to produce only ten writers during the same period (3.3% of outcomes), and Eton only twelve (3.4% of outcomes). Moreover—and this is surely the most telling point of all—whereas at Eton and Westminster, the number of pupils who went into the Military exceeded by six or seven times the number who engaged in writing, at Palgrave the number who engaged in writing was more than twice the number who entered the Military. This fact suggests that the combination of rich verbal pedagogy and deprecation of militarism at Palgrave had its effect.[81]

Several pupils credited their literary or rhetorical skills explicitly to Barbauld's teaching. Taylor and Sayers both, as I have mentioned, especially valued her writing instruction; Taylor also reports of Sayers that he had an "unfaultering memory" and was "one of the finest readers I ever heard: expression of every kind was at his command" (*CW*, p. xi)—surely an effect of the dramatics training at Palgrave. Denman, too, "always attributed to the judicious care of his first instructress much of the retentiveness of his memory, of his fondness for literature, and of the force and clearness of his elocution." According to Lucy Aikin in 1832, Denman "cherishes her memory most religiously. In a great public entertainment where I met him last year, he came up to me and said with a look of delight, 'I dreamed of Mrs. Barbauld only last night!' He has a love and a taste for poetry and elegant literature worthy of her scholar, and I doubt not that she sowed the seed." Even the glamorless Thomas Martineau, who went into the manufacture of bombazine, carried with him a taste for reading, discussion, and "literary talk," and represented to one of his daughters a Palgrave success story: "My father was a plain, business man . . . he had passed some portion of his childhood under the roof of the Barbaulds, and if he did not bring away much learning from them, I like to indulge the belief that from Mrs. Barbauld he acquired the strong political leanings, and the firm principles of Nonconformity, that marked his after citizenship, and certainly descended in no equivocal way to his sons and daughters"—who included not only Harriet Martineau but also James, a distinguished Unitarian theologian and teacher.[82]

The politics of Barbauld's pupils should be a touchstone of her success, given Palgrave's consciously political-ethical program. Almost uniformly, the politics of those whose politics are known were liberal, sometimes radical. Sayers is an exception: he abjured his early liberal beliefs and declined into the "stye of Toryism" (to quote one observer's remark). The Aikins and William Taylor became liberal reformers (Taylor was a member of the Norwich Revolution Society in the 1790s and wrote most of its correspondence and resolutions), Priestley a radical (he embarrassed his politically embattled father by accepting French Republican citizenship in 1792). George Coldham, like Thomas Martineau, was an informed and principled Dissenter. Richard Enfield embraced republicanism; William Firth argued for it in the debates of the Norwich Tusculan Society. Patteson and Charles Marsh (of whom more later) contributed to the radical monthly *The Cabinet,* published at Norwich in 1794-95. That these effects could be produced by schooling nobody doubted: William Taylor once recommended a school on the grounds that it tried "to inspire habits of piety [and] a leaning to republican theories of government."[83]

Their prevailingly left-liberal politics surely help to explain why Palgrave pupils entered government in proportionally lower numbers than Etonians. From 1793 to 1830, the years of reaction and war with France, and of Tory stonewalling of reform thereafter, Palgrave graduates were likely to find themselves not in charge of events—as the

Barbaulds must have hoped they would be—but in opposition. In that political climate, the critical citizenship they had been taught to practice kept them in the minority. (A warning against invidious dualism: Eton produced liberals also—Charles James Fox and Earl Grey among them—and they too were in opposition, although they were also in government. Eton produced, as well, to be sure, the Duke of Wellington, who has no Palgrave counterpart, and whose boast that "the Battle of Waterloo was won on the playing-fields of Eton" perfectly expresses a military-political ethos quite foreign to Palgrave.)

I close with accounts of three pupils in whose lives these mingled themes of outcome can be traced in some detail. Denman is one. His biographer asserts that "the fearless advocate and upright judge of the future owed much, beyond a question, to the simple homely training in truthfulness, honor, and self-denial which he received at Palgrave." His letters to a friend, D'Ewes Coke, in 1810-15, when he was still a practising advocate, exhibit a man forthright in his opinions and engagingly rash in his expression of them. In one, he characterizes himself (not without humor) as "the champion of Liberty, Religion & Truth, against Bigotry, Intolerance, Persecution, Priestly execration, & Judicial Despotism" (6 May [1810]). In another he utters the sentiments of reformist Whigs, who in 1814 were condemned to wait out the costly war against France and lament the shelving of reform at home:

> The time for common place politics I think is gone by; the country must understand the truth, & act upon it. The discontent that prevails among the lowest orders & the lower of the middling must not be inflamed into sedition & violence; but arising as it does from the most wellfounded & rational condemnation of the absurd impolitic & unjust expences which have brought us to a state of unparalleled & inextricable embarrassment, it ought *only* to be allayed by measures of . . . reformation in the public expenditure. And unless ministers & Parliament shew that degree of respect for it, I have no idea that the frame of government can be kept together many years. (28 Sept. 1814)

Denman appreciated the depth of working-class hatred of the war, and in his proposed remedy (government retrenchment and electoral reform) he agreed with middle- and working-class radicals. Denman also watched closely the Luddite trial at Nottingham in 1815 and was happy to see the accused acquitted (23 August [1815]).[84] The capstone of Denman's career was his drafting of the Reform Bill of 1832, the legislation which finally, after fifty years of effort, redistricted Great Britain and extended the franchise. The limitations of that reform notwithstanding, I imagine the Barbaulds would have wept for joy.[85]

What link can be established between Denman's principles and his being taught by Anna Letitia? The moral influence of a teacher is notoriously hard to specify; what works on one pupil slides off another. (Palgrave had its failures, too.) I suspect, however, that Denman would have been impressed by the difference between Palgrave's and Eton's treatments of him: fostered and brought forward at Palgrave, at Eton he experienced actual torture. At Palgrave he would have learned—and felt—principles of fairness and kindness by which to judge, and find wanting, the conduct of Establishment agencies. Obviously, Palgrave did not so "radicalize" Denman as to alienate him entirely from those agencies. Rather, in good citizenly fashion he sought to enter Government—and, when in it, to change it from within.

Denman is interesting in another respect as well, his practical feminism. He believed in women's education and taught Latin to his own daughters (LeBreton, *Memories*, pp. 148-49). He also acted as chief counsel for Queen Caroline in the divorce proceeding brought against her by George IV in 1820. Given that liberal opinion tended to side with Queen Caroline against her generally-detested husband, befriending the queen could signify political liberalism or sympathy with women, or both. It may therefore be a telling fact about Anna Barbauld's influence that not only Denman, but also William Gell befriended Caroline: Gell became, in fact, one of her closest confidants. Other Palgrave pupils, the Baddeleys and Patteson, came to the aid of another mistreated woman, Sarah Spence. Her story is obscure, but it appears that her husband unjustly turned against her; and in 1794 a large number of East Anglians came to her assistance by sponsoring the publication of her *Poems, and Miscellaneous Pieces* (1795).[86] That two or three Palgrave names appear on the list of subscribers may signify not simply that they were East Anglians, but also that Palgrave education induced feminist sympathies in the pupils.

Two other careers in which Palgrave themes best exhibit themselves are those of Basil William Douglas (Lord Daer) and Charles Marsh. Douglas, Barbauld's favorite pupil, went on to Edinburgh University and then to Paris, where, in July 1789, he met Thomas Jefferson and toured the ruins of the Bastille. He became obsessed by transportation projects—canals, roads, and bridges—and is said to have been so public-spirited about them that he willingly gave up his own land if doing so could improve a road. This interest allies him with the Lunar Society of Birmingham, whose members—such people as Josiah Wedgwood, Thomas Bentley, Richard Lovell Edgeworth, and Thomas Day—likewise invested their energies and assets in road- and canal-building schemes. Douglas interested himself also in abolition of slavery, and in employing freed Africans in a colony in Sierra Leone (another imperial enterprise, although, believing as he probably did that "free labor" really is emancipated, he would have perceived it as a project for liberation). He joined political societies considered dangerous by the government: the Society for Constitutional Information (where one of his sponsors was Tom Paine), the Friends of the People, which advocated Scottish independence from England, and the almost working-class London Corresponding Society, whose leaders in 1794 would be tried for treason. Finally, he attended the scandalously radical British Convention, which met in 1793 to propose a new constitution on the model of the United States. Whether he would have stayed radical, or recoiled (as so many did) from the aftermath of the French Revolution cannot be told: he died in 1794. His political radicalism embarrassed even his admiring obituarist: "over his political character, while the minds of men are heated by party spirit, it may be proper to throw a veil. Though his prudence may justly be called in question, yet the purity of his intentions was superior to suspicion. In every enterprize he embarked with invincible zeal. . . ." But Barbauld had no reservations whatever about him: "He was before the finest boy; he is now the finest young man I know."[87]

The other pupil, Charles Marsh, is interesting in part because he is a claimant to authorship of *Reasons for National Penitence* (1794), an eloquent antiwar tract also attributed to Barbauld. If he did write it, his doing so testifies as much to Barbauld's composition teaching as to the political culture of the school: the piece reads like her own antiwar tract, *Sins of Government, Sins of the Nation* (1793). Marsh took up a more radical politics than hers or Denman's, espousing anti-Christian and republican positions in the debates of the Norwich Tusculan Society. Later he went to India as Advocate-General for the Crown. In 1813, back in England and an M.P., he delivered an impassioned speech against a motion to fund Christian missionary activity in India.[88] He was not an anti-imperialist; he spoke from within the assumption that India would remain a British economic property. Nevertheless he distinguishes himself by his sensitivity to the cultural concerns of Indians: he respects what he understands as their different ways of being, and he defends Indian culture against the fanatical missionarizing view that Christianity is needed to "civilize" them:

> Indeed, when I turn my eyes either to the present condition or ancient grandeur of that country; when I contemplate the magnificence of her structures; her spacious reservoirs constructed at an immense expence, pouring fertility and plenty over the land, the monuments of a benevolence expanding its cares over remote ages; when I survey the solid and embellished architecture of her temples; the elaborate and exquisite skill of her manufactures and fabrics; her literature, sacred and profane; her gaudy and enamelled poetry on which a wild and prodigal fancy has lavished all its opulence; when I turn to her philosophers, lawyers, and moralists, who have left the oracles of political and ethical wisdom . . . I cannot hear without surprise, mingled with horror, of sending out Baptists and Anabaptists to civilize or convert such a people. . . . (*Parliamentary Debates*, cols. 1038-39)

The ground of Marsh's speech is the Dissenting demand for freedom of conscience: in language whose adjectival effervescence reminds one a bit of Thomas Denman, he condemns the missionary scheme as "headlong, precipitate, busy, meddling, gossipping, officious, interference with matters, which the laws of God and nature have placed beyond our jurisdiction" (cols. 1026-27). His appreciation of Indian culture surely derives as well from the ethos of geography lessons at Palgrave: "There is no use, but a great deal of harm . . . in an arrogant assumption of superiority" to "laws, government or manners which [are] not fashioned after our partial ideas . . ." (see p. 298). And his defence of Indian cultural autonomy—his acceptance of practical economic imperialism notwithstanding—can also be seen to parallel his teacher's defence, in "On Education," of children's autonomy against the tendency of adults to impose totalitarian moral regimes on them.

In her reflections on women and teaching, *Bitter Milk*, Madeleine Grumet observes that Judeo-Christian culture has always looked to the innocent child to "reform the world." In order to redeem society's failures and evils, the child would have to be so educated as to be kept pure of them—a utopian project, of course, which could never succeed. It could not succeed because the very fact of growing up in "the temporal order, the spatial configurations, social relations, means of production, and patterns of communication . . . in which he lives" transforms the child into the very thing he is supposed to redeem.[89] This irony was appreciated as well by Barbauld. In "On Education" she argues the limited power of any educational project, in the face of the massive influence of a child's surroundings on him. "Education . . . includes the whole process by which a human being is formed to be what he is, in habits, principles, and cultivation of every kind. But of this, a very small part is in the power even of the parent himself; a smaller still can be directed by purchased tuition." Thus, to a rich man who lives in comfort but proposes to raise his son in spartan simplicity and thus reproduce his own younger, innocent self, she suggests that his scheme is impracticable: "Poverty educated you; wealth will educate him. You cannot suppose the result will be the same" (*Works*, II, 306, 307-308).

Barbauld, that is, was no utopian; she fully understood the pressures of reality on people. She and her husband were, however, principled reformers. They would have agreed with Grumet that a potential for change "is lodged . . . in the work that women do daily teaching children in classrooms" (Grumet, p. 19), and they regarded that work as "contribut[ing]," in the words of Rochemont Barbauld, "to the welfare of [their] country." At Palgrave School they tried to realize the potential; their "child redeemer" was to be the Priestleyan "active citizen," nourished by the school's democratic polity and by its inculcation of a citizenly culture. Given the smallness of the school and the short term of the Barbaulds' management, not to mention the immensity of the task of transforming society and their own participation (inevitable, naturally) in the society to be transformed, it would be foolish to claim that they achieved more than a small, qualified, ambiguous success. Yet they appear to have succeeded well enough to make one wonder what turns British history might have taken had Eton, Westminster, and their fellows been remodeled on Palgrave School.

APPENDIX I
PALGRAVE SCHOOL BUSINESS DOCUMENTS

This appendix consists of seventeen documents arranged chronologically. Thirteen are letters from Anna Letitia or Rochemont Barbauld conducting Palgrave School business. They are of two kinds:

1. Letters to tradesmen who supplied the school. Most are to Joseph Johnson, bookseller to the school, ordering books and supplies. Since Johnson was also the guardian of pupil Rowland Hunter, the letters also occasionally report on Hunter and forward requests from him.

2. Letters to parents of pupils or prospective pupils. Of the first, there are only a few from Anna Letitia to Thomas Denman's mother. Their dating (see note to item 15) suggests that she wrote monthly to inform Mrs. Denman of her son's condition and progress. She also wrote an unknown number of letters to William Taylor's parents (they were mentioned by Taylor's biographer, Robberds, in 1843 [I, 9], but I do not know their whereabouts today). Since the school handled some 130 pupils during the Barbaulds' tenure, and the average length of a pupil's stay was two and a half years (a year consisting of ten months), we could infer that Anna Letitia wrote 3,250 notes of the type she wrote to Mrs. Denman, if those notes and their intervals are representative. However, more probably she gave this kind of attention only to her parlor boarders, and then perhaps only until they were able to write letters for themselves (see item 15). I include here, also, part of a letter to her brother which deals with his son George: although not in the main a business letter, this part of it illustrates her pedagogy and concerns as a teacher.

Of letters responding to parents' inquiries about the school or accepting pupils there are only two, both by Anna Letitia, and both to the same mother, Mrs. Thomas Harvey of Norwich.

The letters are seldom dated by year, and dating them is often a matter of conjecture: See the notes to each letter. Here, however, a general note on the school calendar is in order. According to Lucy Aikin there were two sessions: "winter," which ended with "a public examination" in which the boys contended for prizes, and "summer," which ended with the performance of a play and oratorial exhibitions ("Memoir," p. xxvii, and see item 7 below; what Aikin calls the "winter" session Barbauld may have called "Midsummer" [see *Works*, II, 8]). The examination (and therefore the end of winter session) fell in December, just before Christmas (*Works*, II, 21-22); the play and exhibitions (and therefore the end of summer session) occurred in late May. At the end of each session the Barbaulds presented and paid their bills for the session (see the bill for Austin Bulwer [Appendix II] and items 6 and 8 below). Their two vacations lasted no more than a month each: from Christmas to late

January (*Works*, II, 18, 21-22, and 23-25, all letters written by Anna Letitia upon returning to Palgrave from vacation), and most of June (see item 5).[90] The Christmas vacation they spent either in London, mingling with fashionable and literary people (see item 14, anticipating the pleasure the Barbaulds expect to have in Joseph Johnson's company), or at Norwich, visiting friends. Summer vacations were spent in travel, sometimes to revisit Warrington.

1. [Advertisement for Palgrave School, 1774]

PALGRAVE SEMINARY, near DISS,

WILL be opened at the Expiration of the ensuing Midsummer Holidays,

By the Rev. Mr. ROCHMONT BARBAULT,

From WARRINGTON.

Where Youth intended for any of the Professions, or Trade, may be educated accordingly, with the greatest Care and Skill, in the Classics, French, Mathematics, &c.

This Institution is patroniz'd by a set of Trustees, who think themselves very fortunate in having met with a Gentleman so eminently qualified to undertake it, and upon whose Account they have interested themselves in it.

The TERMS are 25 l. per Annum each Pupil, several are already engaged, and no more than Thirty will be admitted.

For any further Particulars, enquire of

PHILIP MEADOWS, Esq; at Diss.
Mr. WILLIAM TAYLOR.
WILLIAM CARTER.
THOMAS NASMITH, of Norwich.

TEXT: *Norwich Mercury*, 21 May 1774. Repeated on 28 May; 4, 11, 18, 25 June; 2 July. On 9 July the text changes from "at the Expiration . . . Holidays" to "Monday the 25th Instant." This revised text was reprinted on 16 and 23 July. The original text, with minor variants, also appeared in the *Ipswich Journal*, 2 July. I have found no advertisement for Palgrave School in the London *Daily Advertiser* or *Public Advertiser* for 19 May-30 June.

Philip Meadows, solicitor, of Diss (1719-83) was the uncle of Philip Meadows Martineau, who attended Warrington Academy 1765-67 (see n. 24), and of Thomas Martineau, a Palgrave pupil. William Taylor, manufacturer of cloth for export, was the father of one of the first Palgrave pupils; so, probably, was William Carter. Thomas Nasmith was a deacon of the Octagon Chapel, Norwich (Octagon Chapel Vestry Book, Norfolk R. O. MS FC13/2); he was probably the Nasmith described as a "considerable carrier from

Norwich to London" (*The Diary of Sylas Neville*, ed. Basil Cozens-Hardy [London: Oxford Univ. Press, 1950], p. 130n). Membership in the Octagon congregation, whose chapel was a showplace among Dissenting meeting-houses, implies good standing in the Norwich middle class.

2. [Announcement of a play performance, 1776]

> The PLAY to be performed by the young Gentlemen educated at PALGRAVE SEMINARY, is fixed for Thursday the 9th instant, being the day after the county election.

TEXT: *Norfolk Chronicle*, 4 May 1776, under "HOME NEWS." The play is not named, but it was probably *1 Henry IV* (see Appendix IV). The date is earlier by almost three weeks than the dates of performances in 1780 and 1782, suggesting to me that the Barbaulds had not yet settled their school routine.

3. [ALB, fragment of a lecture(?) on Islamic history, c. 1780?]

> *Ismael Abulfeda* Prince of Hamah & descended from brother of Saladin b at Damascus 1273 d 1333. Skilled in medicine philosophy & Poetry, but chiefly known by his his^{l} & geograph^{l} works. Designs of publish^{g} his works have failed both in France & England. Michaelis has lately pub^{d} his Geo^{y} of Egypt. He brought his General His. down to latter years of his life, of which only portions have been given to the world, his *acc^{t} of Mahomet* by Gagnier printed Oxford folio 1723. his *his. of the Arabian Caliphs* to 406 Hegira Leipsic 1754, & narrative circumstances relating to Saladin annexed by Schultens to Bohaddin's life of Saladin. He did not possess his dominions till established in them by sultan of Egypt 13 years before his death

TEXT: Warrington Library MS 269, fol. 5 (printed by permission of David Rogers, Area Librarian). Fragment of one leaf, numbered by ALB at top right, "32"; verso endorsed by Lucy Aikin, "Handwriting of Mrs Barbauld." Undated; I infer c. 1780 from her reference to the "lately" published *Abulfedæ descriptio Aegypti, Arabice et Latine*, ed. J. D. Michaelis (Göttingen, 1776). The other works mentioned are *Ismael Abul-feda de vita et rebus gestis Mohammedis*, ed. J. Gagnier (Oxford, 1723), *Abulfedæ Annales Moslemici. Latinos ex Arabicis fecit J. J. Reiske* (Leipzig, 1754), and *Vita et res gestæ . . .*

Saladini auctore Bohadino F. Sjeddadi, nec non excerpta ex Historia Universali Abulfedor, easdem res gestas . . . exhibentia, ed. A. Schultens (Leiden, 1732) (*BLC*).

4. [RB to Joseph Johnson, 30 March 1782]

Palgrave, March 30th 1782

Dear Sir,

I must beg of you to send me by the Norwich-Coach which passes thro' Bury, as soon as ever you can, as we are extremely pressed for time, six copies of Mason's Caractacus; or, if you cannot immediately get half a dozen, as many as you can get now. The rest may be sent afterwards with ~~part~~ ^the whole or part^ of the books ~~am~~ I am going to mention, supposing they cannot be had directly, for otherwise I would desire you to send them along with such of the copies of the play as may be procured at once. I should wish to have Caractacus singly, and such as it was originally printed; but if that cannot be, still I must have that play in one shape or other. The other books I want are as follows——
2 1
3 Bailey's Ovid 6 Castalio 1(Fontenelle)~~s~~ Pluralité des Mondes par) supposing it may be got single, otherwise l'histoire de Charles douze par Voltaire— the volume of Shakespear which, when we were last in Town we desired you to get for us—and, besides the copy of Tremble's book which has been sent to Mr Vaughan's, and which I beg you would send me as soon as returned, another copy, if another can possibly be got. I also wish you would let me know, when you send the parcel, what Roman histories there are in English together with the character they bear, the number and size of the volumes, and the price of each.

Mr Harmer wishes much to have the book I mentioned to you during my last vacation; he is sure there is such a book, and, if you can lay hands upon it, I wish you would put up [*sic*] in my parcel. I safely received last Thursday that which contained Hunter's shirts. He desires me to return his thanks to his sister for them, and to give his love to her and to his uncle.

Mrs Marissal, who desires to be remembered to you, begs you would forward ~~her book~~ ^hither^ along with my books a little parcel which she imagines Mrs Mulso has sent directed for her either to your's or Mrs Kinder's. Pray, if you should see Dr Towers, return our joint, ^(my Letitia's and mine)^ thanks for his book against the Dean of Glocester which came in the last parcel you sent me; and which, from the little I have red of it, seems to me a very judicious performance.

My Letitia desires to be affectionately remembered to you, as does likewise M[rs] Aikin who got safely to us last Thursday-morning, and is pretty wel[l.] I am, Dear sir,

Your affectionate Friend &[c]
Rochemont Barbauld

TEXT: Historical Society of Pennsylvania, MS Etting Collection, 26:8 (published by permission). Addressed (in ALB's hand) "M[r] Johnson N° 72/S[t] Pauls' Church yard/LONDON." Postmarked "DISS" and "1 AP."

For the books mentioned, see Appendix IV. "Mr. Vaughan" is probably Benjamin (1751-1835), Warrington Academy graduate and merchant in London. "Mr. Harmer" is either Robert, merchant, or Samuel, attorney, both of Norwich (C. B. Jewson, *The Jacobin City: A Portrait of Norwich in its Reaction to the French Revolution, 1788-1802* [Glasgow: Blackie & Son, 1975], p. 170). For Hunter, see Appendix III. "Mrs. Marissal" is RB's sister Susannah; for her and her friendship with Mrs. Thomas Mulso, see *Poems of ALB*, pp. 260, 275, 276, and 299. "Mrs. Kinder" is Anna Letitia Jennings Kinder, ALB's sister-in-law, married to a draper. Dr. Joseph Towers, an eminent liberal Dissenter, wrote *A Vindication of the Political Principles of Mr. Locke, in answer to the Objections of . . . Dr. Tucker, Dean of Gloucester* (London, 1782). "Mrs. Aikin" is ALB's mother, Jane Aikin; her husband had died in December 1780.

5. [ALB to Elizabeth Brodie (Mrs. Thomas) Denman, 29 June 1782]

Dear Madam,

It is with pleasure I can tell you that your dear little boy is an excellent traveller. —We got in this morning between nine & ten, after as pleasant a journey as one could expect in travelling all night. My little charge was quite merry & full of talk the first stage or two, then he composed himself very quietly to sleep, of which he took so good a dose that he was quite lively when we got to Palgrave, & has been running about ever since. I never saw a child so easy & contented among strangers, I make the observation *to day*, for *to morrow* I see that none of the family will be strangers to him. Indeed from all we have hitherto seen of Master Denman, we have reason to suppose we shall be happy in our pupil, & give me leave to repeat Madam, that we shall endeavour to make him happy, & justify the great confidence you have placed in us, a confidence we feel very sensibly. Mr Barbauld joins me in respectful compliments to the D^r, your little boy sends his love. I am, dear Madam,

Your obliged & obedient serv^t

Palgrave AL Barbauld
June 29th 1782

TEXT: The Pierpont Morgan Library, New York, MS MA2572 (printed by permission). Addressed "Mrs Denman/at Dr Denman's/old Burlington street/ London." Postmarked "1 JY." Endorsed by a contemporary hand (Mrs. Denman's?), "Mrs Barbauld N° 1." Denman was born 23 Feb. 1779; he is now three and a half. For further letters see items 9, 10, 15, and 16.

6. [ALB to (?William) Manning, of Yarmouth, 24 December 1782]

Sir,
On the other side is a draft for the amount of your bill exclusive of bottles & hampers which I shall return by the Carrier next Thursday. We acknowledge your civility in allowing for the broken bottles & I am sorry I did not mention this sooner, as indeed I ought to have done. Mr Barbauld desires to join with me, Sir in compliments to Mrs Manning & in thanks to her & you for your very obliging invitation. Tho I shall not go further than Norwich this Vacation, Mr Barbauld may possibly spend a day at Yarmouth; & if he does, desires me to say that he shall be happy to pay his compliments to your family. We beg to be remembered to the young ladies.

I am, Sir,
Your most obedient servant
AL Barbauld

Palgrave
Dec 24th

TEXT: Misc. MS 3213, The Carl H. Pforzheimer Collection of Shelley and His Circle, The New York Public Library Astor, Lenox and Tilden Foundations (published by permission). No cover, no address, no postmark; endorsed "A. L. Barbauld/26th Decr 1782" by another (contemporary) hand, but the draft is absent (probably on a leaf that was torn off). The addressee is probably William Manning, described by William Taylor as "an eminent general merchant, then of Yarmouth" (*CW*, p. xiii). The bottles were presumably for school consumption, and contained wine (on 18 May [1785] ALB requested 3 dozen red port from a Mr. Walker via her brother [Hornel Library MS 15/21, fol. 29]). Bottles were customarily shipped in hampers; the Rev. John Seddon, of Warrington Academy, mentions "hampers of wine," 20 Nov. 1758 (Unitarian College Archive B1², John Rylands Library).

7. [ALB to John Aikin, early Jan. ?1783: extract]

> And first to mention what you so kindly enquire after, my health. I am really well, & my spirits are much better than they were a little after our Examination, which, with the packing &c fatigued me much, especially as we had company in the house. Mr Barbauld & I are now endeavoring to recruit a little at Norwich, where our friends are very kind to us, but the long dinners & suppers often give me a different kind of fatigue. George is very well & grows stout. He intends to be quite a Nimrod this Vacation, for he has already been out two or three times on a shooting party with the French Usher. I do not mean however that George has let off a gun, but he enjoys the sport, & picks up the Game, & when he comes home, of his own accord examines his birds by Pennant & Goldsmith— With regard to his progress he certainly does improve but his improvement is slow. He nearly knows the four first rules of Arithmetic, but not quite, his writing you have seen. In French he has made more progress. He is reading Fables choisies & Magazin des Enfans. In speaking he is much improved —He articulates, when he attends to it, very distinctly & I think I told my Mother that he repeated very prettily his piece at the Examination. He had no prize to contend for, for the beginners in French were too much below, & the others too much above him. He wrote indeed the lower English, but could have no hopes of success there. I have for some time made him ^write^ something for me every morning, chiefly to improve him in his spelling, in which, thro inattention, he is very deficient—— That George by no means wants sense particularly that kind of sense which will enable him to conduct himself in business & in the common concerns of life, with even a degree of shrewdness—I am well convinced. But at the present time, I believe, he wants that taste & genius necessary to relish a literary life. He does not kindle at any thing great or beautiful. He has no strong curiosity, or vivid feelings of any kind —He has very affectionate ones however,

he is beloved by the boys, & the family. Charles & he are extremely fond of one another, & I must do him the justice to add that we have not of late seen any thing reprehensible in his behaviour with regard to the fault I mentioned to you. On the contrary I hope he begins to have a respect for truth. I hope the lads will not have a disagreeable Christmas, we shall be disposed to be kind to them so that I daresay they would be out almost every day—

TEXT: E. A. Hornel Library, MS 15/21, fol. 32, undated by ALB (published by permission of James Allan, Hornel Librarian). No cover, no address, no postmark; endorsed in JA's hand, "1780." In Feb. 1787 JA arranged all of ALB's letters to him in date order, and perhaps he endorsed the letter at that time (Hornel Library MS 15/21, fol. 13). A much likelier date, based on ALB's mention elsewhere in the letter of Frances Burney's *Cecilia* (1782) and on comparison with item 6 above, is Jan. 1783.

It says something for ALB's "spirits" that the examination which so fatigued her is the one for which she wrote the sparkling Poem 74 (discussed above, p. 303). The identity of the "French Usher" I do not know. "Pennant" is Thomas Pennant, *British Zoology* (1766; new ed., 1768-70); "Goldsmith" is *History of the Earth and Animated Nature* (1774). George's zoological interests at age 9 are consistent with his becoming, in later life, an estate manager: his only known surviving letter concerns finding a cure for sheep rot. For the French books mentioned, see Appendix IV.

8. [ALB to Joseph Johnson, 3 June 1783]

Such a pacquet! I have hardly the conscience to add to it by a letter which yet I have for a good while intended to write to Mr Johnson

We thank you for the *French books*, which have afforded us great entertainment. We think some excellent, more pretty, & a good many mediocre. We imagine a selection would be better than a translation of the whole & that the version should be very free to adapt it better to the English manners. Lastly we are going to take them to Warrington, where perhaps somebody may take them in hand, if they are not already undertaken by any one. I meant to have asked first whether you did not want the books returned to you, but we have been so busy to the last minute you will excuse the neglect—— We are setting off this morning for Warrington with Hunter who begs his duty to you. I should tell you that he wrote sometime ago to his sister & the letter waited for one of mine so long that it was quite out of date, but pray let him have the credit of it. He is perfectly well now. M[r] Barbauld begs the favour of you to distribute either by the penny post or by your servant as is most convenient to you the inclosed bills, & to set down the postage ^ carriage ^ &c to his account. He returns likewise *Denham's poems* which will not do for a prize book, & *Thomson* because it is not the edition he meant. There is a much smaller *Bell's* I believe. The paper is returned to have the edges cut——

My sister who is much better & Miss Nunes beg their comp[ts] to which M[r] B adds his affec[t] regards with those of Yours &c

Palgrave ALB
June 3[d]

TEXT: Historical Society of Pennsylvania, MS Etting Collection, 3:7 (by permission; previously published in Rodgers, *Georgian Chronicle*, pp. 207-208). No cover, no address, no postmark, no endorsement. I assign the letter to 1783 by reference to a letter from John Aikin to Johnson, 20 June 1783, which reports that he and ALB have discussed *L'Ami des Enfants*, agree that 3 volumes might be extracted from it, and are willing to make the translation; the other 3 French volumes they think not worth the trouble (MS Laing II:647/12, Edinburgh Univ. Library). *L'Ami des Enfants*, by Arnaud Berquin, began appearing in London (in French) in 1782; the anonymous English translation, *The Children's Friend*, which appeared in 1783 is probably not by ALB and her brother. Apparently Johnson never got around to procuring a translation for himself; in 1793 he joined a conger which reprinted a translation by Lucas Williams.

For Hunter, see Appendix III; for Thomson, Appendix IV. Sir John Denham's *Poems* (1668) were reprinted in Bell's *British Poets* (Vol. 26, 1779) (*BLC*). "My sister" is ALB's sister-in-law, Susannah Barbauld Marissal (see item 4). "Miss Nunes" is Eliza Nunes, a friend from Warrington who came to live at Palgrave in 1780 (Hornel Library MS 15/21, fol. 31).

9. [ALB to Elizabeth Denman, June 1783]

ALB sends "a little piece of verse which, with great applause, he spoke at our play." "The dear boy has already begun Latin; he has an excellent memory, and delights in learning."

Extracts printed in Arnould, *Life of Thomas, First Lord Denman*, I, 5.

10. [ALB to Elizabeth Denman, Aug. 1783]

> ALB reports that "'he is going on with his Latin, and reading English books of amusement.'" "'I do not know a sweeter child,' writes Mrs. Barbauld, and she duly records . . . his generosity in sharing his cakes and his other home presents among his little schoolmates."

Extracts and summary in Arnould, I, 5-6.

11. [ALB to Mrs. Thomas Harvey, 22 Aug. 1783]

> Madam,
>
> The very obliging manner in which you have exprest your favourable opinion of me, demands my thanks. I am happy at the same time to assure you that at any time after Christmas next, I shall be ready to accept the trust you are willing to repose in me. Sensible as I am of the attention & tenderness requisite in forming such young minds to the love of science, I yet flatter myself in the present instance with the fairest prospect of success, as I doubt not but to meet in a child of Mrs Hervey's with a subject well prepared for culture, & already inspired with a love of excellence & seeds of taste——
>
> I do not know Madam, whether you are acquainted with the terms for my little parlour pupils, which are forty guineas a year. Entrance as for the others, two guineas. Mr Barbauld desires me to add his respectful Comp^ts^ to the sentiments with which I am Madam
>
> Your obliged
>
> Palgrave & obedient servant
>
> Aug 22^d^ AL Barbauld

TEXT: The Huntington Library, San Marino, California, MS HM32699 (published by permission). Addressed "Mrs Thos Hervey/Norwich/by favour of M^r^ T Martineau." No postmark. Endorsed, by a contemporary hand, "about 1783." The Harveys were a leading Norwich banking family. Mrs. H's brother's wife was a sister of Sarah Siddons, who visited the Harveys; the Harveys also owned a fine collection of paintings (Jewson, *Jacobin City*, pp. 42, 149). Hence ALB's confidence that a son of hers would have acquired "a love of excellence & seeds of taste."

12. [ALB to "Madam" (Mrs. Thomas Harvey), 31 Aug. 1783]

Madam,

I am sorry that by having mistaken your intentions a mistake however I was naturally led into by your letter you have had the trouble of writing twice on the same subject. I am sure I need say nothing by way of explanation of the difference of terms since I doubt not it has already occurred to you that any peculiar attentions require that difference. We shall however be equally happy Madam to receive your son on the common terms provided we have room after the next Vacation. You will wonder at making this a contingence after having agreed to take the young gentleman, but the case is this. Mr Barbauld does not reckon those little pupils I have under my own care as properly belonging to the school & therefore I am not scrupulous of taking them when the number of his young gentlemen is compleat. But the school has been so full this half year that till we know whom we shall lose at the Vacation we dare not absolutely engage for any more. You will be so good therefore Madam to give the same answer to your friends whom we thank for the honour of their good opinion & to assure them we will give them the earliest notice of a vacancy.

I am Madam
Palgrave — Your obliged & obedient servant
August 31st — AL Barbauld

TEXT: Evelyn MS 4, I, 13, British Library (by permission of the British Library Board). No cover, no address, no postmark. Endorsed "1783" by the same contemporary hand as Letter 8's; hence I infer the same addressee.

13. [ALB to Joseph Johnson, c. 22 Sept. 1783]

Dear Sir,

We would not have kept the inclosed if we had sooner had an opportunity of returning it. We think the design a very good one & the manner in which it is executed very entertaining & proper for children particularly in the little stories & facts that are introduced. As to the Linnean definitions a child will hardly enter into them I suppose till ten or twelve years old unless very particularly explained by the person who reads it with him. The stile is not elegant, plain & simple it should be no doubt but it ought to have all the elegance of perfect clearness & correctness. Please to send

1 Enfield's Hymns — Books for Arthur if any have been

3 Greek Testaments sent to Mr Johnson's
2 Dawson's Lexicon Remember the Prize books
~~2~~ ^ 4 ^ Perrins Exercises
1 Martial
12 Thompson's seasons, small & cheap edition, but good type
If you shall have any time within ten days a parcel to Birmingham & Warrington be so good to forward the enclosed by them, if not pray put them in the post ^ except one ^. Hunter is well so is Mr Barbauld who is very much yours as well as
AL Barbauld

TEXT: The Huntington Library, San Marino, California, HM 1605 (published by permission). Addressed "Mr Johnson/No 72 St Paul's Church Y^{d}/London." Endorsed "Wed^{y} Septr 22^{d}" and "Septr 1783" (by a contemporary hand, perhaps Johnson's), and "Mrs Barbauld Em. Authoress," and "113" (by other hands). Postmarked "PENNY POST PAID W WF" and "BROY[?]." I have not identified the book discussed in the text; for the others, see Appendix IV. A parcel for Birmingham would be meant for Joseph Priestley; he was then living there. Possibly his son William was already attending Palgrave (see Appendix III).

14. [RB to Joseph Johnson, 4 Nov. 1783]

Palgrave, November 4th 1783

Dear Sir,

By the very next wagon, if time will permit, be so kind as to send me the following articles———
5 L'Ami des Enfans 4^{e} 5^{e} & 6^{e} volumes—3 Charles douze——1 Racine—6 Perrin's French Grammar—6 Perrin's Exercises—6 Enfield's Exercises—6 Clarke's Esop 6 Bible (the same I had last) 1 Ovid (Bailey's)—6 Entick—4 Phaedrus (Sterling's) [undecipherable insertion]— 1 Gilblas (French)—4 Walkingham's Assistant—36 Copy-books N.B. I must beg of you to tell your stationer that if they are not made of better paper than the last were, I shall never again apply to him for any thing. I must also earnestly desire you to remember that it is high time I should be informed what prize-books I can get. Pray, do not drive me to the very last moment.
Before you close your parcel, be so kind as to send to Mr Kinder's to inquire whether he has any parcels &c. for us or for M^{r} or Mrs Marissal. In case he has any, or you—do me the favour to send them along with the books. Hunter is well and

goes on well. He desires me to give his affectionate duty to you, and begs you would send him a parcel which his mamma, it seems, or his sister (I am not sure which) has sent to you for him. Miss Enfield also desires you would put up along with my books a music-book which she is informed by her mother has been sent to S[t] Paul's Church-Yard some time ago. Pray, when you see M[rs] Newnum (I am not sure whether I spell her name properly) wish her joy in my Letitia's name and in mine. My Letitia desires to be affectionately remembered to you; and we both look forward with pleasure to the time, when, agreeably to your friendly invitatio[n,] we shall again enjoy your company under your hospitable roof. Till then we shall be very busy.

Farewell for the present, and believe me to be, Dear Sir, with sincere regard

Your obedient humble servant
Rochemont Barbauld

TEXT: Warrington Library, MS 269, fol. 18 (by permission of David Rogers, Area Librarian). Addressed "To/M[r] Johnson N° 72/S[t] Paul's Church-Yard/ London"; postmarked "6 NO" and "CYR[?]." Endorsed (probably by Johnson) with additions of prices in shillings and pence: "13.6—" and "2.14" totalled to "16——," and "3.6—" and "2.14" totalled to "6.0—."

For the books, see Appendix IV. RB's impatience ("pray, do not drive me to the very last moment") may testify to the state of his nerves but can be matched among other people who had to deal with Johnson, whose dilatory habits were a constant irritant. For Kinder and Marissal, see above, item 4; for Hunter, see Appendix III. "Miss Enfield" is a daughter of William, Rector of Warrington Academy and a close friend of Aikins and Barbaulds; she was perhaps visiting Palgrave. Mrs. "Newnum" (or Newnham) was an Irish relative of Johnson (Gerald P. Tyson, *Joseph Johnson: A Liberal Publisher* [Iowa City: Univ. of Iowa Press, 1979], pp. 213, 258).

15. [ALB to Elizabeth Denman, 8 Mar. 1784]

Dear Madam,

I am afraid you will think it long since you have heard of your, & permit me to add *my* dear boy, who begs his love & duty to yourself & his Papa & sisters & all his kind friends, & desires me to say that he is learning to write, & when he can write well enough he will not be obliged to me for sending messages but write himself a pretty letter in return for all those

you have sent him. We rejoice in the return of warm weather, for not only he takes great pleasure now in being out & watching the growth of the plants & flowers, but I likewise hope he will feel the effects of the warm weather in losing a cough which has hung upon him for a good while, tho as it has not seemed otherwise to affect his health, or at all to be increasing, I did not trouble you with the mention of it. He has twice had a bottle of pectoral mixture. Pray by the way give me leave to ask if he has had the hooping cough———
I must mention another circumstance which may injure his health. He has a trick of eating the most out of the way & dirty things you can conceive of, such as soot &c. I have more than once spoken to him about it, but I think a letter from you would greatly contribute to strengthen my remonstrances. I the rather mention it as perhaps the D^{r} may think it a sign of worms, tho he has no want of appetite for proper eatables———
I am grieved to think we have passed his birthday without noticing it, for tho I know it is in February I did not remember on what day. Mr Barbauld joins me in respectful compts to the D^{r} & aff^{t} remembrances to the young ladies.

I am dear Madam
Your obliged & obedient
AL Barbauld

Palgrave
March 8th

TEXT: The Pierpont Morgan Library, New York, MS MA2572 (published by permission). Addressed "Mrs Denman/old Burlington street/London." Postmarked "10 MR" and "DISS." Endorsed, by the same hand as on "N°. 1" (item 5 above), "Mrs Barbauld N° 20." My assignment of this letter to 1784 is based on my assumption that the endorsements on it, "N° 1," and "N° 21" (below), are indeed by Mrs. Denman, recording the letters as received. If ALB wrote monthly, letters 2-11 would have been written between Aug. 1782 and May 1783, the letters in Arnould (items 9-10, June and Aug. 1783) would have been 12 and 13, letters 14-19 would have been written between Sept. 1783 and Feb 1784, and "N° 20" would be dated March—as, indeed, it is. ALB's fearing that Mrs. D "will think it long since [she has] heard" does suggest a gap of more than a month since letter 19, but the average interval between letters may nevertheless have been a month. A further reason for dating to 1784 is the sentence rejoicing "in the return of warm weather": "I think that for a very long time people will remember the winter of 1783-4. I have never been so cold. It lasted almost four months, without the ground once being uncovered . . ." (François de la Rochefoucauld, *A Frenchman's Year in Suffolk: French Impressions of Suffolk Life in 1784*, ed. Norman Scarfe [Suffolk Records Society, 30 (1988)], p. 30).

16. [ALB to Elizabeth Denman, 15 Mar. 1784]

Dear Madam,

I am happy to remove the fears of the hooping cough, which I doubt my last letter may have raised, by assuring you that our Apothecary tells me he is certain Master Denman's cough cannot be that, & indeed it is now so much better that I have no doubt it will soon be entirely gone. The oranges & apples I shall certainly procure. I hope I did not alarm you by the mention of the cough, I did not mean to do so, but I own as it had hung on him some time & had I thought sometimes the appearance of a hoop I could not make myself easy without telling you. The parcel committed to Miss Martineau arrived to day by the Norwich carrier & your little boy begs to thank you for the contents. We were much amused with Ruspini's card. You will excuse the shortness of this as I only write to make you easy about the cough I am dear Madam with Comp^ts^ to the D^r^ & the young ladies Your obliged & obedient

AL Barbauld

Palgrave
March 15th

TEXT: The Pierpont Morgan Library, New York, MS MA2572 (published by permission). Addressed "Mrs Denman/old Burlington street/London." Postmarked "17 MR." Endorsed, in the same hand as "N° 1" and "N° 20," "Mrs Barbauld No 21." For date, see item 15, notes. The oranges and apples were probably Dr. Denman's prescription for the cough. "Miss Martineau" is probably a sister of Thomas, the pupil. Ruspini advertised a "dentifrice" and announced that he would make house calls on Tuesdays and Thursdays only (*General Evening Post*, 28-30 Aug. 1777).

17. [ALB to Joseph Johnson, 29 Apr. 1784]

Dear Sir,

You will excuse my giving you this trouble in order to know whether you received a pacquet of letters which I sent off in a parcel my sister conveyed to Mr Marissal's East street, & which was to come to you from thence by the penny post. The parcel got to London April 17th, & there was enclosed in your pacquet a letter for my brother, for Miss Belsham, for Mrs Priestley, & for Mrs Kinder. Now I find my brother had not received his on the 24th. Do me the favour therefore to let me know by a line whether they got to you & whether they have been put into the post. We are very well & have fixed our plan for the Vacation which is if nothing prevents to visit Dr Priestley at Birmingham. He has been in town I hear. Pray tell me if he looks well. Hunter is a good boy & very well. I suppose he stays here during the Vacation———

Please to send by the very next waggon, 2000 quills. 3 doz copy books. The Anatomist. Florizel & Perdita. Farces. A Shakespear's Henry 4th if to be had. & be so good to desire Mr Kinder to enclose in it, two turbans such as I had before. Mr Barbauld joins with me in all cordial greetings. I am dear Sir,

Your obliged &c
AL Barbauld

Palgrave
April 29th

TEXT: Folger Shakespeare Library, MS Y.c. 82(1) (published by permission). Addressed "Mr Johnson/N° 72/St Paul's Church yard/London." Postmarked "1 MA." I assign the letter to 1784 by reference to John Aikin's diary of a visit to Joseph Priestley at Birmingham, 21-25 June 1784, in company with the Barbaulds (Royal Society of London, Yates Album, fol. 24). ALB plans this visit in a letter to John Aikin, 4 June [1784] (Yale Univ. Library).

"My sister" is Susannah Barbauld Marissal; "Miss Belsham" is Elizabeth (1743-1819), ALB's second cousin and closest friend. For Mary Priestley, see *Poems*, p. 219; for Mrs. Kinder, item 4. She was married to a linen-draper (hence the request for turbans). The books are identified in Appendix IV.

APPENDIX II
SCHOOL BILLS

Among the papers of the Bulwer family of Norfolk (Norfolk R. O. MS BUL 4/120) are four bills for the schooling of Bulwer sons. Two of them relate to Palgrave School in 1780 and 1790, one to Bury School in 1786, and one to a school managed by a Mr. Smith in 1783. They shed some light on the costs of schooling (both to parents and to schoolmasters) in East Anglia during and just after the Barbaulds' tenure at Palgrave School. I am grateful to the Norfolk Record Office for permission to print them.

[1.]

W. W. Bulwer Esq. D^r to R. Barbauld
for Master A. Bulwer

1780				
May 31	For ½ a Year's Board & Tuition	£12.	10.	—
——	Pens, Ink Paper & Pencils———	—	2.	6.
——	School Library & Prize Books——	—	2.	6.
		12.	15.	—
	For a single room—————	2.	12.	6
		15.	7.	6

Palgrave, July 31st 1780
Received the contents in full of all demands
Rochemont Barbauld

MS: Norfolk R. O. MS BUL 4/120/1. The last three lines are in RB's hand. On the verso is an endorsement: "July 31st 1780/P^d Mr Barbauld ½ yrs board & tuition for Austin."

For Augustine Bulwer, see Appendix III. The charge for "School Library & Prize Books" may have been customary; at Rugby, pupils paid entrance and departure fees which were used to buy books for the school library (Butler, I, 30).

[2.]

Ma^{r} Edwd Bulwer D$^{r.}$

July 14th 1783	£	s	d
To ½ Year's Board & Education at Chr^{s}	8.	8.	0
Copy Books, Quills, Ink &c	0.	6.	
Pocket money 21 Wks at 3^{d}	0.	5.	3
D^{o} extra	0.	1.	0
Latin Grammar	0.	1.	6
Virgil (in usum Delph.)	0.	4.	0
Vyses' Arithmetic	0.	3.	6
Mr Browne's Bill	2.	4.	6
Extra washing	0.	4.	0
Tradesmens Bills	5.	8.	8
School Candles 1/. Worstead 1/.	0.	2.	0
	£ 17.	8.	6

On account of the great advance in every material article in house-keeping, Mr Smith's Terms, in future, will be Sixteen Guineas p annum, exclusive of Washing.

MS: BUL 4/120/4

[3.]

Mas^{r} Bulwer D^{r} to Phil Laurents [Bury School]
mid^{r} 1786

	£	s	d
To ½ a years, Boarding, washing, &c	12	10	0
Tuition	3	3	0
Extra articles			
To Letters & parcels	0	4	7
mending shoes	0	2	7
To the Taylor for mending cloaths	0	6	4
Hair cutting 6^{d} school-woman & sexton 7^{d}	0	1	1
servants at Xmas	0	10	6
Buckles 5^{s}6^{d} Apothecary 1^{s}6^{d}	0	7	0
Poikilæ Historiæ 3^{s} prayer Book 2^{s}6^{d}	0	6	0
paper 3^{s}6^{d} shoe-maker 11^{s}	0	14	6
weekly allowance	0	12	6
writing master	1	2	0

windows Etc	0	3	11
cash for a Journey 1ˢ postchaise to Scole Jan⁷ˢ Etc:	0	8	0
	20	12	0

MS: BUL 4/120/6

[4.]

W W Bulwer Esqr D^{r} to N: Philips

1790.	For Mr E. Bulwer			
June 14th	To Board & Tuition as usual	£13.	2.	6
	Geog^{y}————	1.	1	—
	To Private Tuition 4 G^{s} P^{r} 2^{r}	8.	8.	—
	Single Room————	2.	12.	6
	Books————	—	13.	6
	Copy Books————	—	1.	—
		£ 25.	18.	6
	Disbursements————	2.	19.	6
	Total	£ 28.	18.	—

MS: BUL 4/120/10

These bills may be compared with one from Harrow in 1779 and one from Eton in 1769. At Harrow, the son of Philip Francis was charged as follows: "Entrance Fee, 9 guineas, Board £25, Schooling 4 guineas, an extraordinary tutor 12 guineas, a bed to himself 5 guineas, Washing 1 guinea, French 4 guineas, Writing 2 guineas, Drawing 6 guineas, Dancing 4 guineas," Fencing 12 guineas, "'besides books, servants and clothes.'" At Eton, a son of Mrs. Dent paid £25 for Board, £3 for Fire and Candles, one guinea for a Study, 10 shillings sixpence for Washing, 5 guineas for a single bed or 10 for a single room, 4 guineas to the Master, 8 to the Tutor, and 9 Entrance Fee (*HSR*, p. xix and n. But the usual fee at Eton was 2 guineas; only wealthy parents paid more, and poor parents might pay no fee at all [*ECR*, p. xxvii]).

Certain charges are constant in all these bills: £25 per year for board is typical (but at Palgrave, unlike Bury, Harrow and Eton, it included tuition), as is five guineas for a single room (at Palgrave) or a single bed (at Harrow and Eton).[91] An entrance fee was probably universal except in cases of poverty. Extra tutelage cost extra; at Palgrave the writing master perhaps submitted his own bill. Supplies were always paid for by the pupils, but books might be provided by the master (Butler, I, 31). The Barbaulds appear to have provided them.

The most striking difference between Palgrave and the other schools lies in the extras. At Smith's and at Bury, pocket money and tradesmen's bills for students were advanced by the school, then recouped from the parent. At Palgrave, apparently students were responsible for arranging these matters with their parents (or else accounts of them were separately kept). Fire and candles, undoubted items of expense to a school, were apparently included in the cost of board at Palgrave: a Palgrave parent's £25 bought more than an Eton parent's did. (To be sure, Eton needed many more candles than Palgrave did.) Drawing, dancing, and fencing, costly items at Harrow, were not offered at Palgrave.

APPENDIX III
PUPILS AT PALGRAVE SCHOOL, 1774-1785: A LIST

The surnames on this list derive chiefly from three sources: 1. The printed playbill for performances of William Mason's *Caractacus* and Samuel Foote's *Taste* "by the Young Gentlemen of Palgrave-School," 27 May 1782 (Norfolk R. O. MC257/63/1/3, 684 × 4: see illustration). 2. A contemporary MS "List of Mr Barbauld's Pupils 25th March 1785" (Norfolk R. O. MC257/63/1/4, 684 × 4, hand unknown, among the papers of Philip Meadows Taylor, a Palgrave pupil after the Barbaulds left). 3. Brodribb, "Mrs. Barbauld's School." Brodribb describes three MS documents then in his possession: a playbill for a performance of Rowe's *Tamerlane* with interludes; one issue of a "weekly chronicle" of the school written by Barbauld; and the parts of "a little masque . . . an ingenious blend of *Comus* and Shakespeare, with possibly a touch of Mrs. Barbauld herself," with the names of the pupils who performed it. The playbill is dated only "Friday, 26 May," but can be assigned to 1780.[92] The chronicle can be dated between July 1782 and ?June 1784, the period when Thomas Denman, who is named in it, attended Palgrave. (Some of the names in it are subjects of riddles, and these are noted in the list below.) The *Comus*-Shakespeare masque must be the earliest of the three documents, for Carter, who played in the masque, was probably one of Palgrave's first pupils; I conjecturally date it 1776.

The playbills and "weekly chronicle" normally give only surnames and no other information; the 1785 "List" gives, in addition to surnames, ages and some family information. I have been able to identify some boys from other sources, listed below. The identities of many, however, can only be conjectured. In conjecturing, I assume that a pupil is likely to be of East Anglian family, and also that if a biographical connection (however remote) can be found between a family name and ALB or her husband, a son of that family is likely to have been the pupil. Different degrees of conjecture are signified by "perhaps" and "possibly."

Other sources of names are cited as they occur. Names asterisked are listed in the *DNB*. Biographical information on pupils comes also from: J. A. Venn, *Alumni Cantabrigienses: A Biographical List of all known Students, Graduates and Holders of Office at the University of Cambridge . . . Part II* (Cambridge: Cambridge Univ. Press, 1940-54) (*AC*); *Alumni Oxonienses: The Members of the University of Oxford, 1715-1886*, ed. Joseph Foster (Nendeln: Kraus Reprints, 1968) (*AO*); *The Eton College Register* (*ECR*); and Joseph Hunter, *Familiae Minorum Gentium*, ed. John W. Clay (London: Harleian Society, 1894-96) (*FMG*).

In the list, pupils' names are followed by their dates of birth and death if known, and then by the dates during—or at a time between—which they are known to have attended Palgrave. In assigning these dates I have assumed that a pupil who took part in the end-of-year exhibitions attended Palgrave for the entire school year preceding the exhibitions.

My estimate of a total pupil population of 131 is based, in the first place, on the numbers we glimpse at different dates (p. 279 above) conjointly with Rochemont Barbauld's stated limit of thirty pupils—which, however, he exceeded in 1779-80, 1783-84, and 1784-85. I can identify six of the original eight pupils, twelve of the sixteen-plus of 1775, and eight of the twenty-nine expected in 1776, leaving a total of at least twenty-seven unknowns. I assume that in the school years 1777-78 and 78-79 the school was full at thirty; I can identify ten of those present in 77-78 and twelve of those in 78-79, leaving thirty-eight unknowns. In 1779-80 the school was overfull at thirty-two; if thirty-two remained constant through 82-83, I can identify all but a total of twenty unknowns in those years. Finally, of the thirty-six pupils in 1783-84, I can identify all but five. The total of unknowns comes to ninety. (That number includes, however, two pupils—Crabb and Gooch—whose dates of attendance I don't know; hence it is really eighty-eight.) I reckon the average length of stay for pupils whose times at Palgrave I know at just under two and a half years. Assuming that the stays of unknown pupils averaged the same as those of known pupils, I divide 88 by 2.45, which yields 35.9. Thirty-six added to ninety-five (the number of known pupils) makes one hundred thirty-one.

Following the list is a Table of Outcomes which shows the numbers of Palgrave pupils who went into various professions, compared with the corresponding numbers of graduates of Eton College and Westminster School.

AIKIN

*Arthur (1773-1854; 1783-85)

ALB's eldest nephew. At Palgrave c. 22 Sept. 1783 (Appendix I, item 13); named on 1785 "List." Afterwards editor, *Annual Review* (1803-09), author, *A Dictionary of Chemistry and Mineralogy* (1807-14, with C. R. Aikin), *A Manual of Mineralogy* (1814), and other works. A founder of modern geology: See H. S. Torrens, "A A's Mineralogical Survey of Shropshire 1796-1816 and the Contemporary Audience for Geological Publications," *British Journal for the History of Science*, 16 (1983), 111-53.

*Charles Rochemont (1775-1847; 1777-85)

ALB's third nephew; adopted as her son, June 1777. At Palgrave examination, Dec. 1778, recited "verses on the boy who would not say

A lest he should be made to say B" (ALB, *Poems*, Poem L6); in Dec. 1779, "repeated very prettily at the examination a little description of Spring I wrote for him, & along with Bunbury [q.v.] a scene from [Young's] Busiris" (ALB to John Aikin, 19 Jan. 1780 [Hornel Library MS 15/21, fol. 31]); recited "L'Allegro," 26 May 1780. Named on 1785 "List." Afterwards trained as doctor and chemist; vaccinator on the "National Vaccine Establishment," 1805 (Rodgers, *Georgian Chronicle*, pp. 164-65); secretary, London Medical and Chirurgical Society (predecessor of Royal Society of Medicine). Author, *A Concise View of all the most important Facts . . . concerning the Cow Pox* (1800).

George (1774-1847; 1782-83)

ALB's second nephew. In Jan. 1783 ALB reports his progress in arithmetic, French, elocution, and writing (Appendix I, item 7). Went into trade, failed; later estate agent to William Dodge Cooper Cooper of Toddington, Beds. (C. R. Aikin, ALs to Susan Aikin, 11 Dec. 1830 [MS, Rodgers]).

AUGHRIM: See GINKEL

BADELEY, Samuel (1770-1854; 1781-85) and Joseph (1782?-85)

"Sons of M^r [Samuel] Baddeley [*sic*] of Walpole [Suffolk]," ages fourteen and eleven (1785 "List"). One, "Master S.," played a Bard in *Caractacus*; he or his brother is mentioned in weekly chronicle. "Master S." must be Samuel, afterwards adm. to Trinity Hall, Cambridge, 1787 (LL.B. 1795) and Lincoln's Inn, 1789, and was Vicar of Ubbeston, Suffolk, 1800-54 (*AC*, I, 111). In 1795 the B family subscribed to Sarah Spence's *Poems, and Miscellaneous Pieces* (Bury St. Edmund's, 1795, p. [v]): Samuel senior; Samuel, "LLB Trinity College Camb."; "S. Badeley, Fig Tree Court, Temple"; and "Joseph Badeley, Esq. Walpole." Joseph is probably the other Palgrave B.

BAKER (1779-82?)

Played Isaac Fungus in scene from Foote's *The Commissary* on *Tamerlane* evening, 26 May 1780. Probably Frederick (or Frederic) Francis (1772-1830), son of Sir George, MD, of London, court physician (*DNB*), friend of Dr. Maxwell Garthshore, whom ALB's brother had served as apprentice (Robert H. Jeffers, "Richard Pulteney, M.D., F.R.S. . . . and His Correspondents," *Proceedings of the Linnean Society of London*, 171 [1959-60]:16), and subscriber to Sarah Spence. Attended Eton, 1782-88; adm. to St. John's College, Cambridge, 1789, and Lincoln's Inn, 1790; migrated to Balliol College,

Oxford, 1791; succeeded to father's baronetage, 1809. Died of a blow from a windmill blade while showing his children how the mill worked (*ECR*, p. 21; *AC*, I, 126)—a scientific interest characteristic of the Palgrave curriculum.

BRETT, Joseph (1774-1835?; 1783-85)

"Son of Mr Jos. Brett [wool-stapler,] Norwich," age ten (1785 "List"). At Palgrave two years under Barbaulds, five under Nathaniel Phillips, their successor. Afterwards adm. to Caius College, Cambridge (1790); B.A., 1796; M.A., 1799. Perpetual Curate of Cringleford, Norfolk, 1805-35 (*AC*, I, 371; *Biographical History of Gonville and Caius College, 1349-1897*, ed. John Venn [Cambridge: Cambridge Univ. Press, 1898] II, 121).

BROWNE

Harry (1775 or 76-before 1847; 1781?-85)

"Son of Mr Browne of Diss," age nine (1785 "List"). Father probably Henry Browne, Clerk to Trustees of Norwich and Scole Turnpike, member of association for prosecuting horse-stealers (*Norfolk Chronicle*, 10 Feb. 1776 and 26 July 1777) and also, with Philip Meadows, large Palgrave landholder (1784 Palgrave survey, Suffolk R. O. HA68:484/691). Norfolk R. O. Persons Index lists a Mary Anne Browne who in Feb. 1847 was widow of Henry, of Diss, "gent."—i.e., probably the Palgrave pupil.

Another B (b. 1771 or 72; 1781?-85), apparently no relation to Harry

"A Ward of Mrs Fox near London. 40 G^s^"; age thirteen (1785 "List"). Fee implies parlor boarder; "ward" may mean that he came from West Indies or North America. Mrs. Fox is not identified. Perhaps this B is John, adm. to Corpus Christi College, Cambridge, 1790, and said to have been "of Jamaica" (*AC*, I, 415).

One of these—or perhaps a third boy?—is the "Brown" who played Caleb in *Taste*, 27 May 1782.

BULWER

Augustine (1762-1831; 1779-80)

Second son of William Wiggett B, of Heydon, Norfolk. A bill for his tuition and expenses, Jan.-May 1780, is given in Appendix II. Adm. to Pembroke College, Cambridge, 1780; B.A. 1784; M.A. 1787; D.D. 1813. Curate, then Rector, of Heydon, 1784-1831; Vicar of Guistwick, 1790-96, and Corpusty, 1790-1831; Rector of Salle, 1796-1818, and Cawston, 1818-31 (*AC*, I, 444).

William Earle (1757-1807; 1774?-76)

Appears on 1785 "List" as addendum in hand of Philip Meadows Taylor: "W. E. Bulwer Norfolk [illegible] Father of Sir E & H Bulwer." Elder brother of Augustine, above. Adm. to Pembroke College, Cambridge, 1776; afterwards of Wood Dalling and Heydon Hall, Norfolk. In 1794 raised corps of six hundred men called Royal Norwich Volunteers (Jewson, *Jacobin City*, p. 45). A colonel in 106th regiment and Brigadier-General, Army; father of Sir Henry, diplomatist, and Edward B-Lytton, novelist (*AC*, I, 445).

BUNBURY, Charles John (1772-98; 1779-80)

Recited, with C. R. Aikin (q.v.), scene from Young's *Busiris*, Dec. 1779; played Selima in *Tamerlane*, 26 May 1780. Elder son of Henry William (1750-1811), of London, cartoonist, friend of Reynolds and Goldsmith (*DNB*); brother of Sir Henry Edward (*Memoir and Literary Remains of Lieutenant-General Sir Henry Edward Bunbury*, ed. Sir Charles J. F. Bunbury [London: Spottiswoode, 1868], pp. 3-7); subject of Reynolds's portrait, "Master B" (1781). Afterwards attended Westminster School, where he was friend of Robert Southey; adm. to Trinity College, Cambridge, 1790; B.A. 1795 (*AC*, I, 446); to India in the 25th Light Dragoons, 1796; d. at Cape of Good Hope.

The Bs were a Suffolk family. A French visitor in 1784, François de la Rochefoucauld, described CJB's uncle, Sir Thomas Charles, as "one of the leading Suffolk gentry, through his family which has always been in Parliament, and through his very considerable wealth: he himself has been a Member of Parliament for the county for thirty years. He was supported by all the Gentry of the County and consequently by the more enlightened part of it"; the farmers and freeholders resented him, however, because he supported the Fox-North coalition, and they blamed Lord North's American war for their troubles (*A Frenchman's Year in Suffolk*, pp. 75-76). CJB's father, the cartoonist, was younger brother to Sir Thomas (Bunbury, *Memoir*, p. 4).

BUNNY (1781-84)

Played Aulus Didius in *Caractacus*, 27 May 1782; named (spelled "Bunney") in weekly chronicle. Probably brother of Joseph and Margaret (or Marriot), of Newbury, Berks., the first of whom married ALB's friend Elizabeth Rigby and the second John Martineau (1758-1834) (David Martineau, *Notes on the Pedigree of the Martineau Family*

[London: privately printed, 1907], Family Tree II; *FMG*, pp. 1107, 1108).

BUTCHER (Robert? b. 1768 or 69; 1782-85)

"Son of M[r] Butcher near Bungay [Suffolk]," age sixteen (1785 "List"); named in weekly chronicle. *Alumni Oxonienses* (p. 202) lists a Mark Grigson B, only son of Robert, of Bungay, "arm.," who matric. at Brasenose College in 1822, age nineteen; probably son of the Palgrave B.

CARTER, Joseph Chamberlin or Chamberlain (1774?-76?)

Played in *Comus*-Shakespeare masque, ?1776. Fellow-pupil and friend of Frank Sayers (q.v.; Taylor, *CW*, p. xiii), hence perhaps one of first eight. Probably son of William, of Norwich, trustee of the school (Appendix I, item 1), and brother of the Miss C who spoke the prologue to a different school performance of *Comus*, Sept. 1781 (Brodribb, p. 735). Joseph "Chamberlain" C, of Norwich, is described as "gent." in *AO*, p. 224, which lists his son, Joseph, as matric. 1806 (age eighteen) at St. John's College, Oxford.

CHOLWICK (1779-80)

Read "part of Young's paraphrase of Job" before *Tamerlane*, 26 May 1780. Possibly William, son of William, of London, "arm."; matric. Oriel College, Oxford, 1784, aged seventeen; B.A. 1789 (*AO*, p. 250).

COCKELL, William (1779-80)

A steward for Palgrave anniversary dinner, 8 Sept. 1794 (*Cambridge Intelligencer*, 23 Aug. 1794); probably the C who played Axalla in *Tamerlane*, and perhaps the WC who died at Trowbridge, Wiltshire, in 1817 (*Gentleman's Magazine*, 87:1 [1817], 644). An Edgar William C (b. 1859) who entered Cambridge in 1878 was son of William, Comptroller-General of Inland Revenue (*AC*, II, 81); perhaps the Palgrave C was their ancestor.

COLDHAM

Sons of George, of Norwich, linen-draper, and Dorothy (Wright) (*FMG*, p. 1131):

George (1766-1815; 1775?-80)

Played Bajazet in *Tamerlane*, 26 May 1780. Afterwards settled at Nottingham; joined law partnership with Henry Enfield, younger brother of Richard (q.v.); Commissioner in bankruptcy; also Town

Clerk for twenty-three years; d. of an accident in a gig (*GM*, 85:2 [1815], 378). "A strict integrity marked his dealings with the world, he had a hand ever open, as well to the alleviation of private distress as to forward every scheme of public utility. . . . His abilities were rather solid than shining. . . . A Dissenter on principle, he was well acquainted with the grounds of dissent. . . . He was . . . a frequent attendant at the Unitarian Chapel on the High Pavement" (*Monthly Repository*, 10 [1815], 597 and 656; Nottingham Archives, MSS DDD 181/5, DDR 2/10, and MR 12/312).

Wright (1770-1816; 1779-84)

Read "Part of Young's Paraphrase of Job" on *Tamerlane* evening. Afterwards alderman of Nottingham (*GM*, 86:2 [1816], 466).

One C (probably George) also played in *Comus*-Shakespeare masque, ?1776; one (probably Wright) was named in weekly chronicle and played a Bard in *Caractacus*. Both afterwards subscribed to William Enfield's *Sermons on Practical Subjects* (London: Johnson, 1798).

CORBETT, Stuart (1774-1845; 1778?-85)

"Grandson of Lord Bute & Nephew of Lady Macartney under whose care he is"; age eleven (1785 "List"). Lady M is Lady Jane, second daughter of John Stuart, third earl of Bute; she married George, earl Macartney, the diplomatist, in 1768 (*DNB*). Stuart was son of Andrew, of Hanover Square, London, "arm." and captain in the Blues, by Augusta (d. 1778), third daughter of John, earl of Bute. Matric. at Merton College, Oxford, 1792 (B.A. 1796, M.A. 1800, D.D. 1816). Vicar of Luton; afterwards Curate of Westley, Rector of Bramwath (or Kirk Bramwith) and Scrayingham, Yorks., Archdeacon and Canon of York (*FMG*, p. 1248; *AO*, p. 297).

COTTEREL (1775-76?)

Played in *Comus*-Shakespeare masque, ?1776. Perhaps Charles Cottrell (1765-1829; the name is spelled variously), son of Rev. Charles Jeffreys C, of Hadley, Midd.; he attended Eton, 1778-85, was adm. to King's College, Cambridge, 1786 (B.A. 1790, M.A. 1793), and Lincoln's Inn, 1789; called to bar, 1795; afterwards King's Advocate in Martinique (*ECR*, p. 131; *AC*, II, 149).

COTTON (1782-84)

Named in weekly chronicle. Perhaps related to Nathaniel, from St.Alban's, who entered Warrington Academy in 1759, went on to

Cambridge, took orders in the Church, and settled in Northamptonshire (Turner, "Historical Account of Students," p. 204). Or perhaps the "Mr." C, of Kesgrave, who subscribed to Enfield's *Sermons on Practical Subjects*; the other C who subscribed, Bayes, would have been too old, although his son William, attorney, afterwards Clerk to Dr. Williams's Trustees, would also have been of suitable age (*FMG*, p. 123).

*CRABB, George (1778-1851)

Named by William Taylor as a Palgrave pupil (Robberds, *Memoir*, II, 421). Son of Zachariah, of Palgrave, gent. (*AO*, p. 311). Afterwards classical master at Thorp Arch School, Yorks.; studied German at Bremen; matric. at Magdalen Hall, Oxford (B.A. 1821; M.A. 1822, with mathematical honors); called to bar at Inner Temple, 1829. Author, numerous books on German grammar and English law; also dictionaries, the best known being his *English Synonymes Explained* (1816; subsequent eds.to 1916 [*BLC*]).

DAWSON (1770 or 71-1793; 1782-85)

"Son of M[rs] Dawson Norwich," age fourteen (1785 "List"); also named in "weekly chronicle." A John D sat on Norwich Common Council (*Norfolk Chronicle*, 6 Apr. 1776): his father? The pupil is perhaps George, adm. pensioner at Emmanuel College, Cambridge, 1789, of whom nothing else is recorded (*AC*, II, 255). D's death is mentioned by C. R. Aikin to ALB in Dec. 1793 (MS, Rodgers): "Did you hear of the death of Dawson your former pupil. He paid dear for his imprudences. His mother, Mrs Barnard, is very much afflicted. . . . He left a widow & infant child, but not in very easy circumstances."

*DENMAN, Thomas (1779-1854; 1782-84)

Entered Palgrave 29 June 1782, age three-and-a-half; son of Thomas, physician, of London. Recited verses written for him by ALB, Christmas 1782 and ?May 1783 (ALB, *Poems*, Poems 74 and L7). Began Latin by June 1783; going on with Latin and reading English books for amusement, Aug. 1783; learning to write, 8 Mar. [1784]; named in "weekly chronicle" ("a man behind a cave"); not on 1785 "List." (See Appendix I, items 5, 9, 10, 15-16.) Lucy Aikin ("Memoir," pp. xxix-xxx) lists D among ALB's parlor-boarders, for whom she wrote *Hymns in Prose*. (Actually *Hymns* was published [1781] before D arrived at Palgrave, but his age and the fact that ALB wrote regular letters to his mother do suggest that he was her parlor-boarder.)

From Palgrave, D went to school in Kensington and then (1788-95) to Eton, where he "was branded with a red-hot poker for refusing to make a speech for the amusement of the elder boys." Adm. St. John's College, Cambridge, 1796 (B.A. 1800, M.A. 1803); adm. Lincoln's Inn, 1798; called to bar, 1806. Occasional contributor to *Monthly Review*, *Critical Review*, and *Edinburgh Review*. M.P. for Wareham, 1819-20; for Nottingham, 1820-26, 1830-32; solicitor-general to Queen Caroline, 1820-21 (he is pictured foremost among the Queen's counsel in Sir George Hayter's painting, *The Trial of Queen Caroline*, 1820-23 [National Portrait Gallery, London]); Attorney-General, 1830, in which capacity he drafted Reform Bill of 1832 and defended it in committees; knighted, 1830; Chief Justice of King's Bench, 1832; created Baron Denman, 1834; Speaker, House of Lords, 1835. From 1843, he wrote and agitated against slavery and the international slave trade (*ECR*, p. 156; *DNB*). See also Outcomes.

DOUGLAS family, of St. Mary's Isle, Kirkcudbright, Galloway, earls of Selkirk

Four sons of this family attended Palgrave (family information from *The Scots Peerage*, ed. Sir James Balfour Paul, VII [Edinburgh: David Douglas, 1910], 521-23):

Basil William, Lord Daer (1763-94; 1774?-82)

Eldest surviving son of Dunbar Hamilton (later Douglas) of Baldoon, fourth earl of Selkirk. Schoolmate of Sayers and Taylor (Taylor, *CW*, p. xxvii), and thus perhaps one of first eight pupils. A parlor-boarder, and ALB's favorite pupil (Aikin, "Memoir," p. xxxi). She praises him to William Taylor: "Your friend Mr Douglas is grown a very fine youth & is in every sense one of the first if not the first among his fellow students" (10 Apr. [1782], in Rodgers, *Georgian Chronicle*, p. 206). Afterwards attended Edinburgh University. For his subsequent career, see Outcomes.

John, also Lord Daer after his brother's death (1765-97; 1779?-82)

Second surviving son. Played Elidurus in *Caractacus* and Carmine in *Taste*, 27 May 1782. Adm. Faculty of Advocates, 1787.

Alexander (1767-94; 1779?-82)

Fourth surviving son. Played Evelina in *Caractacus*. Joined 22nd Regiment, 1785; went to West Indies with 38th Foot, 1793; d. of yellow fever.

*Thomas, fifth earl of Selkirk (1771-1820; 1779?-85)

"[Fifth surviving] son of the Earl of Selkirk," age fourteen, paying thirty-five guineas (1785 "List"). Parlor-boarder? Close friend of

Arthur Aikin (Aikin to Martha Aikin [his mother], 30 Sept. 1797 [ts by Conant Brodribb]). Attended Edinburgh University. From 1802, devoted himself to founding and directing British colonies in Prince Edward Island and Manitoba. A close friend of Sir Walter Scott. Author of numerous publications arising from his colonial activities: e.g., *Observations on the Present State of the Highlands of Scotland, with a View of the Causes and Probable Consequences of Emigration* (1805, characterized by *DNB* as "a strikingly clear, well-written work") and *A Sketch of the British Fur Trade in North America* (1816).

One D read Collins's "Ode to Fear" on *Tamerlane* evening, 26 May 1780.

DUMBLETON (b. 1778 or 79; 1784-85)

"Son of a Gent. in Surry. Mr Richd Gurney knows him"; age six, forty guineas (1785 "List"). Parlor-boarder. Richard Gurney was the eldest partner in the Norwich Quaker banking firm of that name (Jewson, *Jacobin City*, pp. 103, 114).

DUNHAM, Charles (b. 1771 or 1772; 1784-85)

"Son of M^{r} [Charles] Dunham [?partner in Dunham and Yallop, Norwich, goldsmiths, haberdashers, and cutlers (Jewson, *Jacobin City*, pp. 119-20)] of Bridgham Norfolk," age twelve (1785 "List"). At Palgrave one year under Barbaulds, five under Philips. Adm. Caius College, Cambridge, 1790, "age 19." (*AC*, II, 355; *Gonville and Caius College*, II, 121).

DYSON, E. and J.

Probably sons of Edward, of Diss (1726-1812), whose daughter, Elizabeth, married Meadows Taylor, a member of Rochemont Barbauld's congregation (*FMG*, p. 1131; *Christian Reformer*, 24 [1838], 432), and who was perhaps connected with the Norwich Bank (*Norfolk Chronicle*, 18 May 1776). If so, they are:

Edward (1766-1803; 1779-84?)

Played Dervise in *Tamerlane*, 26 May 1780.

John (d. 1833; 1779-84?)

Played Zama in *Tamerlane*. Perhaps the "Mr Jno Dyson" shown on 1812 map of Palgrave (Suffolk R. O. B150/1/3.12) as a P householder.

One D played Brush in *Taste*, 27 May 1782; one is named in weekly chronicle ("hang yourself child"). A Dyson is mentioned by C. R. Aikin to ALB, 6 Dec. 1792, as now at Ostend on his way home, to his mother's relief; CRA describes this D as a "farmer" (MS, Rodgers).

In 1833 CRA described as "my old friend" a Dyson who was planning to quit London for Westmoreland with his sister (Rodgers, *Georgian Chronicle*, pp. 265-66), but biographical details mentioned by CRA appear incompatible with those in *FMG* above.

ELLIOTT (b. 1772 or 73; 1784-85)

"Under the care of Crisp Molineux Esq[r]."; age twelve (1785 "List"; see also MALBROUGH, below). Molineux (1730-92), of Garboldsham in Norfolk, M.P. for Castle Rising and King's Lynn, owned plantations in St. Kitt's, West Indies (*Caribbeana*, ed. Vere Oliver, III [1913-14], p. 2). Probably, therefore, E came from West Indies. Perhaps William Thomas, son of Thomas E, of Christ Church Parish, Barbados, bap. 24 Jan. 1773 (*Barbados Records: Baptisms 1637-1800*, ed. Joanne M. Sanders [Baltimore: Genealogical Publishing Co., 1984], p. 308)—who happens to be the only male Elliott (with single or double *t*) listed in *Barbados Records* whose birthdate fits the age of the pupil. Perhaps also he is the unidentified E who attended Eton, 1785-90 (*ECR*, p. 178).

ELSDEN

H. (1779-80)

Played Omar in *Tamerlane*, 26 May 1780.

Another E, presumably brother to the above (1779-80)

Read "Harpagon's Soliloquy" by Moliere on the same evening. (That this is a different E is inferred from the use of the fore initial to distinguish the first E.) Perhaps related to Robert, member of Independent Church at Hadleigh, who d. 1832 (Suffolk R. O. Biography Index).

ENFIELD, Richard (1768-91; 1780-?)

Son of William, rector and tutor at Warrington Academy 1770-83 and close friend of ALB and John Aikin, who published William's posthumous *Sermons* (1798) by subscription. Expected to enter Palgrave after summer vacation, 1780 ("You will have an extremely promising lad in Dicky": John Aikin to ALB, 18 Feb. 1780 [Hornel MS 15/21, fol. 6b]). On 3 Feb. 1779 Aikin, at Warrington, asks ALB on William's behalf what "Greek grammar Dicky is to learn" (Rodgers, *Georgian Chronicle*, p. 205): he is being groomed for Palgrave? Afterwards law clerk to William Roscoe, of Liverpool; appointed Town Clerk of Nottingham. Converted to republicanism by reading Paine's *Rights of Man* (John Aikin, "Biographical Account of the Author of these Volumes," in Enfield, *Sermons*, I, xxi; RE to Roscoe, 14 May 1791, Liverpool R. O. 920 ROS 1446).

FIRTH, William (b. 1768; 1775?-82)

Friend of Frank Sayers at Palgrave (Taylor, *CW*, p. xiii, who identifies him as "Serjeant," i.e., serjeant-at-law). Played in *Comus*-Shakespeare masque, ?1776; probably the F who played Mirvan in *Tamerlane*, 1780, and Druid in *Caractacus*, 1782. Son of William, of Norwich ("a considerable manufacturer" [Jewson, *Jacobin City*, p. 29] and member of Norwich Common Council, 1776 [*Norfolk Chronicle*, 6 Apr. 1776]); adm. Trinity Hall, Cambridge, 1792, age twenty-four (*AC*, II, 499). Member of Norwich Tusculan Society, 1793-94: argued against evidence of Christianity, and in favor of republicanism ("Minutes," Norfolk R.O. MS N N A S G2).

FOY (b. 1772 or 73; 1784-85)

"From New York," age twelve, "Ward of Brooke Watson's" (1785 "List"). The only Foy shown in 1790 New York census is Martin "Foye," of New York City, age and occupation not given (*Heads of Families at the first Census of the United States taken in the year 1790: New York* [Washington: Government Printing Office, 1908], p. 136). A Sydney Hollis F, from Dorsetshire, entered Warrington Academy 1774 but d. after graduation from Cambridge (Turner, "Historical Account of Students," p. 528). Sir Brook Watson (1735-1807), merchant, was M.P. for the City of London, 1784-93, and a director of the Bank of England (*DNB*).

GELL

"Sons of a Gent of large fortune near Ashburn in Derbyshire"; ages ten and eight; forty guineas each (1785 "List"). Parlor-boarders. Sons of Sir Philip, of Hopton, Derbs. (1723-95), at one time Sheriff of Derbs., and his wife Dorothy (b. Milnes, of Chesterfield, Derbs., 1758-1808) (*FMG*, p. 46):

Philip (1775-1842; 1780?-85)

Afterwards M.P. for Malmesbury, 1807-12, and Penrhyn, 1812-18; D.C.L., Oxford, 1824 (*AO*, p. 516).

*William (1777-1836; 1780?-85)

Lucy Aikin ("Memoir," pp. xxix-xxx) lists WG among pupils for whom ALB wrote *Hymns in Prose*; if true, he must have entered Palgrave aged about three. Afterwards adm. Emmanuel College, Cambridge (B.A. 1798, M.A. 1804 [*AC*, III, 32]). Explored topography of Troy and Pompeii. Published *The Topography of Troy* (1804), *The Geography and Antiquities of Ithaca* (1807), *An Itinerary of Greece* (1810), etc. Chamberlain to Princess (afterwards Queen) Caroline, 1814-20; lived in Italy after 1820 (*DNB*).

GINKEL, Frederick-William, Lord Aughrim (1766-1810; 1776?-)

Parlor-boarder (Aikin, "Memoir," p. xxxi, which spells it "Aghrim"). Son of 5th earl of Athlone; a Dutch family whose ancestor (the first earl) commanded King William's army in Ireland, 1691, and returned to Holland. In summer ?1776 the Barbaulds visited Holland (*Works*, II, 9), perhaps to collect this pupil. He afterwards succeeded his father as 6th earl (Burke, *Dormant, Abeyant, Forfeited, and Extinct Peerages of the British Empire*, new ed. [London: Harrison, 1883], pp. 231-32).

GLANVILLE

John (b. 1778 or 79; 1784-85)

William (b. 1777 or 78; 1784-85)

"From Jamaica"; ages six and seven; "Wards of M[r] Barbauld" (1785 "List"). Probably sons of W. J. Glanville, who was adm. Warrington Academy 1766 and afterwards went to West Indies (Turner, "Historical Account of Students," p. 268). A John, of St. Kitt's, d. 30 July 1798; another John, of Tobago, witnessed a will in 1812 (*Caribbeana*, III [1913-14], pp. 108, 295).

GOOCH, William (1762?-1814)

Son of Henry, mercer, of Gorleston, Suff. At Palgrave four years under Barbaulds. Adm. Caius College, Cambridge, 1787, age twenty-six; B.A. 1792, ordained, 1792; d. at Douglas, Isle of Man (*Gonville and Caius College*, II, 116).

?HARVEY (1784?)

A son of Mrs. Thomas H, of Norwich, perhaps enrolled in Jan. 1784; see Appendix I, items 11-12.

HAYWARD (1782-84)

Named in "weekly chronicle" ("the food for horses and what has a guardian"). Many Hs lived in Suffolk in the sixteenth, nineteenth, and twentieth centuries (Suffolk R. O. Biography Index); one who would have been of suitable age is Henry, d. 1818, aged forty-eight, buried in St. Matthew's Churchyard, Ipswich (BM Add. MS 24886, fol. 59[v]).

HUNTER, Rowland (1773 or 74-after 1836; 1781-85)

"From Liverpool, sent by Mr Johnson Bookseller St Pauls Churchyard"; age eleven (1785 "List"). Grandnephew of Joseph Johnson, of London, publisher and bookseller to Palgrave School. Named in weekly chronicle; for other mentions, see Appendix I, items 4, 8, 13-14, 17. Afterwards heir and successor to JJ, publishing Maria Edgeworth and providing ALB with books as late as 1824 (ALB to RH, 5 March [1824]; Evelyn MS 4, I, 15, British Library; Tyson, *Joseph Johnson*, pp. 185, 212; Marilyn Butler, *Maria Edgeworth: A Literary Biography* [Oxford: Clarendon Press, 1972], pp. 490-91). Subscribed to Enfield's *Sermons* (1798).

ISAACSON (1779-80)

Played Gruel in scene from *The Commissary* on *Tamerlane* evening, 26 May 1780. "Isaacson's heirs" were Palgrave householders, 1812 (map of Palgrave, Suffolk R. O. B150/1/3.12). Or perhaps John (b. 1768), son of Rev. John, of Lydgate, Suffolk, adm. St. John's College, Cambridge, from Bury St. Edmunds School in 1786 and afterwards rector of Little Bradley and Lydgate, Suffolk, 1794-1828 (*AC*, III, 529).

JEFFERY (1779-80)

Recited (with Scott, Jr.) Pope's Third Pastoral on *Tamerlane* evening, 26 May 1780. Possibly from Norwich: a John J (1647-1720) had been an archdeacon there. Possibly J. Jeffery, Esq., M.P., who subscribed to Enfield's *Sermons* (1798).

KERRISON (1775-76?)

Played in *Comus*-Shakespeare masque, ?1776. Probably Thomas Allday (1768-1818), only son of Roger, banker, Grand Master of Norwich Masonic Society, Alderman, Justice of the Peace for Norfolk, Treasurer to Corporation of Guardians for the Poor, and Mayor of Norwich 1778-79 (often mentioned in *Norfolk Chronicle* for those years). Afterwards attended Eton (1782-84), adm. Pembroke College, Cambridge, 1786 (B.A. 1793, M.A. 1796); Alderman (1803) and Mayor (1806) of Norwich (*ECR*, p. 309; *AC*, IV, 31); probably entered his father's bank, Kerrison and Son (Jewson, *Jacobin City*, p. 88).

KING, Isaac (1776-1832; 1782-85)

"[Son of Isaac,] Of High Wycomb [Bucks.]"; age nine (1785 "List"). Also named in weekly chronicle. Afterwards attended Westminster School; adm. Trinity College, Cambridge, 1795; LL.B. 1801, Trinity

Hall. Chaplain to Prince Regent; Perpetual Curate of Aston Clinton with St. Leonards, and of Lee, 1801-17; Vicar of West Wycombe and Rector of Halton, 1805-26 (*AC*, IV, 42). "He was a great promoter of National and Sunday Schools, and of religious societies in connection with the Church of England" (*The List of the Queen's Scholars of St. Peter's College, Westminster, . . . Collected by Joseph Welch*, new ed. [London: G. W. Ginger, 1852], p. 444).

LA FITE (b. 1773 or 74; 1784-85)

"Son of Madame La Fite french Governess to some our princesses"; age eleven (1785 "List"). Frances Burney, when at Court, knew a Mme. de la Fite, who is identified by Austin Dobson as Marie-Élizabeth Bouée, dame de la F (1750-94), French woman of letters, who had been "invited by Queen Charlotte to England as a reader to the Royal Family, perhaps especially the Princess Elizabeth" (*Diary and Letters of Madame d'Arblay* [London: Macmillan and Co., 1904], II, 263n). She is mentioned by Theophilus Lewis Barbauld (RB's father), 15 July 1777, as manager, with her husband, of a periodical (*Bibliothèque des Sciences et des Beaux Arts* [*BLC*]); he speaks as if he knows her (BM Add. MS 48706, fol. 18). The only difficulty with this identification is that the son of Mme. de la F in 1785 seems to have been a little boy (*Diary and Letters*, II, 306, 334). Perhaps she had two sons.

LAWTON (1784-85)

"Son of a Widow Lady near London, under the care of Captn Tipple of Stuston near Palgrave"; age thirteen (1785 "List"). Probably Joseph Thomas, posthumous son of Thomas, of Walthamstow, Essex (near London). Afterwards attended Westminster School; adm. Trinity College, Cambridge, "age 18," 1793; B.A. 1798, M.A. 1805; Rector of Elmswell, Suffolk, 1810-63, and Perpetual Curate of Walsham-le-Willows, 1813-52. Died 1863, "aged 88" (*AC*, IV, 117).

LE BLANC

C. (1781-82)

Played Boy in *Taste*, 27 May 1782.

Another, presumably brother to above (1781-82)

His existence inferred from the fact that the one L is distinguished by his fore initial. Possibly Thomas (1774-1843), of Cavenham, Suffolk, who was adm. Trinity Hall, Cambridge, 1789, age fifteen, from Charterhouse School, who went on to be Fellow (1800-15) and Master

(1815-43) of Trinity, and Vice-Chancellor of the University (1824-25; secured purchase of present Botanical Garden by the University), and who is said to have been "a devotee of port, and a martyr to gout" (*AC*, IV, 128).

LEWIS (1779-80)

Performed "a song by Marloe" on *Tamerlane* evening. Several Ls lived in Suffolk, nineteenth century (Suffolk R. O. Biography Index). This is probably one of three who subscribed to William Enfield's *Sermons* (1798): Israel, of Hampstead; J. W., of Parliament Street, London; and Leyson, of Hampstead, who subscribed for ten copies and whose wife's sister married a Martineau (*FMG*, p. 1108).

LOWNDES (1778?-82)

Played Zachary Fungus in scene from *The Commissary* on *Tamerlane* evening; is distinguishing himself in poetry (ALB to William Taylor, 10 Apr. [1782; Rodgers, *Georgian Chronicle*, p. 206]; ALB's mentioning him to WT implies that he was contemporary with WT); played title role in *Caractacus*, 27 May 1782. Perhaps son of Thomas, who entered Warrington Academy in 1763, from Liverpool (Turner, "Historical Account of Students," p. 264). A Thomas, of Hampstead, subscribed to Enfield's *Sermons* (1798). Or perhaps elder son of Robert, of Chesterfield, Derbs.: Milnes (1764-1800), who matric. at Christ Church College, Oxford, July 1782 (B.A. 1786) and became barrister at Middle Temple, 1794 (*AO*, p. 877). Like the Gells (q.v.), he was descended from the Milneses of Chesterfield, although through a different line of the family (*FMG*, p. 53).

MALBROUGH (b. 1772 or 73; 1784-85)

"Under the care of Crisp Molineux Esqr."; age twelve (1785 "List"). See s.v. ELLIOTT.

MARSH

*Charles (1774?-1835?; 1782?-85)

"[Youngest] son of M^{r} E[dward] Marsh [manufacturer,] Norwich"; age ten (1785 "List"); steward for Palgrave anniversary dinner, 8 Sept. 1794 (*Cambridge Intelligencer*, 23 Aug. 1794). His sister was second wife of Peter Finch Martineau (*FMG*, p. 1108). Entered New College, Hackney, 16 Sept. 1790 (Hackney College "Minutes," Dr. Williams's Library); admitted pensioner, St. John's College, Cambridge, Oct. 1792, did not graduate. Contributor to Norwich *Cabinet*, 1794. Adm.

Lincoln's Inn, 1791; called to bar, 1797; went to India, 1804, as King's Advocate in Admiralty Court and Advocate-General, Madras; M.P. for East Retford, 1812-18 (*AC*, IV, 327). Also subscribed to Enfield's *Sermons* (1798). See Outcomes.

S. (1779?-82)

Played Puff in *Taste*, 27 May 1782. Perhaps Samuel Graeme M (1768-1803), son of Samuel, of London, adm. Pembroke College, Cambridge, 1785 (B.A. 1789, M.A. 1792), and afterwards Vicar of Manewden, Essex (*AC*, IV, 329).

"Wm" (b. 1769 or 70; 1779?-85)

"Son of Mr Robt Marsh Norwich"; age 15 (1785 "List"). Played a Bard in *Caractacus*. Father probably the RM who subscribed to Warrington Academy (WA "Register of Admissions," Manchester College, Oxford) and Enfield's *Sermons*. "Wm" is probably William Heath M (d. 1849?), of Norwich, who matric. at Corpus Christi College, Cambridge, 1787 (B.A. 1791, M.A. 1794) and was afterwards Curate of Swanton-Abbot, Norfolk (1792), Rector of Wramplingham (1800-11) and Eccles (1810-12), and Vicar of Calthorpe (1801-48) (*AC*, IV, 330).

S. or Wm. played Prince Henry in scene from *Henry IV, Part 1* on *Tamerlane* evening; one M is named in weekly chronicle ("a fen").

MARTINEAU

Thomas (1764-1826; 1774?-79?)

Son of David, surgeon; father of Harriett (1802-76) and James (1805-1900). Harriett implies that he was a Palgrave contemporary of William Taylor (*Autobiography*, I, 302), which may make him one of the first eight pupils. His brother, Philip Meadows M (1752-1829), entered Warrington Academy in 1765 (Turner, "Historical Account of Students," p. 265). The Ms were a leading Norwich Huguenot family, related by marriage to the family of Rev. John Taylor, Warrington tutor, whose grand-daughters the Rigby sisters were close friends of ALB. See Outcomes.

John Scott (1779-1845; 1784-85)

"Son of Mr Peter [Finch] Martineau, [banker]"; age six (1785 "List"). Afterwards employed at The Amicable Insurance Office (Martineau, *Notes on the Pedigree*, Family Tree IV; *FMG*, p. 1108).

MASON (1782-84)

Named in weekly chronicle. Perhaps John Ball M, of Beeston, near Nottingham, who subscribed to Enfield's *Sermons* (1798); or perhaps William, of Somersham, Hunts., whose daughter married William

Orton Aikin (son of ALB's nephew George), 1825 (Joseph Foster, *Collectanea Genealogica* [London: privately printed, 1887], I, 44; *FMG*, p. 183).

MEREWETHER (1782-84)
Named in weekly chronicle ("a sheep and nothing else").

MOORE (1782-84)
Named in weekly chronicle. Possibly J. M, of Altringham, Cheshire, or P. M. M, bookseller in Dublin, who subscribed to Enfield's *Sermons* (1798); or possibly Richard, of Kentwell Hall, Long Melford, Suffolk (1769-1826), who matric. at Caius College, Cambridge, in 1786 and of whom little else is known (*AC*, IV, 456).

MOORE, Edward, Lord (1770-1837; 1784-85)
"Son of the [sixth] Earl of Drogheda"; age fifteen; forty guineas (1785 "List," which spells "More"). Lucy Aikin ("Memoir," p. xxxi) identifies him—also spelled "More"—as a parlor-boarder. Afterwards succeeded his father as seventh Earl and second Marquess of Drogheda (*Burke's Peerage*, 1959 ed., p. 708).

MOTT, Thomas Vertue (1761-88; 1777-79)
Son of Thomas Vertue, surgeon, of Burnham Market, Norf. Schools: King's Lynn, six years; Sutton near Woodbridge, one year; Palgrave, two years, under Barbaulds [present at the wedding of Sarah Rigby to Caleb Parry, 1778]. Adm. Caius College, Cambridge, 1779, age eighteen; B.A. 1783, M.A. 1786. ?Author, *The Stranger's Visit, and other Poems*, 1813 (*Gonville and Caius College*, II, 102; I thank Prof. David Weinglass for bringing this pupil to my notice).

MOUNSEY (1779-82)
Played Haly in *Tamerlane* and Mrs. Mechlin in scene from *The Commissary*, 26 May 1780; spoke prologue and played Arviragus in *Caractacus* and Lady Pentweazel in *Taste*, 27 May 1782. Possibly related to the Monsey family of Norfolk, to which Messenger M, physician (1693-1788), belonged (*DNB*).

OLIVE (1781-82)
Played Vellinus in *Caractacus*, 27 May 1782. Possibly James (1775-1822), son of Gregory, of London, merchant, who was adm. Glasgow University, 1790, and afterwards matric. at Emmanuel College,

Cambridge (LL.B. 1803); minister, St. Paul's, Portland Square, Bristol, 1814-22 (*AC*, IV, 588).

PAGE (1782-84)

Named in "weekly chronicle." P a leading family name in Suffolk (Suffolk R. O. Biography Index); Ps also served on Norwich Common Council, 1776 (*Norfolk Chronicle*, 6 Apr. 1776). A Mr. P, surgeon, practicing at Woodbridge in Suffolk, 1785 (Bodleian Library MS Beckford c. 32, fol. 82), was probably a cousin of the Palgrave pupil, for C. R. Aikin reports meeting a Woodbridge cousin of P at Norwich —and also meeting the pupil himself "at our Palgrave meeting this year" (CRA to ALB, 6 Dec. 1792 [MS, Rodgers]).

PAKENHAM, Thomas (1774-1835; 1784-85)

"Son of Lord Longford"; age eleven; paying forty guineas (1785 "List"). "I found my Son very much improved in the year he was with Barbauld to whom he seemed very much attachd" (Lord L to Lord Mulgrave, 5 Sept. 1785 [ts communicated by Dr. P. O'Brien]). Afterwards succeeded to earldom of Longford (*Burke's Peerage*, p. 1403). In 1816 Maria Edgeworth reported to ALB his approaching marriage; ALB remembered him as especially suited "to appreciate and enjoy domestic pleasures" (LeBreton, *Memoir*, pp. 173-74, 179). The Ps, of Co. Westmeath, Ireland, were a military and diplomatic family, several of whose members appear in *DNB*; their representative today still holds the title of Lord Longford.

PARTRIDGE (1779-80)

Read Virgil's Pollio on *Tamerlane* evening. An important Suffolk family: possibly James, of Stratford (1774-1858), or Robert, of Shelley Hall (1773-1854) ([Charles Partridge,] "A History of the P Family," Suffolk R. O.). However, there were also Norfolk Ps: a Robert served as Sheriff (1780) and Mayor (1784) of Norwich (*GM*, 87.i [1817]:282).

PATTESON, William (b. 1777 or 78; 1784-85)

"Son of Mr [Jacob] Patteson [or Pattison] of Witham Essex"; age seven (1785 "List"; *FMG*, p. 15). Henry Crabb Robinson's early friend, whom he identifies as a Palgrave pupil and contributor to the Norwich *Cabinet*. (HCR, "Reminiscences," I, 54-55 [MS, Dr. Williams's Library]). Also a subscriber to Sarah Spence's *Poems*, 1795.

PHIPPS, Hon. Augustus (1775?-80?)

A parlour-boarder, according to Lucy Aikin ("Memoir," p. xxxi). The Ps, barons and earls of Mulgrave, were a leading family in government and the military; several appear in *DNB*. They were related distantly to ALB through her maternal grandmother's connection to the Anneseley family (*DNB*, XLV, 229; LeBreton, *Memoir*, p. 6). The pupil must have been Augustus (1762-1826), fifth son of Constantine, Lord Mulgrave: at Harrow, 1774-75 (*HSR*, p. 113); adm. St. John's College, Cambridge, 1782 (M.A. 1785), and Lincoln's Inn, 1780; Commissioner of Excise, Stamps, Taxes and Inland Revenue, 1792-1822, and Commissioner of Excise for United Kingdom, 1823-26 (*AC*, V, 117).

PRICE (b. 1770 or 71; 1784-85)

"Son of Mr Price London," age fourteen (1785 "List"). A Robert P, only son of James, merchant, of London, was adm. Pembroke College, Cambridge, 1792, "age 18" (B.A. 1796); afterwards Curate of Croydon, Surrey (1802-16), Prebend of Durham (1804), and Vicar of Shoreham, Kent (1816-42) (*AC*, V, 194). Possibly he was the Palgrave pupil, despite the discrepancy in reported ages.

PRIESTLEY, William (1771-before 1835; 1784-85)

"Son of D^r. Priestly," age fourteen (1785 "List," which spells his name thus). ALB's friend Joseph P, chemist and theologian, had three sons, of whom the second, William, was fourteen in 1785. Afterwards accepted French citizenship (1792) and practised as barrister in Paris; joined his parents in Pennsylvania; died in Louisiana as planter (R. J. Hetherington, MS Priestley pedigree, BRP XII.75 × 12, Girton College Library)

PURDIE (1775?-80)

Played Falstaff in scene from *Henry IV, Part 1* on *Tamerlane* evening; played also in *Comus*-Shakespeare masque, ?1776. Possibly (?) John Purdy (1773-1843), hydrographer, son of a Norwich bookseller (*DNB*).

SAY (see also SAYS) (1779-80)

"S junior" played Humphry Gubbins in scene from Steele's *Tender Husband* on *Tamerlane* evening, 26 May 1780; in the same scene, "S senior" (an elder brother, or his father?) played Tipkin. Possibly William (1768-1834), son of William, land-steward, of Lakenham, Norfolk, and afterwards prolific engraver of mezzotints (*DNB*); or possibly Henry, son of William, of Downham Market, Norfolk, "arm.," who matric. at Trinity College, Oxford, 1786, age seventeen (B.A. 1790, M.A. 1795), and was Rector of North Peckenham with Houghton-on-the-Hill, Norfolk, 1794-1855 (*AO*, p. 1259).

*SAYERS, Frank (1763-1817; 1774-78)

One of first eight pupils (Aikin, "Memoir," p. xxv); left early in 1778, "soon after fourteen" (Taylor, *CW*, p. xiii). Played Prospero in *The Tempest*, Caled in *The Siege of Damascus*, the King in *Henry IV*, and Mark Antony in *Julius Caesar* (*ibid.*, p. xi). Son of an insurance broker; came to Palgrave from Yarmouth. Afterwards studied medicine at London and Edinburgh (M.D. 1789), then took up philology and literature: *Dramatic Sketches of the Ancient Northern Mythology* (verse tragedies and poems, 1790), and other works. Admired by Sir Walter Scott and Robert Southey (*DNB*). William Enfield (to whose posthumous *Sermons* S subscribed) described him as "a man of sterling merit. Besides his poetical talents which are very considerable, he possesses a sound judgment and correct taste. . . . He is a man of enlarged views and a liberal spirit" (WE to Ralph Griffiths, 1 Mar. 1793; Bodleian Library MS. Add. c. 89, fol. 70). See also Outcomes.

SAYS (1782-84)

Named in "weekly chronicle." Possibly, however, "Says" is plural for "Say senior" and "Say junior" (q.v.).

SCOTT

John (b. 1770 or 71; 1779?-85)

"Son of M^{r} [John] Scott of Norwich, [merchant]," age fourteen (1785 "List"; *FMG*, p. 1108); his sister was first wife of Peter Finch Martineau (see MARTINEAU, John Scott). Probably the S who played Novice in *Taste*, 27 May 1782, and the S named in "weekly chronicle"; perhaps the S who played Prince of Tanais in *Tamerlane*, 26 May 1780 (unless that S was his father or elder brother, and "Scott junior" [below] was John). JS witnessed the siege of Willemstadt in Holland, 1793 (C. R. Aikin to ALB, March 1793 [MS, Rodgers]).

"Scott junior" (1779-80)

Recited Pope's Third Pastoral on *Tamerlane* evening (and thus may be younger brother to John).

STEELE (1779-80)

Played title role in *Tamerlane*. Assuming that only an older pupil would play a title role, I suppose he was b. ca. 1765. I have not, however, found any Steele or Steel who seems a plausible candidate for the pupil.

*TAYLOR, William, Jun. (1765-1836; 1774-79)

One of first eight pupils (Aikin, "Memoir," p. xxv); his memoir of Sayers (in his edition of S's *Collective Works*) provides some of the best information on the school. Son of William, of Norwich, manufacturer of cloth for export and sponsor of the school (Appendix I, item 1); cousin of Dysons (q.v.; *FMG*, p. 1131). Left Aug. 1779, to learn family trade. Visited Germany 1781-82, met Goethe, became enthusiast (perhaps the first in England) of German language and literature; translated Gottfried Bürger's ballad "Lenore" (1790, publ. 1796); also translated Goethe and Lessing. Later an associate of Southey and Coleridge. Active in Norwich radical circles, 1790s. Wrote prolifically for *Monthly Magazine*, *Annual Review*, and *Monthly Review*; credited with broadening English literary taste by reference to European trends. Noted stylistic innovator (Sir James Mackintosh said that he wrote not English but "Taylorian" [*Memoirs*, I, 443]); author of *English Synonymes Described* (1813) and *Historic Survey of German Poetry* (1828-30). Both he and his father subscribed to Enfield's *Sermons* (1798).

TEMPLETOWN: See UPTON

THOMPSON (1779-82)

Read "a Fable of Dodsley" before *Tamerlane*, 26 May 1780; played a Bard in *Caractacus*. A "Miss Thompson" visited Palgrave, Nov. ?1783 (ALB to Miss Dixon, 18 Nov. [?1783], Hackney Borough Archives MS 3734): his sister? Perhaps he is Culpho T, "near Ipswich," who subscribed to Enfield's *Sermons* (1798).

TROWER (b. between 1771 or 72 and 1774 or 75; 1784-85)

Three brothers, ages ten to thirteen, "sons of Mr Trower of Clapham" (1785 "List"). Perhaps relations of James and Richard, who attended Eton in 1765-71 and 1765-68 respectively, from Jamaica (*ECR*, p. 528).

UPTON, Hon. John Henry (1771-1846) and Hon. Fulke Greville, afterwards Howard (1773-1846; each 1781?-85)

"Sons of Lord Templetown" (Clotworthy U, 1721-85, created Baron T 1776), ages thirteen and twelve; forty guineas each (1785 "List"). Parlour-boarders. One played Lord Duke in *Taste*, 27 May 1782; one is named in "weekly chronicle." Their mother, Elizabeth, was a close friend of Frederica Locke, of Norbury Park, close friend in 1790s of Frances Burney (Burney, *Letters and Journals*, ed. Joyce Hemlow [Oxford: Clarendon Press, 1972-84], III, 241n).

JH succeeded as second Baron T, 1785; attended Eton, 1787-88; was M.P. for Bury St. Edmunds, 1803-12; created viscount, 1806 (*ECR*, p. 512).

FG entered St. Peter's College, Westminster, 1787; elected to Cambridge but instead went to Christ Church College, Oxford, 1791; entered 1st Foot Guards as ensign, 1793; promoted to lieutenant and captain, 1794, and lieutenant-colonel, 1804; served in Holland till 1807; full colonel, 1813, and aide-de-camp to the Duke of York; M.P. for Castle Rising, 1808-32 (*List of the Queen's Scholars of . . . Westminster*, p. 434; *AO*, p. 1459). In 1802 Charles Burney declared him "a charming young man, & an excellent Musician" (Burney, *Letters and Journals*, V, 325n).

WAKKER (1779-80)

So given by Brodribb, p. 735; perhaps really WALKER? Played Sir Harry Gubbins in scene from *The Tender Husband* on *Tamerlane* evening. A Walker attended a meeting of Norwich Tusculan Society, 11 Apr. 1794, voting in favor of republicanism over monarchy (Norfolk R. O. MS N N A S G2), and several Walkers subscribed to Enfield's *Sermons* (1798): Rev. George, of Nottingham; Sayer, M.D., of Charterhouse Square, London; and William, of Leeds.

*WELD, Isaac (1774-1856; 1784-85)

"Son of [Rev.] M[r] [Isaac] Weld of Dublin," age eleven (1785 "List"). Related by marriage to John Taylors of Norwich (*FMG*, p. 1104). Afterwards removed to Norwich. Travelled in North America, 1795-97, meeting George Washington; published *Travels through the States of North America*, 1799, to international acclaim. Afterwards travelled in Ireland, publishing *Illustrations of the Scenery of Killarney* (1807) and *Statistical Survey of the County of Roscommon* (1838) (*DNB*).

WORTH (b. 1771 or 72; 1784-85)

"Son of M[r] W Worth Norwich," age thirteen (1785 "List"). Both W. and W. C. Worth, of Norwich, subscribed to Enfield's *Sermons* (1798).

WRIGHT, Thomas (b. 1768; 1781-85)

"Son of [Francis,] the late Town Clerk of Norwich," age sixteen (1785 "List"). Probably the same W who played Alderman Pentweazel in *Taste*, 27 May 1782, and who is listed in "weekly chronicle." Afterwards matric. at Christ's College, Cambridge, Michaelmas 1785; B.A., 1789 (*AC*, VI, 597).

TABLE OF OUTCOMES

To estimate the influence of Palgrave School on its pupils I have tried to compare their after-careers with those of graduates of Eton College and Westminster School. To do this, I have depended on biographical information given in the *Eton College Register* and *The Westminster School Register from 1764 to 1883*, ed. G. F. Russell Barker and Alan H. Stenning (London: Macmillan, 1892). I chose Eton for its obvious prominence and Westminster because its school register, although inferior to the magisterial *ECR*, is adequate to the purpose; also because some Palgrave boys went on to Eton and Westminster. The *ECR* and *WSR* cover, of course, much longer periods than the eleven years of the Barbaulds' work at Palgrave. To achieve parity between them and Palgrave, I counted only Etonians and Westmonasterians at least half of whose periods of attendance fell in the years 1774-85.

An "outcome" is a career. In assessing Palgrave's influence on its pupils, I am interested in careers which were likely to carry cultural or political weight; because my assessment is comparative, I am also interested in careers entered by graduates of all three schools. Perusal of some hundreds of pages of school and university registers (*AC*, *AO*, *ECR*, *WSR*) has driven home to me the stereotypicality at this period of careers in Church, Army or Navy (Military for short), Law and Government (national or local)—and also the frequency with which graduates of British schools and universities ended up in India or some other imperial venue. (The historian of imperialism would do well to study school registers.) So dominant are these careers that it was apparently difficult for the editors of the published registers to identify graduates who did anything else. The outcomes on which I compare Palgrave with Eton and Westminster, then, are careers in the Church, the Military, Law, Government, and Empire. "Empire" includes, besides colonial administrators, all who engaged in official foreign relations on behalf of Britain. Empire and Government do not overlap—except, of course, when individuals happened to serve (say) as ambassadors (thus being counted under Empire) and also to serve, at some other time, as (say) members of Parliament (thus being counted under Government).

The two remaining categories on the Table, Authors and *DNB*, owe their presence to the fact that Palgrave graduates published, and got listed in the *Dictionary of National Biography*, in what seemed to me rather high proportions. Comparison with Eton and Westminster graduates in the first category appears to confirm my hunch, as does comparison with Eton graduates in the second (I did not collect comparable information on Westmonasterians).

As I imply above, the ratio of outcomes to individual pupils is not necessarily one-to-one: a single pupil may enter a profession (say, Law) in practising which he has occasion to publish books (and thus becomes an author); since I am interested in each of these outcomes, I count this pupil twice. Graduates of Eton and Westminster are treated, of course, in the same way. By this method, I count thirty-nine Palgrave pupils (about one-third of the total number) whose lives yield fifty outcomes; 316 Etonians (of 688 who attended Eton contemporaneously with the Barbaulds' time at Palgrave) who yield 355 outcomes; and 249 Westmonasterians (of 743 ditto) who yield 306 outcomes.

The names of Palgrave pupils appearing in the *DNB* are asterisked in the list of pupils, above. Under the other categories on the Table I count the following pupils:

Church: S. Baddeley, Brett, A. Bulwer, Corbett, Gooch, Isaacson, King, Lawton, S. Marsh, W. Marsh, Olive, Price.

Military: W. E. Bulwer, Bunbury, A. Douglas, Pakenham, F. G. Upton.

Law: G. Coldham, Cotterel, Crabb, Denman, J. Douglas, Enfield, Firth, C. Marsh, Priestley.

Government (national or local): G. Coldham, W. Coldham, Denman, Enfield, P. Gell, Kerrison, C. Marsh, Phipps, F. G. Upton, J. H. Upton.

Empire: Cotterel, T. Douglas, C. Marsh.

Authors: A. Aikin, C. R. Aikin, Crabb, Denman, T. Douglas, W. Gell, C. Marsh, Patteson, Sayers, Taylor, Weld.

TABLE OF OUTCOMES

School	Church	Military	Law	Gov't	Empire	Authors	*DNB*
Palgrave (n=50)	12 (24%)	5 (10%)	9 (18%)	10 (20%)	3 (6%)	11 (22%)	10 (20%)
Westminster (n=306)	99 (32.4%)	61 (19.9%)	39 (12.7%)	73 (23.8%)	24 (7.8%)	10 (3.3%)	—
Eton (n=355)	110 (31%)	85 (23.9%)	33 (9.3%)	93 (26.2%)	22 (6.2%)	12 (3.4%)	35 (9.9%)

APPENDIX IV
TEXTS USED AT PALGRAVE SCHOOL

This list of texts is derived from the school's business correspondence (Appendix I), from playbills for its dramatics evenings in 1780 and other years (described by Brodribb), and from other incidental sources. Unless otherwise identified, bibliographical details come from the *British Library Catalogue of Printed Books* (*BLC*).

ADDISON, Joseph, and Richard Steele

The Spectator. ALB recommends it to her pupil Lydia Rickards, c. 1794 (E. C. Rickards, "Mrs. Barbauld and Her Pupil," *Murray's Magazine*, 10 [1891], 709). It was so commonly used as a school text that it is likely to have been used at Palgrave.

AIKIN, John, MD

Heads of Chemistry (printed by William Eyres of Warrington for the use of Aikin's class at Warrington Academy). On 4 Feb. 1782 Aikin asked Joseph Johnson to send a copy to RB (MS Laing II, 647/11, Edinburgh Univ. Library).

ANACREON

ALB reports that C. R. Aikin could have recited an ode, 20 Jan. 1779 (*Works*, II, 22).

THE ANATOMIST

The Anatomist; or, the Sham Doctor. A Farce. [in one act. Adapted from Edward Ravenscroft's comedy.] London: T. Davies [1762?]. Also a 1763 ed. in three acts (*BLC*). ALB orders, 29 Apr. 1784 (Appendix I, item 17).

BERQUIN, Arnaud

L'Ami des Enfants (see Appendix I, item 8). RB orders five copies each of Vols. 4-6, 4 Nov. 1783 (Appendix I, item 14).

BIBLE

RB orders six copies, 4 Nov. 1783 (Appendix I, item 14).

CASTALIO

RB orders six copies, 30 March 1782 (Appendix I, item 4). Probably *Youth's Scripture Remembrancer: or Select sacred Stories, by way of familiar Dialogues, in Latin and English: with a short Application to each Story. The original by S. Castalio: to which are added some explanatory Remarks. By Dr. Bellamy.* London: J. Robinson, 1743.

CLARKE'S ESOP

Fabulæ Æsopi selectæ . . . With an English Translation . . . by H. Clarke. (Advertised as published "This Day" by W. Strahan, *Public Advertiser*, 28 May 1774.) RB orders six copies, 4 Nov. 1783 (Appendix I, item 14).

COLLINS, William

"Ode to Fear" recited by a Douglas, 26 May 1780.

COMUS-SHAKESPEARE MASQUE

ALB compiled a masque from Milton's *Comus* and Shakespeare, in which Carter, Coldham, Cotterel, Firth, Kerrison, and Purdie performed (Brodribb, p. 735); I conjecturally date it 1776.

DAWSON, John

Lexicon Novi Testamenti alphabeticum, nunc primum plenè editum: omnes voces . . . in sacro textu occurrentes, cum earum resolutione grammatica, completens. In usum Scholarum (*BLC*, which lists only a 7th ed., London: Rivington, 1784). ALB orders two copies, c. 22 Sept. 1783 (Appendix I, item 13).

DODSLEY, Robert

A fable (not identified) recited by Thompson, 26 May 1780.

ENFIELD, William

Exercises in Elocution; selected from various authors . . . intended as a sequel to The Speaker. Warrington, 1780. RB orders six copies, 4 Nov. 1783 (Appendix I, item 14).

Hymns for Public Worship. Warrington, 1772. ALB orders, c. 22 Sept. 1783 (Appendix I, item 13).

"ENGLISH HISTORY IN QUESTION AND ANSWER"

ALB recommends this title to her pupil Lydia Rickards, c. 1794 (Rickards, p. 709); probably, then, she used it at Palgrave. Probably *A New History of England. By Question and Answer. Extracted from the most celebrated English Historians* (advertised in *Public Advertiser*, 27 Dec. 1768); *BLC* lists nearly the same title, *The History of England . . . Historians, particularly M. Rapin de Thoyras*, in a 25th ed. (London, 1811), attributing it to John

Lockman. The *DNB*'s entry on Lockman (1698-1771) does not list this title.

ENTICK, John (1703?-73)

RB orders six copies of an unspecified title, 4 Nov. 1783 (Appendix I, item 14). Entick wrote several school texts; his most popular was *A New Spelling Dictionary* (1764; new ed. 1774 [*Public Advertiser*, 20 June 1774]), but also widely used were *A New Latin and English Dictionary, designed for the Use of Grammar Schools and Private Education* (1770, new ed. 1782) and "Entick's New Spelling Book, intitled, The Child's Best Instructor" (*PA*, 27 July 1774). For political implications of using Entick, see n. 71 below.

FABLES CHOISIES: See La Fontaine.

FARCES

ALB orders, 29 Apr. 1784 (Appendix I, item 17). Probably *A Collection of the most esteemed Farces, and Entertainments performed on the British Stage* (1782); it contains *The Anatomist*, *Florizel and Perdita* (below), and *Taste*.

FLORIZEL AND PERDITA

Florizel and Perdita: or, The Sheep-Shearing. A dramatic Pastoral, in two acts. In prose and verse; altered by Maurice Morgann from Shakespeare's *Winter's Tale*. ALB orders, 29 Apr. 1784 (Appendix I, item 17).

FONTENELLE, Bernard Le Bovier de

Entretiens sur la Pluralité des Mondes (1686). RB orders, 30 Mar. 1782 (Appendix I, item 4).

FOOTE, Samuel

The Commissary (1765, rptd 1779). One scene performed, 26 May 1780. *Taste* (1752; 5th ed., London, 1781). Performed 27 May 1782.

GOLDSMITH, Oliver

History of the Earth and Animated Nature (1774). George Aikin referred to it in studying birds (see Appendix I, item 7).

GREEK GRAMMAR

ALB writes poem in answer to its question, "What do the futures speak of?" (*Poems*, p. 282).

GREEK TESTAMENTS
ALB orders three, c. 22 Sept. 1783 (Appendix I, item 13).

HUGHES, John
The Siege of Damascus (1720; rptd in *Bell's British Theatre*, 1776; in *New English Theatre*, 1776; and singly, 1778). Performed 1776 or 1777, Sayers playing Caled (Taylor, *CW*, p. xi).

LA FONTAINE, Jean de
Fables Choisies, mises en vers par M. de la Fontaine (1668; many rpts in French, most recently 1782). George Aikin reading it, ?Jan. 1783 (Appendix I, item 7).

LE SAGE, Alain René
Gil Blas (1721-24; rptd in French, London, 1777). RB orders in French, 4 Nov. 1783 (Appendix I, item 14).

"MAGAZIN DES ENFANTS"
George Aikin reading it, ?Jan. 1783 (Appendix I, item 7). Not identified.

MARLOWE, Christopher
An unidentified song performed by Lewis, 26 May 1780.

MARTIAL
ALB orders, c. 22 Sept. 1783 (Appendix I, item 13). No ed. specified, but James Elphinston's *Martialis Epigrammata in XII Libros digesta ut Uni* had been published in May or June (*Stationers' Register*, 31 May 1783).

MASON, William
Caractacus. RB orders six copies, 30 Mar. 1782 (Appendix I, item 4); performed 27 May 1782.

MILTON, John
"L'Allegro" recited by C. R. Aikin, 26 May 1780.
Comus presented Sept. 1781 (Brodribb, p. 735).

MOLIERE
Harpagon's "soliloquy" from *L'Avare* recited by Elsden, 26 May 80.

"BAILEY'S OVID"

RB orders three copies, 30 Mar. 1782, and one copy, 4 Nov. 1783 (Appendix I, items 4 and 14). I have not identified this text; the *BLC* shows no such edition.

PERRIN, John (or Jean Baptiste)

Entertaining and Instructive Exercises, with Rules of the French Syntax. 2nd ed., 1773; 4th ed., 1774 (advertisements, *Morning Chronicle*, 6 Jan. 1773, and *Public Advertiser*, 20 June 1774). ALB orders four copies, c. 22 Sept. 1783; RB orders six copies, 4 Nov. 1783 (Appendix I, items 13-14).

Grammar of the French Tongue (1768; 6 eds. by 1789). RB orders six copies, 4 Nov. 1783 (Appendix I, item 14).

PHAEDRUS

Phædri Fabulæ: or Phaedrus's Fables . . . in a method intirely new; . . . by John Stirling, 9th ed. (London: Rivington, 1771). RB orders four copies, 4 Nov. 1783 (Appendix I, item 14).

PLINY

Selecta Quædam ex C. Plinii Secundi Historia Naturali; ad usum Scholarum accomodata, ed. John Aikin (London: Johnson, 1776). "Mr. Barbauld thanks you for your elegant Pliny, which he intends to make a school-book immediately after the vacation" (ALB to John Aikin, ?spring 1776 [*Works*, II, 9]). Aikin recommended Pliny despite its difficulty because it "teach[es] things as well as words" (Lucy Aikin, *Memoir of Dr. John Aikin*, I, 28).

POPE, Alexander

Third Pastoral recited by Scott Jr. and Jeffery, 26 May 1780.

RACINE, Jean

RB orders one copy, 4 Nov. 1783 (Appendix I, item 14)—presumably of R's *Oeuvres*, and probably in French.

"ROMAN HISTORY . . . IN QUESTION AND ANSWER"

ALB recommends this title to her pupil Lydia Rickards, c. 1794 (see above, "ENGLISH HISTORY"); hence I guess that she used it at Palgrave also. Probably *A New Roman History, by Question and Answer* (attributed to John Lockman by *BLC*, which lists a "4th" ed., 1778; but *Public Advertiser*, 27 Dec. 1768, announces the "7th" ed.).

ROWE, Nicholas

Tamerlane (1702; rptd in play collections, 1776 and 1778; also singly, 1776). Performed 26 May 1780.

SHAKESPEARE

Henry IV, Part 1. A scene performed on 26 May 1780; the whole play in preparation, spring 1776 (*Works*, II, 9), Sayers playing the King (Taylor, *CW*, p. xi); ALB orders *Henry IV*, 29 Apr. 1784 (Appendix I, item 17).

Julius Caesar. Sayers played Mark Antony in a performance of it (Taylor, *CW*, p. xi).

A Midsummer Night's Dream. ALB used bits from it to replace the Ceres-Juno masque in *The Tempest*, May 1777 (*Works*, II, 16).

The Tempest. Performed May 1777 (*Works*, II, 16), Sayers playing Prospero (Taylor, *CW*, p. xi).

Unspecified. RB reminds Johnson to send "the volume of Shakespear" he asked for when last in town, 30 Mar. 1782 (Appendix I, item 4).

STEELE, Richard

The Tender Husband (1705; rptd. 1779). One scene performed on 26 May 1780.

THOMSON, James

ALB returns wrong ed., wants Bell's, 3 June 1783; ALB orders *Seasons*, twelve copies, c. 22 Sept. 1783 (Appendix I, items 8 and 13). T's *Poems* composed Vols. 57-58 of Bell's *Poets of Great Britain* (1782).

"TREMBLE'S BOOK"

RB orders two copies of an unspecified title, 30 Mar. 1782 (Appendix I, item 4). I have not identified it.

VIRGIL

"Pollio" (Fourth Pastoral) recited by Partridge, 26 May 1780.

VOLTAIRE

Histoire de Charles XII (1731; a new ed., in French, Dresden, 1761). RB orders one copy as alternative to Fontenelle, 30 Mar. 1782, and three copies, 4 Nov. 1783 (Appendix I, items 4, 14).

WALKINGHAM [WALKINGAME], Francis

The New Tutor's Assistant: being a Compendium of Arithmetic, and a Complete Question-Book, the Rules are versified; a Plan entirely new, and much better adapted to Young Minds than that of Prose (*Stationers' Register*, 6 Feb. 1781; *BLC* shows it as a new ed. of a 1751 work). RB orders four copies, 4 Nov. 1783 (Appendix I, item 14).

YOUNG, Edward

Busiris (1719; rptd in *Bell's British Theatre*, vol. 16, 1776). C. R. Aikin and Bunbury recited a scene from it, Dec. 1779.

A Paraphrase on Part of the Book of Job (1719). Recited by Cholwick and W. Coldham, 26 May 1780.

NOTES

1. *European Magazine*, 54 (Dec. 1808), 478. Opening date from Appendix I, item 1, notes. For Palgrave School after the Barbaulds, see S. W. Rix, "Honest Tom Martin's House," *Notes & Queries*, Ser. 2, Vol. 10 (4 Aug. 1860), 86-87.
2. The story is told cogently by Mitzi Myers in a long series of essays on Maria Edgeworth and her contemporaries. See especially "Impeccable Governesses, Rational Dames, and Moral Mothers: Mary Wollstonecraft and the Female Tradition in Georgian Children's Books," *Children's Literature*, 14, ed. Margaret Higonnet and Barbara Rosen (New Haven: Yale Univ. Press, 1986), 31-59; "Little Girls Lost: Rewriting Romantic Childhood, Righting Gender and Genre," *Teaching Children's Literature: Issues, Pedagogy, Resources*, ed. Glenn Edward Sadler (New York: MLA, 1992), pp. 131-42; and "Of Mice and Mothers: Mrs. Barbauld's 'New Walk' and Gendered Codes in Children's Literature," *Feminine Principles and Women's Experience in American Composition and Rhetoric*, ed. Louise Wetherbee Phelps and Janet Emig (Pittsburgh: Univ. of Pittsburgh Press, 1995), pp. 255-88. An example of the rumor-mills miscalled "histories" is Geoffrey Summerfield, *Fantasy and Reason: Children's Literature in the Eighteenth Century* (Athens: Univ. of Georgia Press, 1985), reviewed as it deserves by Myers, *Children's Literature Association Quarterly*, 12 (1987), 107-10. Another fine exponent of the progressive pedagogy embodied in *Lessons for Children* is Sarah Robbins, "*Lessons for Children* and Teaching Mothers: Mrs. Barbauld's Primer for the Textual Construction of Middle-Class Domestic Pedagogy," *The Lion and the Unicorn*, 17 (1993), 135-51.
3. Lucy Aikin, "Memoir," in *The Works of Anna Letitia Barbauld* (London: Longman, 1825), I, [v]-lxxii (hereafter cited as "Memoir"); C. W. Brodribb, "Mrs. Barbauld's School," *Contemporary Review*, 148 (1935), 731-36 (hereafter cited as Brodribb). On the fate of the Aikin-Barbauld family papers, see *The Poems of Anna Letitia Barbauld*, ed. William McCarthy and Elizabeth Kraft (Athens: Univ. of Georgia Press, 1994), pp. 203-204 (hereafter cited as *Poems*), and [C. W. Brodribb,] "'Life, I Know Not': Mrs. Barbauld: 1743-1825," *TLS*, 19 June 1943, p. 298.
4. Norfolk Record Office MC257/63/3-4, acquired in 1989 among papers of Philip Meadows Taylor. I am grateful to Mr. Paul Rutledge, Archivist, for bringing them to my attention and for permission to reproduce the playbill and publish the list (Appendix III).
5. W. A. C. Stewart, *Progressives and Radicals in English Education, 1750-1970* (New Jersey: Augustus M. Kelley, 1972).

6. RB to Philip Meadows, 10 Nov. 1773 (Norfolk R. O. MC257/59/23; for Meadows, see Appendix I, item 1 note). On eking out ministerial incomes by schoolteaching, see John Seed, "Unitarian Ministers as Schoolmasters, 1780-1850: Some Notes," *Transactions of the Unitarian Historical Society*, 17 (1979-82), 170-76. That the Palgrave School anniversary in 1794 was held at the Scole Inn (*Cambridge Intelligencer*, 23 Aug. 1794) suggests how central a gathering-point Scole was. Actually, the school drew more than twice as many pupils from Norfolk as from Suffolk; however, in 1811 Norwich had nearly three times the population of Ipswich (*Monthly Magazine*, 40 [Jan. 1816], 487), so, proportionally, pupils may have come equally from both counties.
7. The house is described by Rix (n. 1) as a large double-roofed building "with central entrance and thirteen windows in front, looking towards the village church, upon a tongue of greensward, and its northern end adjoining the road to Botesdale and Bury St. Edmund's." In the middle window, above the entrance, was a medallion bearing the arms of Archbishop Sancroft (it can be seen in the daguerrotype). Jerom Murch, who claimed to remember the house, placed it "very near the village church" (*Mrs. Barbauld and her Contemporaries* [London: Longmans, Green & Co., 1877], p. 68n), but an 1812 map of Palgrave shows "Palgrave School" on the southwest edge of the village close to the Bury road, where Rix locates it (Suffolk R. O. B150/1/3.12, first noticed by Betsy Rodgers, *Georgian Chronicle: Mrs. Barbauld and her Family* [London: Methuen, 1958], p. 197). By 1856, when Philip Meadows Taylor, a later Palgrave pupil, revisited it, it was called "Barbauld House" (Taylor to Isaac Weld, 6 May 1856 [Norfolk R. O. MC257/63/1/5]). For Thomas Martin, see the *DNB*. Archbishop Sancroft, a Nonjuror in 1689, died a Jacobite (*DNB*); that his arms adorned a school kept by Protestant Dissenters is ironic. The daguerrotype belongs to Mr. Simon Martyn, who has kindly permitted me to reproduce it here.
8. See William Taylor, *Collective Works of the late Dr. Sayers; to which have been prefixed some Biographic Particulars* (Norwich: Matchett and Stevenson, 1823), p. x (hereafter cited as Taylor, *CW*). "We supped two & twenty in the parlour, the upper house, the lower house in the hall" (ALB to her brother John Aikin, 18 May [1785], thanking him for a gift of a mackerel which "feasted the whole house" [Hornel Library MS 15/21, fol. 29]). 21 occupants are shown in the 1841 Census (the earliest to give complete data), at which time the house held a girls' school: the headmistress, a "governess," fourteen pupils ages six to seventeen, four servants, and one person of unknown occupation (Public Record Office, HO107/1024).

9. RB to Philip Meadows, 10 Nov. 1773 (n. 6).
10. For ALB's accounts see Anna Letitia LeBreton, *Memoir of Mrs. Barbauld* (London: George Bell, 1874), p. 49; for their fate, n. 3 above. For making beds and scolding maids see ALB to John Aikin, [?Feb. 1776], *Works*, II, 7-8; for her overseeing provisions, Appendix I, item 6. Thomas James, ex-master of Rugby, counselling his former pupil Samuel Butler in 1798 about setting up as schoolmaster, warns that "Mrs. Butler . . . must oversee all provisions," as Mrs. James did (Samuel Butler, *The Life and Letters of Dr. Samuel Butler* [London: Jonathan Cape, 1924], I, 37; hereafter cited as Butler).
11. James advised keeping "at least" two maids and a boy-servant for a school of four pupils; at Rugby, a much larger school, however, he had only eight servants (Butler, I, 17, 37). No more than five of the 1841 occupants of "Barbauld House" were servants (n. 8 above). For ushers, see ALB, *Works*, II, 6, and Appendix I, item 7.
12. Butler, I, 17-18. James reckoned the cost of beds for four "decent private pupils" at £100.
13. See RB to Philip Meadows (n. 6). For Palgrave's actual charges, see Appendix I, items 1 and 11, and Appendix II. Starting number of pupils from Aikin, "Memoir," p. xxv.
14. See Appendix I, items 11-12 for terms on which ALB took her pupils. That she took them after the success of *Lessons* and wrote *Hymns in Prose* for them is claimed by Aikin, "Memoir," pp. xxix-xxx; and see p. xxxi also. On parlor boarders at Daventry, see John Williams, *Memoirs of the late Reverend Thomas Belsham* (London, 1833), p. 255; on Christ's Hospital, Leigh Hunt, *Autobiography* (London: Smith, Elder, 1891), p. 61 (hereafter cited as Hunt). The Palgrave pupils listed in 1785 as paying forty guineas are Browne, T. Douglas (for some reason he paid only thirty-five), Dumbleton, the Gells, Moore, Pakenham, and the Uptons; only Dumbleton and William Gell were under age ten.
15. *The Eton College Register, 1753-1790*, ed. Richard Arthur Austen-Leigh (Eton: Spottiswoode, Ballantyne & Co., 1921), p. xxvii (hereafter cited as *ECR*). A bill for A. Bulwer (Appendix II, item 1) shows a charge for a single room even though he was not a parlor boarder, and thus further complicates the housing question.
16. The 1785 calculation omits C. R. Aikin, who, as the Barbaulds' adopted son, would not have paid. Accounts kept in 1786-91 by Meadows Taylor, a school sponsor, suggest something of its value to the sponsors. In 1786-89 he held a one-fifth interest in "Palgrave School House & Land," worth £130 (£160 in 1789); in 1790-91 he appears to have held a one-eighth interest, worth £100 (Norfolk R. O. MC257/69/3). These figures

imply a property valued at £650 in 1786-88 and £800 in 1789-90. The same documents indicate that Palgrave School paid £6 rent per month in those years.

17. See, for starting number of pupils, n. 13; for 1775, Anne Thackeray Ritchie, *A Book of Sybils* (London: Smith, Elder, & Co., 1883), p. 21 (date inferred); for 1776, ALB to John Aikin, *Works*, II, 8 (date inferred); the 1780 number is my count based on Appendix III. Some pupils in these counts were ALB's "little parlour pupils," not counted by RB as "properly belonging to the school" (Appendix I, item 12); hence his top number of thirty could be exceeded.
18. *The Harrow School Register, 1571-1800*, ed. W. T. J. Gun (London: Longmans, Green & Co., 1934), p. xv (hereafter cited as *HSR*).
19. The age distribution of the forty-one boys present in March 1785 agrees closely with this distribution based on the known ages of all boys, whenever present. Three were age six; six were age seven to nine; seventeen were age ten to twelve; thirteen were age thirteen to fifteen; and two were age sixteen.
20. I allude to Johnson's wisecrack—numbingly repeated by writers on ALB and her school—about ALB's having dwindled from precocious genius to mistress of "an infant boarding-school" (Boswell, *Life of Johnson*, ed. G. B. Hill, rev. L. F. Powell, 6 vols. [Oxford: Clarendon Press, 1934], II, 408). The remark shows that Johnson knew the school only from *Lessons for Children*. Boswell got this anecdote from Charles Burney and entered it—wrongly—under 1775, before *Lessons* existed. Johnson cannot have said it before the middle of 1778.
21. Of twenty-one more whose origin is conjectural, nine may have come from Suffolk, five from Norfolk, and one from Essex. So many came from Norfolk that they split into rival factions, the "Norwichians" and the "Yarmouthians"—i.e., from Great Yarmouth (Taylor, *CW*, p. xii).
22. Andrew "Aitkin," "merchant burgess of Kirkcudbright," gave a bond to Lady Lucy Douglas, Countess of Nithsdale, in 1691 for livestock bought from her husband (*Kirkcudbright Sheriff Court Deeds, 1676-1700* [Edinburgh: Oliver & Boyd, 1953], entry 1980). Andrew's relation to ALB's grandfather, who left Kirkcudbright for London in the 1690s, is not clear, but probably he was a cousin. For Lady Lucy Douglas and her brother, see William Fraser, *The Douglas Book* (Edinburgh, 1885), II, 424, 425; her brother, the 1st earl of Selkirk, was great-great grandfather of Basil William Douglas (perhaps one of the first eight pupils) and his brothers. For Maxwell Garthshore, see *DNB*.
23. LeBreton, *Memoir*, p. 49; the text actually says "county," probably an error for "court" (cf. Aikin: RB's connections were "among courtiers and

supporters of the establishment" ["Memoir," p. xxxiii]). For Theophilus Lewis Barbauld (below), see [ALB,] "Memoir of the Rev. R. Barbauld," *Monthly Repository*, 3 (1808), 706, and his petition to Lord Bute, 1763 (MS North. b. 4, fols. 195-200, Bodleian Library), which presumably got him his later appointment as preacher at the French Chapel, St. James's.

24. For Rigby and Martineau, see "V. F." (William Turner), "Historical Account of Students educated in the Warrington Academy," *Monthly Repository*, 9 (1814), 205, 265; for the Rigby sisters, see *Poems*, p. 231.

25. Anna Maxwell, *Hampstead* (London: James Clarke & Co., nd), pp. 142-43. I count as sons of professional families the Aikins, H. Browne, Denman, Enfield, LaFite, T. Martineau, Mott, Priestley, and Weld, and as sons of commercial families Brett, the Coldhams, Dunham, the Dysons, Firth, Gooch, Hunter, Kerrison, C. Marsh, J. S. Martineau, Sayers, Scott, and Taylor. Sons (or at least close relatives) of titled families were Aghrim, Corbett, the Douglases, Moore, Pakenham, Phipps, and the Uptons, and I count as sons (or close relatives) of landed gentry the Baddeleys, the Bulwers, Bunbury, Elliott (a planter), the Gells, the Glanvilles (planters), and Malbrough (a planter). See Appendix III.

26. "A Catalogue of Books parcel of the Library of Philip Gell of Hopton . . . ," Gell Papers, Derbyshire R. O. D258, Box 32/27. Dorothy Gell's library included ALB's *Poems* (5th ed., 1776), *Miscellaneous Pieces in Prose* (2nd ed., 1775), and *Devotional Pieces* (1775), and John Aikin's *Essays on Song-Writing* (2nd ed., 1774, now in my possession). She also made literary attempts herself: Among the Gell papers (Box 36/10b) is a fiction in her hand, titled "A Book Without a Title. Written by Nobody—Of whom nothing is expected," and her copy of Aikin's *Essays* is overwritten with copies of poems in her hand. For enthusiastic responses by young women to ALB's poems, see *Poems*, p. xxxii.

27. *An Address to the Opposers of the Repeal of the Corporation and Test Acts* (1790, in *Works*, II, 365). In 1803, reviewing a book on education, ALB suggested another reason: "It is because in many towns the established clergy want either the sobriety, the literature, or, at least, the necessity for exertion, requisite to make them able to fulfil, and willing to undertake the task" (review of *An Essay on Education* by the Rev. William Barrow, *Annual Review*, 1 [1803], 578; I attribute this piece to ALB on grounds of sentiments and style). In a letter to her brother, 19 July [1788], revolving options for Arthur Aikin's further education, she even considers one of the "great schools"—"if they took the proper care of their morals & improvement" (Hornel Library, MS 15/21, fol. 27).

28. *Memoirs of the Life of Gilbert Wakefield, B.A.* (London: Joseph Johnson, 1804), I, 214-15. Likewise, ALB explains why Rochemont Barbauld was

sent by his father to Warrington: it was "a seminary . . . frequented by many of the establishment, whose parents wished to avoid both the expense and the hazard to morals of a university education" ("Memoir of the Rev. R. Barbauld," p. 707).

29. See S. J. Curtis, *History of Education in Great Britain*, 7th ed. (London: University Tutorial Press, 1967), pp. 127-29; Curtis mentions Eton and Christ's Hospital in particular, and remarks that many rich parents "dreaded sending their sons to the larger boarding-schools." See also H. C. Barnard, *A History of English Education from 1760*, 2nd ed. (London: Univ. of London Press, 1961), pp. 18-19; Barnard cites an article in *Punch*, "as late as 1854," about an "aggravated assault at Rugby by two ruffians of 16 or 17 on a little fellow, apparently between 10 and 11." The sadistic Bowyer is remembered by Hunt, pp. 60-64. At Eton there were two rebellions within fifteen years (*ECR*, p. xxiii); in the 1830s conditions there were severely scrutinized in the press (Brian Simon, *Studies in the History of Education, 1780-1870* [London: Lawrence & Wishart, 1960], pp. 101-102).

30. *Works*, II, 311-12. The line attributed to the parent, "Tell me . . . when you have broken any thing, and I will not be angry with you," closely follows two speeches in Edgeworth's "The Little Dog Trusty" (*The Parent's Assistant*, 2nd ed. [London: Joseph Johnson, 1796]): "let us run and tell my mother. You know she bid us always tell her directly when we broke any thing" (Frank to his brother Robert, p. 4); and "tell me the truth—I shall not be angry with you, child" (their mother to Robert, p. 6). ALB regards skeptically this tale's optimism about the harmony of interests between parents and children. Her views also, of course, smack of Rousseau on lying (see *Emile*, trans. Allan Bloom [New York: Basic Books, 1979], pp. 101-103); yet ALB could have learned everything Rousseau teaches about lying from her father, the Rev. John Aikin. When master of Kibworth School, he scrupulously refrained from questioning his son John about the doings of John's schoolmates—with the result that John "more than once" stood guard while his schoolmates stripped Mr. Aikin's apple trees (Lucy Aikin, *Memoir of John Aikin, M.D.* [London: Baldwin, Cradock and Joy, 1823], I, 3).

31. See her "Memoir of the Rev. Mr. Barbauld": "Children . . . were always attracted to him and he delighted to entertain them with lively narratives suited to their age, in which he had great invention" (p. 708).

32. On punishments at Warrington, see the Academy's published "Report" for 1769 (Manchester College Library, Oxford); jobations are lengthy rebukes. ALB expresses distaste for flogging in her review of Barrow (n. 27 above). Fines almost replaced corporal punishment at William

Gilpin's Cheam School as early as 1752 (Stewart, *Progressives and Radicals*, pp. 4-5); hence they would not have been a very radical innovation at Palgrave.

33. On Williams's juvenile court, see Stewart, p. 25; Gilpin had used student juries at Cheam also. John Aikin's juvenile trial is "Trial of a Complaint made against sundry Persons for breaking the Windows of DOROTHY CAREFUL, Widow, and Dealer in Gingerbread" (*Evenings at Home*, 3 [London: Joseph Johnson, 1793], 64-82); it carries a footnote, "This was meant as a sequel of that very pleasing and ingenious little work, entitled *Juvenile Trials*, in which a court of justice is supposed to be instituted in a boarding-school, composed of the scholars themselves, for the purpose of trying offences committed at school." I have not traced the title to which Aikin refers. I thank Betty Rizzo for bringing this subject to my attention. There is, however, no reference to juvenile trials in any document of Palgrave School known to me.
34. Butler, I, 8. The system was long credited to Dr. Butler of Shrewsbury, but Samuel Butler traces it to Rugby under Dr. James, who served from 1778 to 1794 (*ibid.*).
35. J. Fitch "on the Discipline of Schools," *Monthly Magazine*, 47 (1819), 21.
36. *A Legacy for Young Ladies*, ed. Lucy Aikin (London: Longman *et al.*, 1826), pp. 70-72 (hereafter cited as *Legacy*). Undated, but evidently written for Palgrave, and probably about one of the Gells, who came from Derbyshire.
37. I do not know whether the rules were drawn up in consultation with the pupils, as at David Williams's school (Stewart, p. 25), or presented to them as accomplished facts. At the experimental Hazelwood School the boys themselves drew up the rules—but that was in 1819 (Stewart, p. 59). It should not be assumed that pupils always wanted this power; Williams's boys tried to shift responsibility for discipline back to him (Stewart, p. 25).
38. John Milton, *Of Education*, in *The Prose of John Milton*, ed. J. Max Patrick (Garden City: Doubleday Anchor, 1967), p. 233; RB, *The Duty of Promoting the Welfare of the Rising Generation: . . . a Sermon preached at St. Thomas's, Jan. 2, 1792. For the Benefit of the Charity-School, in Gravel-Lane, Southwark* (London: H. Goldney, 1792), p. 13; Joseph Priestley, "An Essay on a Course of Liberal Education for Civil and Active Life" (1765), in *Theological and Miscellaneous Works of Joseph Priestley*, ed. J. T. Rutt (1831-32, rpt. New York: Kraus, 1972), XXIV, 11, 10. On educational tradition in English Protestant Dissent, see J. W. Ashley Smith, *The Birth of Modern Education: The Contribution of the*

Dissenting Academies, 1660-1800 (London: Independent Press, 1954; hereafter cited as Smith, *Birth*).

39. RB lists the subjects he will teach in his letter to Philip Meadows (n. 6 above). John Aikin's asking Joseph Johnson to send RB a copy of his *Heads of Chemistry* (Appendix IV) may mean that Palgrave boys studied chemistry; other evidence of science teaching there is noticed below. On science in the Dissenting academies, see H. McLachlan, *English Education under the Test Acts: Being the History of the Nonconformist Academies, 1662-1820* (Manchester: Manchester Univ. Press, 1931—hereafter cited as *EETA*); McLachlan notices experimental apparatus owned by academies. Trap-ball and swimming at Palgrave come from Taylor, *CW*, p. xiii (Sayers excelled at both); in *Lessons for Children*, (London: Joseph Johnson, 1794), II, 62-63, Charles is promised a garden if he is good. Gardens were progressive: Gilpin's pupils at Cheam grew them (Stewart, p. 6), as do Rousseau's Emile (*Emile*, p. 98) and Maria Edgeworth's Harry and Lucy (*Early Lessons*, new ed. [London: R. Hunter, 1815], II, 236-38).

40. Jennings's curriculum is known from a description by Doddridge, on which are based the accounts in *EETA*, pp. 135-38, and Smith, *Birth*, pp. 112-19: it included geometry and algebra, geography, French, civil history, oratory, and several sciences as well as Hebrew, Latin and Greek. Doddridge's offering at Northampton Academy in the 1730s and 40s was almost the same (*EETA*, p. 147). For the Rev. John Aikin's Warrington pedagogy, see "V. F." (William Turner), "Historical Account of the Warrington Academy," *Monthly Repository*, 8 (1813), 166 (encouragement of inquiry) and 169 (English literature).

41. Smith, *Birth*, claims Priestley as an innovator in history teaching (pp. 153ff, 243). But the Priestleyan emphases and de-emphases listed in my text were typical of Dissenting teaching. Doddridge, before Priestley, conceived of education as preparation for one's life calling (Smith, *Birth*, pp. 133-34); Jennings, before Priestley, demoted Latin (ibid., p. 119) and promoted French—even though he could not pronounce French (*EETA*, p. 135). The Rev. Philip Holland, a Dissenter who kept a school at Bolton in Lancashire, described his aims in 1758: "it is my endeavour to adapt [my instruction] . . . to the particular employments the young gentlemen are design'd for." If for "one of the learned professions," he grounds them thoroughly in Latin and Greek and gives them "a taste" for English authors, "a general knowledge" of history and geography, and some mathematics. But with "those who are intended for the counting-house [i.e., trade]," he reduces classics to "a by-business" and emphasizes instead English authors, geography, history,

and maths (Francis Nicholson, *Memorials of the Family of Nicholson of . . . Liverpool and Manchester*, ed. Ernest Axon [privately printed, 1928], pp. 69-70). These are the same options that RB offered.

42. James's lesson plans for the fifth and sixth forms are Butler, I, 25-29; extracts of his plans for lower forms are pp. 33-36. Aside from their heavy Classicism, his plans are strikingly repetitive: the same *kind* of activity takes place in every hour of the day, every day of the week. In various ways James acknowledges that this is boring for the master. What must it have been for the pupils?

43. Lamb's bromide against modern education in the form of the "Barbauld Crew" (delivered in a letter to S. T. Coleridge [*The Letters of Charles and Mary Anne Lamb*, ed. Edwin W. Marrs, Jr. (Ithaca: Cornell Univ. Press, 1975-78), II, 81-82]) has been fetishized by the folklorists of children's history, but in its time it made no sense. The real "blights and blasts" of infant happiness were masters like Bowyer at Christ's Hospital and innumerable petty tyrants and timeservers at village schools around England, who perpetuated an antique curriculum from sheer lack of imagination. See Simon, *Studies* (n. 29 above), pp. 97-98, for the curriculum at Louth School in the 1810s: a sixteenth-century scholar would, he says, have been at home in it. The *Westminster Review* in 1825 —forty years after the Barbaulds left Palgrave—attacked this hoary curriculum in language reminiscent of Milton's two hundred years earlier (ibid., pp. 99-100).

44. "On the Classics," in *Legacy*, pp. 73, 78. (I suspect that this is a lecture from Palgrave; it refers to Sir William Jones's translations of Oriental texts, which appeared in the early 1770s, as if they are recent [p. 84].) For Taylor's comments see J. W. Robberds, *Memoir of the Life and Writings of the late William Taylor of Norwich* (London: Murray, 1843), I, 7; for Denman, Appendix III; for C. R. Aikin and Anacreon, Appendix IV.

45. *Annual Review*, I, 577 (n. 27 above). For ALB's view that learning languages per se is an error, see her *Female Speaker* (London: Johnson & Co., 1811), p. iv. She speaks there about women, but would have said it of men as well. Her brother, John Aikin, agreed with her: Teaching boys what they need to know as men "will scarcely include Latin verses and the study of Greek dialects" (Aikin, *Memoir of Dr. John Aikin*, I, 59; see also Appendix IV below, s.v. Pliny). ALB expressed the same opinion with regard to French: reading, not speaking it is what matters ("On Female Studies," *Legacy*, pp. 48-49). Utilitarianism also marked the language teaching of William Gilpin at Cheam and of David Williams (Stewart, pp. 6, 28).

46. ALB mentions writing geography lectures (*Works*, II, 8); LeBreton speaks of "charming lectures on History and Geography" (*Memoir of Mrs. Barbauld*, p. 49), and Brodribb mentions lectures on Roman history (p. 733). A fragment of what looks like a lecture on Islamic history is Appendix I, item 3. The poems on flora, fauna and humans are *Poems*, Poems 75 and 161-64. An example of verses written to be recited at exams is Poem 74, discussed below; see also *Poems*, p. 206. "On the Uses of History" (*Legacy*, pp. 117-64) is addressed to "My dear Lydia" (Lydia Rickards, a pupil in the 1790s). In composing her own teaching texts ALB was not unusual: her grandather Jennings composed some of his, and so did James at Rugby.
47. *EETA*, pp. 136, 158; Smith, *Birth*, p. 137.
48. Her citing editions of Abulfeda (Appendix I, item 3) seems to follow in the spirit of John Jennings, who cited the best authorities on every subject (*EETA*, p. 136).
49. *Lessons for Children*, 4 (London: Johnson, 1792), 31-51. The technical name for this way of teaching geography is "the concentric system" (Stewart, p. 28).
50. I take the phrase "interrogative system" from *Monthly Magazine*, 41 (1816), 224-25, an article in praise of up-to-date schooling signed "Common Sense." Question and answer—in dialogue between Mamma and Charles—is one of the main methods of *Lessons for Children*. The use of verse to help children remember information was considered a novelty: See Appendix IV, WALKINGHAM.
51. In *Annual Review*, I, 577, ALB lists "all those novelties which so many ingenious people have deemed improvements in the mode of instruction," but which Barrow, whose book she is reviewing, condemns; one is "the . . . system of *playing children into learning*."
52. "Aunt Julia's Recollections," William Smith Papers, MS Add. 7621 (Box 1), Cambridge Univ. Library. Julia Smith was the youngest daughter of William Smith, MP, of Parndon, a Barbauld friend. On rote learning, see Simon, *Studies*, p. 98.
53. *Lessons for Children*, 3 (London: Johnson, 1803), 32-38 (the caterpillar lesson); the experiment on the cat first appears in the 1808 ed. (II, 7) but would have been familiar at Warrington in the 1760s. The members of the Lunar Society (who included Joseph Priestley) encouraged science experiments by their children (Simon, *Studies*, pp. 50-56).
54. ". . . The striking feature of the arguments used by tutors in favour of the study of Natural Science is the rarity with which the utility of the subject was mentioned, and the constancy with which its use as leading

to an appreciation of the presence of order and design in the universe was emphasized" (Smith, *Birth*, p. 248).

55. For the impact on her of Priestley's sermon, see *Poems*, Poem 2 and its notes. Her essay, "Thoughts on the Devotional Taste, on Sects, and on Establishments" is the preface to *Devotional Pieces* (1775).
56. *Hymns in Prose for Children* (London: Joseph Johnson, 1781), pp. 65, 70-71.
57. For ALB's "wise book," see Aikin, "Memoir," p. vi; for Priestley, *Theological and Miscellaneous Works*, I, i, p. 5n.; for Wakefield, [Anna Letitia LeBreton,] *Memories of Seventy Years* (London: Griffith & Farran, 1883), p. 179.
58. *Matins and Vespers*, 2nd ed. (London: printed for the author, 1824), sig. A2. Eight-year-old Harriet Martineau was deeply affected by *Hymns*: "parts of them . . . I dearly loved: but other parts made me shiver with awe" (*Autobiography* [London: Smith, Elder & Co., 1877], I, 34). One "S. A. A.," writing in 1869, says *Hymns* so moved her in childhood that when she first visited London she "made a pilgrimage" to Stoke Newington to see where ALB lived ("Notable North Londoners. No. VI. Anna Laetitia Barbauld," *The North Londoner*, 13 March 1869, p. 83).
59. [LeBreton,] *Memories*, p. 43. For a sample of ALB's print hand, see reproduction of "Animals, and their Countries." (I am grateful to Conant Brodribb for permission to reproduce it.)
60. For ALB's reading prowess see n. 57. Mitzi Myers notes that learning to read at two was not altogether "atypical," to judge "from memoirs and the assumptions of many late-eighteenth-century reading texts" ("Of Mice and Mothers," p. 39). C. R. Aikin turned two in August 1777; in January 1778, ALB responded to a test printing of *Lessons*, Vol. I (*Works*, II, 19).
61. I don't claim that the infant Denman knew he was quoting *Macbeth*; the quotation is for the amusement of older hearers. But in learning the lines he was learning a melange of literary styles, from folk tale to high drama. Another noteworthy irony: Leigh Hunt, like Charles Lamb, decried the children's books of his youth for tediousness and moralizing, and professed to see in Percy's *Reliques of Ancient English Poetry* portents of "a nobler age" to come (Hunt, p. 46). In her lines for Denman, ALB takes the Babes in the Wood from Percy's *Reliques*; once again, Palgrave pedagogy satisfies the demands of Romantics under whose influence historians of children's literature routinely denigrate ALB.
62. Poems perhaps written for the "chronicle" are Poems 67 ("Petition of a Schoolboy to his Father"), 77 (a logogriph, not easy to solve), 78 ("Written on a Marble"), 147, 148 and 158 (all riddles). I know of no

predecessors to ALB's chronicle; a newspaper was produced at Daventry Academy, 1785-96 (*EETA*, pp. 164-65), and the first school journal at Eton began in 1786 (*ECR*, p. xxxiii). ALB's contributions to her brother's *Evenings* are named in Aikin, "Memoir," p. xxxvi n.

63. Hunt, p. 71. Dissenting academies emphasized English composition: at Warrington it made part of the weekly exercises in the first and second years (*EETA*, p. 211); at Jennings's academy it included such sophisticated practices as "translations" of one writer's style into that of another (Smith, *Birth*, p. 114); at Northampton, Doddridge laid as much stress on it as Priestley was later to do (ibid., p. 138). However, it was not neglected at Rugby: James required "at least one English exercise a week" (Butler, I, 36).
64. Robberds, *Memoir of . . . William Taylor*, I, 8; Taylor, *CW*, p. xii.
65. On oratory at the academies see Smith, *Birth*, pp. 237-39; again, Jennings and Doddridge stand out (p. 137). James, at Rugby, did assign Latin declamations (Butler, I, 30), but probably for the Latin, not the oratory. At Warrington, by contrast, the tutors and students gathered every Saturday for oral compositions, speeches, and readings of scenes in plays (*EETA*, p. 216).
66. A preference for the vernacular, of course, marked Protestantism from its beginning, and was always associated with the (at least, theoretical) aim of textual democracy. Smith regards "the study of English Language, inextricably bound up with English Elocution" and English literature, as the great curricular innovation of the Dissenting academies (*Birth*, p. 237).
67. In ?spring 1776 ALB made costumes for a performance of *I Henry IV* (*Works*, II, 9); the *Caractacus* playbill denies admittance "behind the Curtain." Whether plays were actually performed or just read aloud varied from one academy to another. Jennings's pupils performed Rowe's *Tamerlane* (Smith, *Birth*, pp. 112, 240), but Doddridge seems to have disapproved of plays (ibid., pp. 138, 240). At Warrington, plays—or at least, scenes from plays—were read, but performance appears to have been frowned upon; ALB's ode "To Wisdom" is supposed to have been written in protest against the tutors' banning a play (see *Poems*, p. 253).
68. Brodribb, pp. 734-35. For the date of this event, see n. 92.
69. "Preface of the Editor," *Theatre of Education. Translated from the French of the Countess de Genlis* (London: T. Cadell and P. Elmsly, 1781), I, 4.
70. In *Lessons for Children*, the hare is IV, 84-88, the robin-torturing boy II, 73-77. "The River and the Brook" is *Legacy*, pp. 209-10. The two playlets are *Evenings at Home*, I, 32-42 and 102-105. Lucy Aikin considered all of *Evenings* a political text: "the morality which they

inculcate is not that of children merely, but of men and of citizens; . . . it engages the youthful feelings in the cause of truth, of freedom, and of virtue" (*Memoir of Dr. John Aikin*, I, 159).

71. The Barbaulds' choice of other people's texts could be political also, as I suspect it was with a text by John Entick. Entick, a schoolmaster and miscellaneous writer, joined the political opposition to Lord Bute in 1762 and was a celebrated victim of the government's use of general warrants for search and seizure (*DNB*). Which of his popular school texts the Barbaulds ordered for Palgrave I do not know; perhaps it was his *New Latin and English Dictionary* (see Appendix IV). The latter was scorned by Thomas James: "Entick is not fit for even second form boys; look out *correptus* or any participle, and you will perceive that there is nothing to lead the child to *corripior*. Then the several meanings of words are not differently distinguished by figures . . . but are all huddled together" (Butler, I, 39). I suspect that James's finical Latinity registers "aristocratic" contempt for Entick's "popular" utilitarianism. Again, the author of two other Palgrave texts, John (Jean Baptiste) Perrin, joined the agitation for Irish independence in the 1780s and 90s (*DNB*, s.v. Perrin, Louis).

72. Rodgers, *Georgian Chronicle* (n. 7 above), pp. 206-207, dated by reference to Taylor's movements in Germany. *Caractacus* is quoted from the 1796 ed. (London: George Cawthorn), pp. 37-38 (later citations in text); for ALB's similar language in "Corsica," ll. 126-30, see *Poems*, p. 24. In February 1782 Parliament passed Gen. Conway's motion to petition the King to cease offensive operations in America; the motion begged Lord North's ministry to admit that its American policy had failed (John Aikin, *Annals of the Reign of King George the Third* [London: Longman, 1816], I, 302-303; J. Steven Watson, *The Reign of George III, 1760-1815* [Oxford: Clarendon Press, 1960], p. 218).

73. On the oppositionist ideology of "patriots" see John Brewer, *Party Ideology and Popular Politics at the Accession of George III* (Cambridge: Cambridge Univ. Press, 1976), pp. 249-57 and 201-207 (their views of the government's American policy).

74. Indeed, the analogy had been mooted in the press in 1777. The *London Evening Post*, 8 and 18 Nov. and 2 Dec., printed letters to Lord George Germain, then secretary of state for the American colonies, questioning his pursuit of a military solution: "I will venture to foretell, that your scheme of subjugating America, will end in the independence of the Colonies, [and] the almost final ruin of this country. . . ." The letters are signed "Caractacus" and "Caradoc" (a variant of the same name). The *LEP* was a vigorously anti-ministerial paper.

75. During a debate on American policy, also in 1777, following the disaster at Saratoga, C. J. Fox "asked what we had got by trying the force of arms?" (*Edinburgh Advertiser*, 21-25 Nov. 1777).

76. Linda Colley, *Britons: Forging the Nation, 1707-1837* (New Haven: Yale Univ. Press, 1992), pp. 168-69; Appendix I, item 7. "Written on a Marble" is *Poems*, Poem 78. "Things by their Right Names" was published in *Evenings at Home*, (London: Johnson, 1792), I, 150-52, just before the opening of war with France, and could therefore have been written against the current war fever; but Charles's appearance in it suggests an earlier date, perhaps c. 1780.

77. "Aunt Julia's Recollections" (n. 52 above); Robberds, *Memoir of . . . William Taylor*, I, 8.

78. For Taylor's contributions to the magazines, Robberds, I, 155ff; ALB's letters to Taylor are Robberds, II, 378-79 and I, 92-93 (which also notice *Lenore* and Scott). Taylor's spoof is "Hudibras Modernized," *Iris* (Norwich), 19 Nov.-10 Dec. 1803 (attributed to him by Robberds, II, 3-4); ALB expostulated in a letter "To the Editor of the *Iris*" (i.e., Taylor), 24 Dec. 1803, signed "A Lover of Truth" (attributed by Robberds, *ibid.*), and Taylor retracted in the same issue.

79. Madeleine R. Grumet, *Bitter Milk: Women and Teaching* (Amherst: Univ. of Massachusetts Press, 1988), p. 55. RB's surviving MS diary (Rodgers) covers only the first three weeks of June 1802.

80. The four to eight Palgrave pupils who were sons of West Indian planters and may have followed their fathers' occupation could thereby be counted among contributors to empire, but I have excluded them because I have no numbers for their counterparts at Eton and Westminster.

81. The superior verbal proficiency—or at least proclivities—of Palgrave pupils is further evidenced by the greater proportion of them who went into the Law (a highly verbal profession, after all): almost twice as many entered Law from Palgrave as from Eton.

82. Sir Joseph Arnould, *Life of Thomas, First Lord Denman* (Boston: Estes & Lauriat, 1874), I, 6; Lucy Aikin, *Memoirs, Miscellanies and Letters*, ed. Philip Hemery LeBreton (London: Longman *et al.*, 1864), p. 274; Anna LeBreton recalled Denman's often saying "that if he had any eloquence it was owing to his readings with Mrs. Barbauld" (*Memories of Seventy Years*, p. 148). For Thomas Martineau, see James Drummond and C. B. Upton, *The Life and Letters of James Martineau* (London: James Nisbet, 1902), I, 8.

83. On Sayers's Toryism, see *Memoirs of the Life of . . . Sir James Mackintosh*, 2nd ed. (London: Moxon, 1836), I, 443; on Taylor, Robberds,

I, 67ff. Taylor's commendation of a republican school is quoted in Seed, "Unitarian Ministers as Headmasters," p. 173.

84. See Arnould, I, 6. Denman's letters to Coke are Derbyshire Record Office MS D1881, Box L1 (published by permission). With his letter of 28 Sept. 1814, cf. D. Reid, president of a local radical committee in Scotland: "the real and only source of the calamity may be fairly ascribed to the enormous load of taxes with which the country is so unmercifully burdened" as a result of the "bloody, expensive, unjust and unnecessary" war with France (P. Berresford Ellis and Seumas Mac A'Ghobhainn, *The Scottish Insurrection of 1820* [London: Victor Gollancz, 1970], p. 103). After 1830, it must be noted, Denman as Attorney-General was obliged to prosecute people whom formerly he would have defended.

85. Linda Colley skillfully appreciates the ambiguities of the 1832 Reform Act, concluding that, with all its limitations, it made Great Britain for a time "one of the most democratic nations in Europe" (*Britons*, pp. 345-49).

86. Gell's correspondence with and about Queen Caroline is Derbyshire R.O. MS 3287. On Spence, see *The Feminist Companion to Literature in English*, ed. Virginia Blain *et al.* (New Haven: Yale Univ. Press, 1990), p. 1013; her subscribers' list also includes Sir George Baker (probably the father of Baker the pupil) and ALB herself. The list is a virtual who's who of national as well as East Anglian liberalism.

87. Douglas is not in the *DNB*. Information about him comes, in order, from Christina Bewley, *Muir of Hunterhill* (Oxford: Oxford Univ. Press, 1981), p. 29; *The Papers of Thomas Jefferson*, ed. Julian Boyd (Princeton: Princeton Univ. Press, 1958), XV, 280; ALB, *Works*, II, 81-82; National Library of Scotland MSS 1036, fols. 113-14 (Douglas on building a canal, 1788), and 19,969, fol. 89 (W. Mure to Thomas Telford, 7 Feb. 1833, on D's enthusiasm for roads and bridges); Edinburgh Univ. Library MSS Dc.6.111, fol. 8 (obituary of D, "extracted from the *Dumfries Journal* of 18th Nov. 1794"), and Laing III.352/1 (D to Dr. Walker, 21 Oct. 1791, on Sierra Leone settlement); Public Record Office TS 11/962, fol. 66 (his membership in the SCI); Ellis and Mac A'Ghobhainn, *Scottish Insurrection*, pp. 56-57 (Friends of the People: they print a letter from D to Charles Grey declaring that "Scotland has long groaned under the chains of England and knows that its connection there has been the cause of its greatest misfortunes"); *Selections from the Papers of the London Corresponding Society, 1792-1799*, ed. Mary Thale (Cambridge: Cambridge Univ. Press, 1983), p. 8; Henry Roscoe, *The Life of William Roscoe* (Boston: Russell, 1833), I, 83 (on D at the British Convention);

[LeBreton], *Memories of Seventy Years*, p. 66 (ALB's comment from an undated letter, c. 1790).

88. Marsh is in the *DNB*. Additional information from: *Reasons for National Penitence* (London: Robinson, 1794), Beinecke Library Mhc8.1794.M35 (flyleaf attribution to "my old fellow-student at New Coll[ege]. Hackey—C. Marsh"); Norfolk R. O. MS N N A S G2, 27 Sept., 11 Oct. 1793, 3 and 17 Jan., 11 Apr. 1794; *The Parliamentary Debates from the Year 1803 to the present Time* (London: Longman *et al.*, 1813), XXVI, 1018-51. Unfortunately, the motion to authorize missionaries was carried (col. 1082).
89. Grumet, *Bitter Milk*, pp. 155-56.
90. ALB speaks in 1779 of having a full house ten months in the year (*Works*, II, 22). In a letter to John Aikin (15 Apr. [1780], Hornel Library MS 15/21, fol. 28) she says the Palgrave vacation ends on "June" 1, but this must be a slip for "July." A problem in the calendar as I've described it is that the two terms are unequal in length, and I do not know how to reconcile this anomaly with the information I have.
91. A school kept by the Rev. Stephen Freeman for "young gentlemen" charged, in 1796, twenty-five guineas per year (*Cambridge Intelligencer*, 2 Jan. 1796). Another took boys under seven for ten guineas, ages seven to twelve for twelve guineas, and ages twelve to fourteen for fourteen guineas (*ibid.*, 9 July 1796; these fees are probably for the term, not the year). A boarding school in Surrey in 1774 charged twenty guineas for pupils under nine, twenty-five for ages nine to twelve, and thirty for twelve and up (*Ipswich Journal*, 28 May 1774). William Enfield announced in 1783 a plan to take private pupils at fifty guineas (Warrington Academy Papers, Unitarian College Archive Cupboard A, John Rylands Library). For other fees, 1779-1812, see Seed (n. 6), p. 172: twenty-five guineas is a typical fee.
92. In only two years during the Barbaulds' time at Palgrave did 26 May fall on a Friday: 1775 and 1780. On 19 Jan. 1780 ALB told her brother that they were thinking of doing *Tamerlane* that spring (Hornel Library MS 15/21, fol. 31).

ACKNOWLEDGMENTS

In addition to persons acknowledged elsewhere in the text, I am grateful to Mitzi Myers and Rosanne G. Potter for helping me to think about issues involved in this research; to Carren O. Kaston for archival assistance; to John Goodchild, Archivist, Wakefield Library, for advice on genealogy; to Lady

Rodgers for kind permission to quote from manuscripts in her possession; to the Hon. Thomas Pakenham and the Marquis of Normanby for permission to quote from Lord Longford's letter to Lord Mulgrave (Appendix III); and to the libraries and archives named in Appendix I for permission to print materials in their holdings.

JOHNSON'S BOOKMAN

ISOBEL GRUNDY

Studies in Bibliography. Edited by David L. Vander Meulen, Volume Forty-Eight, Bibliographical Society of the University of Virginia, 1995.

All but two of the articles in this volume were gathered as essays in honor of J. D. Fleeman, "on the completion of his life's principal endeavour, a comprehensive bibliography of Samuel Johnson." Most of them treat Johnson or his contemporaries. David Fleeman followed the project with interest, read several of the essays, and "guided the choice of the photograph that appears as frontispiece." He died less than six weeks after that photograph was taken. It shows him sitting before the open *Dictionary*, wearing the inevitable bow tie, with a faint smile, and the inward look of sickness or of age.

This therefore is a complicated occasion, mingling joy and sorrow in the manner of Johnson's "real state of sublunary nature." How grievous, how shocking, and above all how sad that David Fleeman should be dead, when he was born only in 1932, abounded in intellectual vigor, and was still pursuing research in the month before his death. How right and pleasing, however, that he should be honored in this fine volume.

David taught me bibliography in 1966-67. It is surprising to remember that this was the first year of his teaching career at Oxford, as well as the first time that he and Michael Turner (following in the footsteps of Herbert Davis) had taught the use of a hand printing press to those graduate students who chose to learn it. (He later examined my D.Phil. thesis. The gimlet quality of his blue eyes, already well known to me, was never more in evidence than at the moment when he elicited the fact that where I had written *London* I still did not realize that I ought to have written *The Vanity of Human Wishes*.) He was able to impart intensity of interest to lectures dealing in ems and formes, printers' devils and printers' pie. In no time at all we were all harboring dreams of acquiring our own printing presses at some time in the future, and in the present were devoting our evenings to pulling copies for the Bodleian from a collection of old but recently acquired blocks. Though I am still nothing more than a dabbler in bibliography, I never work out the designation for an unnumbered preliminary page or hold a book up to the light to look for chain lines without silent gratitude to David Fleeman.

This volume has done him proud. David Fairer's memoir rightly hails him as "the leading British Johnson scholar" and as "someone who has worked with absolute devotion in his cause." It relates the early stages of his travel over Johnson's mind. By working on Johnson's proof-changes he discovered behind the reported dictator a "scrupulous craftsman and conscientious refiner of language." By working on later editors' punctuation of the texts, he discovered "the revelation of lifting away those thick overlays [fifty additional commas in the first chapter of *Rasselas* alone] from Johnson's sentences for ever."

This memoir records his scholarly achievements, and itemizes much of the punishing workload which he shouldered while these achievements were accomplished: for his college he was Librarian and later Vicegerent, for the university Proctor, for the scholarly community Editor of *Notes & Queries* and General Editor of the Oxford Bibliographical Society. He held office in the Johnson Societies of London and of Lichfield, the Johnson Club, and as a Governor of Johnson's House. In addition he carried out B.A. examining responsibilities, and faculty responsibilities, and the endless hours of tutoring. It is sobering to reflect that in today's results-oriented academy, someone showing such early promise as a researcher is likely to be offered at least some slight alleviation of these burdens, some slightly improved allocation of time if not leisure for their own work, than ever came David Fleeman's way. Sobering, too, is the reflection that his achievements were registered in the time-devouring practice of bibliography, in which no-one can ever estimate how long the job will take, against a background of tasks that have an unfortunate tendency to nurture habits of haste and flap.

Fairer comments appreciatively on the generosity with which Fleeman functioned not only as a mine of information and understanding, but also a model of collegiality in communicating these to fellow Johnsonians. These qualities emerge even more clearly in one of the tributes brought out by the occasion of his death. Daisuke Nagashima, founder and secretary of the Samuel Johnson Club of Japan, has printed a booklet which he understands as "returning the gifts of my benefactor": David Fleeman's correspondence with him from 1982. (The slim folio volume also includes other letters, three newspaper obituaries, a moving little tribute by Hitoshi Suwabe, and Fleeman's own "Johnson at Oxford and in the Cotswolds in the 1730s?" This last is a model of speculative or inconclusive biographical writing, of how to present possibilities without any irritable reaching after truth and certainty.)

Most impressive are David Fleeman's letters. They begin when Professor Nagashima was a tyro Johnsonian, asking questions to which the easy answers were supplied with exemplary patience and courtesy. They continue into matters of more difficulty: questions, suggestions and interpretations fly back and forth. It is, it must be confessed, heartening to find that Homer could nod, and that when caught nodding he was so disarmingly frank about it. Fleeman had no fear of using the word "error" or the word "unaccountable" of his own work: like Johnson, he knew just how unattainable for fallible humanity is the condition of bibliographical freedom from error. His letters often apologize for being delayed by various kinds of academic business, but they are the richer for their maturing process, for long pondering the issues involved. The correspondence is indeed a monument.

Fleeman was a great representative of the new school of editing. Fairer lists among his strengths "his keen awareness that a literary text is a flux upon which any printed edition is bound to impose an artificial fixity," his "healthy awareness of textual pragmatics," his suspicion "of tidy impositions of consistency upon a text that may reflect its author's uncertainties or oversights." In Fleeman's own words, a text "is not a single entity but is rather a process. Its witnesses are merely markers on the continuum of that process." To Nagashima he writes, "You see that what looks like a debate between two practical procedures, quickly turns into a metaphysical disagreement about the nature of literary texts and their relation to readers." In late years his work on the afterlife of Johnson's texts in the nineteenth century drew him into recognition of "la sociologie du texte," and the light cast on cultural attitudes by the style in which a period handles its texts.

In a field where progress has usually been measured by the abandonment of one dogma for another, David Fleeman did not shun controversy. Nagashima prints the letters from a debate with Donald Greene in 1989 which echoed Fleeman's disagreement with the editors of the Yale Edition of Johnson. Fairer alludes to this controversy, but is silent about the one with George Watson which caused Fleeman's magisterial Johnson entry in the *New Cambridge Bibliography of English Literature* to appear unsigned—the only major article to be so unattributed. Whereas Fleeman's anti-Yale defense or apology is fully presented in his *Review of English Studies* piece on the Yale *Rambler* volumes (August 1971), his anti-*CBEL* defense or apology lies buried in letters in the *TLS*.

James McLaverty's essay, "Pope in the Private and Public Spheres: Annotations in the Second Earl of Oxford's Volume of Folio Poems, 1731-1736," stands first in the *SB* volume after Fairer's memoir. And well it may: McLaverty has incurred a debt of gratitude from the eighteenth-century scholarly world of almost staggering proportions. Having already converted David Foxon's 1975-76 Lyell Lectures into *Pope and the Eighteenth-Century Book Trade* (1990) he has now taken on the completion of Fleeman's monumental Johnson bibliography: an intellectual biography, as he saw it. (This work had reached proof before Fleeman's death, but to say this is not to say it was finished.) To perform one such labor of personal and scholarly love shows praiseworthy generosity; to perform two is beyond praise. It is thanks to James McLaverty that we possess or shall possess the final works of these two great bibliographers.

This issue of *SB* will be a source-book worth having. Besides Fairer's "The Publications of J. D. Fleeman," it offers a number of valuable texts. First comes "On the Character and Duty of an Academick" by Johnson himself, which was first published in *Hospitality* . . ., 1793, by John Moir, who had followed Johnson as tenant of 8, Bolt Court, but never reprinted since. Was Johnson an "academick"? The *Dictionary* interprets this word as meaning a student, and all words of this root as associated closely with universities. He may, nevertheless, have written this character with his mind on self-admonition if not self-portraiture. This addition to his *oeuvre* is a kind of expansion and theorization of Chaucer's thumb-nail sketch of the Clerke of Oxenford: "gladly wolde he lerne, and gladly teche." Its rediscovery is perfectly timed, for it could stand as a picture of the career of either Johnson or Fleeman.

Also gathered here are Lord Oxford's annotations to poems by and about Pope (Bodley M 3.19 Art); four unpublished poems by John Hoadly; the substantive additions and revisions, probably made by Johnson, in the "Grammar" at the head of his *Dictionary*, dating from between the earlier and the fourth editions; William Warburton's letters to David Dalrymple, Lord Hailes; and Walter Scott's notes to Boswell's *Journal of a Tour to the Hebrides*. Useful descriptions of books are given here as well: of Crousaz's *An Examination of Mr Pope's Essay on Man*, translated by Elizabeth Carter, and of Richard Hurd's editions of Horace.

These various richly detailed expositions of texts and of issues in the transmission of texts have posed a multiple challenge to the expertise of the University Press of Virginia. In the main they have responded splendidly, though here and there minor errors can be spotted. (Some may not be the fault of the printers: "the Lifeguards at Windsor" appear bafflingly in the text when Fairer mentions Fleeman's work as teacher—of writing not riding—to Household Cavalry recruits. Lifeguards may do for patrolling pools or beaches, but the various branches of the Guards, so admired by Boswell, are more tenacious of their capital letter.) On one page, 69, an especially difficult layout—main text made up of bibliographical description and notes, with further footnotes of a more usual kind subimposed on that—has proved too much, and blocks of type are misplaced.

Individual essays are richly informative and provocative. The state of sublunary nature is brought vividly before us. Eighteenth-century individuals scurry about their day-to-day book business: writing, publishing, annotating, subscribing, contriving to send their letters gratis, putting their knowledge on display or carefully concealing their sources. McLaverty's piece on Lord Oxford's annotations catches a privileged reader in the process of puzzling out responses to Pope, alert for personal references though sometimes slow to perceive the full extent of character-assassination of those personally known to him. It throws out more than one startling suggestion for further research: for instance, the idea that Pope himself might have had a hand in publication of *Verses Address'd to the Imitator of Horace*.

O M Brack in "Samuel Johnson and the Translations of Jean Pierre de Crousaz's *Examen* and *Commentaire*" sets out to identify the various components of the controversy surrounding these works, to sort out the chronological and textual relationship of the various translations, and to define Johnson's role. It has already been established, though only belatedly, that Elizabeth Carter translated the *Examen* and he the *Commentary*. Brack adds, persuasively if not conclusively, arguments that Johnson (in addition to translating the *Commentary*) had a close advisory relationship to Carter's translation of the *Examen*. Feminist scholars can hardly delight in this account of Carter's extreme timorousness, her readiness to turn to male colleagues for advice and authorization; but it carries conviction.

Keith Maslen identifies unrecorded writings by the Rev. John Hoadly, son of the famous bishop, a man haunted by unfulfilled yearning to "break through the prudery of my profession and . . . produce a play upon the Stage." Donald D. Eddy tracks Richard Hurd's editions of Horace through the Bowyer ledgers, both as first step in a bibliography of Hurd and in order to demonstrate the uses of printing ledgers. Hugh Amory patiently disentangles facts behind the raw data of the subscription list to Fielding's *Miscellanies*, 1743. He not only cracks the anonymity of several names whose possession is shared among several individuals, but also puzzles over patterns of behavior: the way some subscribers never bother to collect their copies (like today's harassed supporters of charities whose newsletters they have no intention of reading); the failure of any subscribers' copies of Fielding's *Miscellanies* to come down to us.

James E. Tierney investigates the abuse of the privilege of postal franking. Everybody who was anybody, it seems, had access to the free postage which was supposed to be used for parliamentary business only. Privileged persons endorsed sheets wholesale for the use of their friends. John Baskerville in Birmingham suggests that Robert Dodsley, so conveniently situated for access to MPs, might like to send him half a dozen franks. Indeed, anyone writing to Dodsley and requiring him, in the absence of a frank, to pay for his own mail, felt called on to apologize. When the Postmaster-General, alarmed at the loss of revenue, attempted to secure reforms, and when Edward Cave, supervisor of franks, began checking the contents of letters in transit by the use of a candle, the result was not that any of this free-loading was curtailed, but that the Post Office had its knuckles rapped and MPs' privileges were ratified.

The last four essays in the volume ride rapidly away in various directions from Samuel Johnson and his circle. Pamela Dalziel unravels Thomas Hardy's characteristically devious and concealed involvement with competing dramatized versions of *Far From the Madding Crowd*. B. J. McMullin discusses "signing by the page," the practice where "page $8r may in certain circumstances be signed '$15'": signed, that is, with the number of the page within the gathering. This kind of signing seems to have been used to guide the placing of a cancellans or new page designed to replace a cancellandum or rejected page. But whether it was meant to convey information exclusively to members of the trade, or how, if it was meant to convey something to book-owners, they might be in a position to act on its instructions, remain questions still awaiting answers.

These articles nicely round out the tribute to Fleeman. It is fitting that his interest in Hardy (memorably reflected in at least one supervisory relationship) should be represented in this honorific volume. More peculiarly, Hardy's deliberate covering of his spoor, his "profound ambivalence," exemplifies this volume's concern with things that cannot be certainly known or confidently interpreted. Similarly, McMullin's essay serves as a reminder not only of Fleeman's command of bibliography in its most uncompromisingly technical aspects, but also, again, of the contingency and imperfection of historical knowledge, of the use of bibliography as a probe into the unknown.

This volume also boasts G. Thomas Tanselle's J. Ben Lieberman Memorial Lecture at the University of Virginia, which then became the inaugural lecture at the Toronto Centre for the Book. This is a brilliant piece and, again, apposite. Tanselle opens with G. M. Trevelyan's plangent remark about the "other men and women, as actual as we are to-day," who once walked our own familiar spot of ground, but who are now vanished. (Trevelyan is clearly important for Tanselle, since he occurs again later in a footnote; for a student of literature other echoes inhabit the garden, and Trevelyan's words are almost drowned by other, more distant voices: "vanished as if it had never been"; "hu seo þrag gewat.") Tanselle moves quickly from what we sense of the past to what went on in the print-shop: to human artifacts as objects which both reflect and stimulate thought, to printed ephemera, and thence with a lurch into the gulf of incomprehension that separates printing historians from analytical bibliographers. He cites Fleeman, but cites him as example of the bibliographic crime of grasping at certainty which ought to remain elusive: specifically, of supposing that Bowyer's ledgers must be accurate, and therefore susceptible of being used to bear a weight of succeeding evidence.

This sounds, once again, the *leitmotiv* of the volume: "the limited nature of all bodies of evidence, however broadly conceived." It also exemplifies the bibliographer's typical method of proceeding by demolishing the arguments of some earlier bibliographer, and exemplifies it by using the man this volume honors as its whipping-boy. This is ironical, certainly, but not on the whole to be regretted. Fleeman dogmatizes, and is contradicted. He would, I think, have enjoyed this.

The book's last essay, by Ralph Hanna III, is, unfortunately but at least without parallel, badly written. (The sentence opening its second paragraph uses words that imply logical sequence where no logical connection exists: "Although 'The Papelard Priest' is certainly Smith's major find in Additional 45896, he fails to investigate either of the other two poems in any great detail.") It is, however, thoughtful and closely in tune with companion essays on very different topics. Variants among MSS make Mr Hanna's thoughts turn to the detail of road-maps on one hand and to "broader literary and cultural problems" on the other. Not, perhaps, stylistically, but in substance, every contribution in this volume speaks the same language.

Most centrally for Fleeman, the volume demonstrates that his tradition of work on Johnson as meticulous refiner lives on. Gwin J. Kolb and Robert DeMaria, Jr, use a welter of tiny detail to establish exactly what Johnson changed in the Preface, Grammar, and History in his unabridged *Dictionary*. Anne McDermott writes from the electronface of the Johnson Dictionary Project, which aims to produce the critical edition which traditional means have failed to encompass, and for which the first-edition and fourth-edition texts have now been readied for publication on CD-ROM. McDermott chooses to investigate *Dictionary* use of the *Memoirs of Martinus Scriblerus*: a text which "seems to run directly counter" to Johnson's "stated principles of selection," and which he always mentions scornfully or dismissively, in spite of his high opinion of Dr Arbuthnot and his frequent use of the *Dunciad* and *Gulliver's Travels*.

This summary of Johnson's responses to *Scriblerus* sounds uncannily like a summary of his responses to Swift. Working on his use of frivolous sources (and the *Dictionary* frequently catches Swift committing the sin of wit as well as exhibiting serious and specialized knowledge of many topics) can be counted on to be *fun*. McDermott has taken the lovely passage in which Martin learns through play—"*Marbles* taught him *Percussion* and the *Laws* of *Motion*," etc.—and has hunted its words, both childish and scientific, through the *Dictionary*'s pages. This has provided her with interesting findings of at least two kinds. Some serve to confirm a general thesis: Johnson apparently liked the passage, since he uses it under *birdcage, percussion, vice* (an instrument, not a moral infringement) and *whirligig*. Others are troublingly inexplicable. Johnson twice takes the Scriblerian "*Axis* and *Peritrochia*" and alters it to "axis in peritrochio." Why? McDermott does not know.

While footnote material has steadily accreted around Johnson's writings and around Boswell's writings about Johnson, a few categories of accretion have been sliced away and forgotten. Walter Scott's commentary on Boswell's *Journal of a Tour to the Hebrides* has great merit as local history. It is founded on close personal knowledge, selected and arranged with the novelist's eye to human interest. Scott touches his reader's capacity for nostalgia when he notes that the coaching inn where Johnson first stayed in Edinburgh has been downgraded to use by carriers, and has then lost even that function. But no more than Johnson himself will he permit nostalgia to rule. His final word on the inn is, "It was a base hovel."

Ann Bowden and William B. Todd show Johnson and Scott as occupying relative positions as commentators on the Highlands that may come as a salutary surprise. Scott is the more down to earth, Johnson the more romantic. When Johnson expresses regret that arms have been allowed to rust in a manner unworthy of martial forebears, Scott finds him guilty of forgetfulness or insensitivity over the current legal position: carrying arms for a Highlander was matter for transportation at a second offense, and a gathering of twelve men and a bagpipe was classed as a rebellion. Scott brings a later witness to attest that conditions in an English jail would be preferable to those in huts where Johnson was entertained on Inchkenneth. Yet he has an eye for romantic color as well as historic harshness: landing on Skye in 1829 he asked his party what they had in their minds, and "all answered separately" that their minds were full of Johnson's Latin ode on Skye addressed to Hester Thrale. Scott also defends the delicacy of cold sheep's head against Johnson's disdain: defends it "*totis veribus*," with words borrowed from Johnson himself and with a fine sense of the mock-heroic.

Proper bibliographers approach such minutiae with gusto. This number of *SB* abounds in reference to valuable scholarly endeavor by Johnson and his contemporaries; and it exhibits working models (as it were) of such endeavor in our own day. Here and there reference is made to publishing activity in an inferior tradition. Warburton writes more in sorrow than in anger of "a miserable & enormous heap of stuff, called *Biographia Brit*. for the composition of which, the undertaking Booksellers called from the way side the lame & the blind &c." We all know of modern reference books just like that. Warburton's further bad news is that "it suited the People"; his good news is that it had some decent contributions "by means of Mr Cambel a man of sense & industry who has written much for the Booksellers." That too has a modern ring. Mere publishers' projects, compiled by the sorriest hacks, often sell well; but the feeblest publisher's project is likely to incorporate somewhere the work of people of "sense and industry."

Warburton, who died just too soon to participate in the multiple controversies surrounding *The Lives of the Poets*, would have had something to say about the topic of Thomas F. Bonnell's contribution. This is the relation between *Bell's Poets* and the edition for which Johnson wrote his lives. The stage of the relation under scrutiny here is not the lead-in to Johnson's contribution, when the conger of London booksellers were spurred by Bell to enlist Johnson and to set ten printshops at work, but the aftermath of Johnson's lives. It is a familiar story how Johnson's delays caused his lives to appear in four volumes separate from the volumes of poetry, becoming in effect appendices rather than prefaces. It is less well known how blatantly and persistently Bell, hastening to provide equivalent lives for his own collection, levied unethical contribution from the work of his competitor. One of Johnson's first-issued lives, that of Denham, was only five months old when it was stripped out and recombined with material from "*Biographia Brit*" (material written by the lame and the blind, no doubt) for a new appearance in Bell's edition. Indeed, while the biographical section of Bell's Denham weaves both sources together, the critical section is *almost* pure Johnson.

This blatant plagiarism brought threats of legal action, and Bell found it prudent to withdraw. Withdrawal meant first cancelling his pirated life from copies of Denham's poems, then hastily compiling, printing, and stitching in a new one. In the new version the narrative and ideas lifted from Johnson remain in substance unchanged, but they are rewritten to conceal—more or less—their indebtedness. Interestingly enough, one of the commonest modes of concealment is elaboration and verbosity. Johnson writes "learned some attention to money"; his plagiarist writes "acquired a greater degree of economical prudence." Johnson writes "law and poetry"; his plagiarist writes "the study of the law and the cultivation of his poetical talents."

This larceny in "Denham" passed unnoticed, so Bell went on to carry out the same kind of depredations on Johnson's later lives, usually, Bonnell says, "in lazy paraphrases which, far from disguising their origin, preserved much of Johnson's syntax and word choice." While in the halls of literature pundits debated the question of a permanent critical placing for Johnson's lives, below-stairs as it were they began their afterlife as raw material for the constant recycling and fresh serving which still makes up the greater proportion of literary productive processes.

Fleeman's *alma mater* was St Andrew's, whose motto, *Ever to be best*, strikes a more pugnacious note than those of Oxford or Harvard. It fits well with Johnson's commitment to competitive striving, as well as with David's hope to deserve the scholar's epitaph, "He got it right." That epitaph belongs unquestionably to him. Yet at the same time "Fleeman's Johnson is not a man of monumental certainties, but a mind at work, facing practical difficulties and encountering challenges at every stage." To get it right is not something that can be done once and for all, but a continuing process. No sooner is the great work in the bookshops than further "enormous heaps of stuff" require to be ordered and interpreted. The extraordinary gifts brought to their scholarly labors by a Fleeman or a Johnson are only a glorified version of what Warburton celebrated in John Campbell: "sense and industry." May present and future scholars continue to bring their gifts to bear as fruitfully as those honored in this volume and those who worked on it!

REVIEWS

MARSHALL WAINGROW, ed. *James Boswell's "Life of Johnson": An Edition of the Original Manuscript, in Four Volumes (Vol. 1: 1709-1765).* Edinburgh, New Haven, and London: Edinburgh Univ. Press and Yale Univ. Press, 1994. Pp. xlii + 518.

This volume represents a major achievement, of the first importance in eighteenth-century scholarship. Its appearance marks a crucial stage in the continuing investigation of Boswell's biography of Johnson and of his creative process more specifically. With the preparation of the first volume (covering the years 1709-1765) of *James Boswell's "Life of Johnson": An Edition of the Original Manuscript*, Marshall Waingrow has added his second major contribution to these studies, after his invaluable *Correspondence and Other Papers of James Boswell relating to the Making of the "Life of Johnson,"* which appeared in 1969. The edition of the manuscript will eventually run to four volumes, with Bruce Redford assuming the editorship after volume 1.

The present volume is one of the editions of the Boswell papers at Yale, known collectively as *The Yale Editions of the Private Papers of James Boswell*. The publications of the Boswellian materials (chiefly manuscripts, journals, letters, and other papers) found earlier this century in Malahide Castle in Ireland and Fettercairn House in Scotland, along with others, principally from Auchinleck, the Boswell estate, have proceeded in two forms: the "trade" edition, a selection of material, chiefly from the journals, without extensive annotation or careful reproduction of details of the manuscripts, intended for a wide reading public; and the "research" edition, intended to serve scholars demanding high standards of textual scholarship and scholarly annotation. Chiefly for financial reasons, the edition proceeded with the "trade" (or "reading") volumes of Boswell's journal, completing their publication in 1989 with *Boswell, The Great Biographer, 1789-1795*, edited by Marlies Danziger. Of the more substantial "research" volumes, five volumes of Boswell's correspondence have appeared. Most recently, vol. I (1766-1767) of *The General Correspondence of James Boswell, 1766-1769*, edited by Richard C. Cole, was published in 1993. The three-volume *Catalogue of the Papers of James Boswell at Yale University*, by Marion S. Pottle, Claude Colleer Abbott, and Frederick A. Pottle, was published in 1993. As many as twenty-five volumes remain to be published, including the much-needed "research" editions of the complete journals.

The case of the manuscript of the *Life* is of special importance, needless to say. What discoveries does this volume reveal? What conclusions does it make possible for the first time? What material does it provide not previously known? What mysteries and controversies does it finally clear up? In fact, what strikes one upon an examination of the edition are not the presence of spectacular revelations, but rather the emergence of fuller contexts in which to study the complex procedures of Boswell's work, especially in relation to his own life, preoccupations, and practices. No doubt, with time, these contexts will divulge new secrets and enable new conclusions. But the process for the scholar and critic will be an appropriately painstaking one.

It is astonishing to consider the layers and collections of texts now available which constitute stages and parts of the composition of the *Life*. These materials, most of which were found in the discoveries mentioned above, are well summarized by Marshall Waingrow in the Introduction to the current volume. They include: Boswell's journals, in which he kept memoranda of his meetings with, and discussions concerning, Johnson; his letters to, about, and from Johnson; diverse Johnsoniana recovered in the aforementioned finds: "chronological lists of publications, anecdotes, memoranda, and items to be quoted—that Boswell had put together in preparation for writing the *Life*" (p. xx); the previously recovered fragment of the first proof-sheets and complete set of revised proofs (now in the Hyde Collection); Boswell's so-called Note Book covering the years 1776-77, in the Hyde Collection, published by R.W. Chapman as *Boswell's Note Book 1776-1777* in 1925. And for secondary external reference, there exist several collections of Boswelliana and Johnsoniana, for example. The manuscript of the *Life* brings these disparate strands and fragments together. Waingrow skillfully makes the appropriate, often ingenious, connections clear. The edition displays Boswell as the superb assembler, stretching his account in relation to his documents across chronological divisions, assembling his "narrative" (I use that term in a very general sense) according to where his papers lead him. It is a fascinating view into the process of assembling fragments into a picture with some appearance of coherence. In fact the Boswell that emerges from this manuscript and its transcription and annotation, with its hundreds of disparate "Papers Apart," strikes me as nervous and at times even compulsive in his attempt to establish coherence (in this case actual physical evidence), against the uncertainty or difficulty of establishing identity and personal history by other means. One might insist that the *Life* as published already gives one the opportunity to perceive and examine these characteristics of Boswell's practice. But never before have we had the opportunity to observe so clearly the emphasis upon the process of assembling, creating, shaping, writing—the process, that is, with which the book seems, at times, to be preoccupied, perhaps even at the expense of its ostensible biographical subject.

This is an edition of the manuscript Boswell sent as printer's copy to the press for the first edition of the *Life*. (The fascinating story of the discovery in three parts of the manuscript of the *Life*, along with the larger story concerning all of the Boswell papers, can be found in David Buchanan's *The Treasure of Auchinleck* [1974] and Frederick Pottle's *Pride and Negligence* [1982].) And extremely complex this printer's copy is. In fact, as evidence of the methods used in preparing printer's copy and the process of printing proof, the manuscript and related papers are of considerable interest. It consists of a manuscript of more than a thousand leaves and virtually the same number of supplementary pages, called by Boswell "Papers Apart." These separate materials were included in the original printer's copy, keyed in by directions to the compositors to "Take in" certain texts. In Waingrow's words, these "'Papers Apart' included books and magazines from which passages were to be reprinted, original letters and copies, Johnsoniana from contributors, portions of Boswell's journal, drafts of texts composed in advance of the composition of the *Life*, articles on special subjects composed when the *Life* was in progress, and various additions and revisions written on separate sheets where there was no space for them on the pages of the main manuscript" (p. xxii). The task of the compositor in setting, and the corrector in correcting this manuscript in preparation of proof must have been extremely difficult and open to mistakes of all kinds. Boswell's diligence in checking proofs, and the preparation of the absolutely necessary "revises" or additional proofs, and his considerable help from Edmond Malone, were all necessary, no doubt, to prevent the many likely mistakes from creeping in. However, neither Boswell nor Malone checked the proofs against the manuscript (p. xxiv).

Since we have such a small part of the manuscript photographically reproduced—only two pages, astonishingly little for such a complex work as this—the accuracy of the verbal recreation of the physical document is of extreme importance. One feels the lack of access to the physical document all the more keenly when one discovers that each of the three illustrations provided (one of the first proof-sheet of the title-page of the *Life*, the other two pages of the manuscript) are misdescribed in their captions, apparently errors which occurred at the press, as an errata list from the editorial staff points out. Waingrow, faced with a very difficult task, decides to opt for a transcription of the manuscript "designed primarily to convey the *substance* of Boswell's composition" (p. xxvi). This method ignores many physical features of the manuscript, and informs us that this is not really a documentary edition of the manuscript, one which would attempt to recreate in symbolic form the physical characteristics of the manuscript itself; rather it is an edition of the *text* of the manuscript, but one which nevertheless retains Boswell's frequent revisions and changes. This is clearly a compromise position, one which will disappoint those of us hoping to recreate as much of the physical manuscript itself, for various purposes—further evidentiary interpretation, bibliographical analysis, etc. Some would insist that the physical qualities of the manuscript necessarily affect or even help to constitute the "substance" of Boswell's "composition." The editor mentions his desire "to preserve something of the aura of the manuscript *qua* manuscript" (p. xxviii), a very odd thing to say in the introduction to an edition *of a manuscript*. Instead of an "aura," many readers would wish for a faithful recreation of the manuscript itself, so far as that is possible. Although Waingrow's discussion of the manuscript text and his reasons for treating it in the ways he does are for the most part persuasive, the occasional inattention to the physical elements of the manuscript inevitably introduces doubt concerning the reliability of his judgments concerning the relative importance of other aspects of the manuscript.

As with any other edition of a manuscript, this is not an "objective" document; in fact, it is subject to interpretation, or re-interpretation (something readers often forget) and the "primary" evidence presented has itself been considered and interpreted by the eyes of the "critical" editor. This is to be taken for granted. But Waingrow's decision to ignore aspects of the physical form of the manuscript is unfortunate. I see no intellectual or methodological justification for such a position. And the tell-tale expression that he has "silently corrected"—in this case "misplaced punctuation" (p. xxviii)—is most disappointing to find, especially in an edition of a manuscript, which should record all text verbatim. Needless to say, as in all cases, those working closely with the texts of Boswell will have to consult the original manuscript for themselves. But without Waingrow's work to guide them, they may have a very hard time proceeding through such a tangle.

Waingrow's discussion of the problems of transcribing the text is illuminating and very interesting, particularly concerning Boswell's persistent rewriting. He concludes that "Strictly speaking, there is no 'first draft' distinguishable in the manuscript as a text separate and distinct from all other elements in the process of composition. There is instead a succession of first drafts of the basic narrative together with revisions and additions, themselves composed in one or more drafts. Only one discrete text is contained in the manuscript and that is the final text or printer's copy" (p. xxv). The manuscript itself is described as follows. What Boswell calls his "rough draught" he wrote on one side of the paper, numbered that page, and left the other side blank. This left a blank page on which he could write additions and revisions for the text facing it—that is, the text on the recto. The "rough" draft was written rapidly, and Boswell chose to incorporate printed and written texts when possible, and to have others perform any necessary transcription. He left blanks for names and references he could not immediately supply, filling them in at later stages, and included alternative words and phrases, some resolved in revision, others not until later. He also wrote queries and memoranda in the margins concerning other material and rewriting of the existing text.

"Boswell's method of revision," Waingrow explains,

> was to write between the lines of his draft, in the margins, or on the (originally) blank verso of the preceding page, depending on the nature or extent of the revision and the space then available for it. Proportionately few verso pages have remained blank; many contain more than one revision, and those usually written at different times; some are completely filled; and a few overflow to Papers Apart. A system of connecting signs was employed to guide the compositor across the terrain (and through the occasional thickets) to the final text. (p. xxiii)

At this point, Waingrow approaches the difficult task of determining the stages of revision. He asserts, on the basis of an examination of the text, the position of the elements on the page, changes in the impression of the quill, and differences in the tone of the ink, not to mention a thorough acquaintance with Boswell's general practice in the manuscript, that "many revisions were made in the act of writing the basic draft itself, and for that matter in the writing of subsequent additions" (p. xxiii), so that, in other words, there are no clearly distinct stages of revision. Boswell and Malone, however, did revise portions of the copy in preparation for the press. But Boswell's own revisions were "made in *ad hoc* fashion" (p. xxiii). He continued revising while writing later parts, acquiring additional materials for parts of the narrative already written, and he continued to revise after the copy was at the press.

It is clear that it is the *process* of revision which interests the editor, as the multi-layered manuscript demands. Waingrow sensibly determines that he must "display that process . . . in the text itself" rather than in textual notes, and so devises a system of transcription which enables him to present "a representation of the manuscript's evolutionary growth from the perspective of successive cross-sections. Through an analysis of *observable* sequences of elements of the composition I have constructed a text in which the elements are juxtaposed so as to show the temporal order in which the elements in a given segment of the manuscript were composed" (p. xxvi). The system, making use of square brackets and other notations, effectively provides what its editor claims.

It is not only the method of transcription, but also Waingrow's clear annotation, in which he cites and analyzes relevant passages from the journal and from other sources, which provides the reader with a sense of a process of composition, across time. In fact, one of the most interesting effects of examining this edition is the way in which Boswell's preoccupation with himself, particularly but not only in relation to Johnson, and with his writing—evident throughout the *Life*—are exemplified or played out on the pages of the edition of the manuscript. His preoccupation with explaining to the reader his methods of recording Johnson—that is, the recording itself tending to become the subject of parts of the *Life*—is seen in action in the manuscript (and in this edition), as Boswell tries alternatives, rewrites passages, includes extraneous material, all in an effort, we might say, to create or at least to capture Johnson, or to translate "him" into the medium of the print biography. The journals, whatever their other interest for us, lacked this dynamic primarily because they were not intended to be published, and so did not attempt the difficult task of constructing the picture of Johnson (and Boswell in relation to him) for posterity. But the subject of Boswell in relation to Johnson, and of the writing of this book, becomes vividly apparent with the edited manuscript.

For example, in the portion of the *Life* covering 30 July 1763 (pp. 320f.; Hill-Powell, I, 457f.), Boswell describes in detail what he and Johnson did and what Johnson said on this day's journey to Greenwich. The *Life* has, in part, what follows:

> I was much pleased to find myself with Johnson at Greenwich, which he celebrates in his "London" as a favourite scene. . . . Afterwards he entered upon the business of the day, which was to give me his advice as to a course of study. And here I am to mention with much regret, that my record of what he said is miserably scanty. I recollect with admiration an animating blaze of eloquence, which rouzed every intellectual power in me to the highest pitch, but must have dazzled me so much, that my memory could not preserve the substance of his discourse; for the note which I find of it is no more than this: —"He ran over the grand scale of human knowledge; advised me to select some particular branch to excel in, but to acquire a little of every kind." The defect of my minutes will be fully supplied by a long letter upon the subject, which he favoured me with, after I had been some time at Utrecht. . . . He asked me, I suppose, by way of trying my disposition, "Is not this [i.e. Greenwich Park] very fine?"

Boswell responds, "Yes, Sir; but not equal to Fleet-street," and follows with a defence of what "many of my readers may censure [as] my want of taste" by citing "a very fashionable Baronet" with similar views. Boswell then complains that, in their return on the river to London, "the night air was so cold that it made me shiver. I was the more sensible of it from having sat up all the night before, recollecting and writing in my journal what I thought worthy of preservation; an exertion which, during the first part of my acquaintance with Johnson, I frequently made. I remember having stayed up four nights in one week, without being much incommoded in the day time." Nearing the close of his account, Boswell includes a paragraph on Johnson's pleasure at hearing "of my family, and of its hereditary estate" and "my description of the romantick seat of my ancestors."

Clearly here as elsewhere in the *Life*, Boswell is pleased to find himself accepted by the great man as a companion; and the record of his own activities takes over from the presentation of his putative subject. Boswell, as viewed by Johnson, and his process of recording or remembering in order to write the biography become, in fact, at this point, the focus of the biography. Boswell's rather childish joy at Johnson's supposed pleasure concerning his aristocratic family and paraphernalia is the predictable conclusion to such a reversion.

Waingrow's edition brings together evidence, most already published in the journal, which makes it possible for us to see vividly the process of Boswell's often plaintive, often irritating, attempts to present himself. For example, the exchange about Greenwich and Fleet Street, we are reminded, derives not from the journal record, but from a marginal note in the journal, "Park not equal to Fleet Street." In the next footnote, we learn that "the very fashionable Baronet" (Sir Michael Le Fleming) in fact could not have made the remarks he is quoted as having made until considerably after the journal entry was written (despite a brief marginal note in the journal concerning them added later). In fact, it is reasonable to suspect that Boswell includes the exchange with Johnson (from memory, we must assume, if it occurred at all, since it is not recorded in the journal) comparing Greenwich Park and Fleet Street in order to insert the paragraph which makes clear his intimacy with fashionable aristocracy and sophisticated urban tastes. Waingrow records that Boswell wrote into his journal only the following brief comments concerning the return from Greenwich: "We then came up by water. I was a little discomposed even by this small excursion, and felt warm comfort at being again in London." A later marginal note in the journal states: "Shivering on the water." As Boswell writes this portion of the *Life*, his activity of recording, transcribing, and recasting emerges and *becomes* his subject.

With the information and disparate texts brought together in this edition, the process of moving from action to sensation to annotation, to expansion, to recreation (of self, as well as scene) is vividly effected. And the alternative and cancelled wordings, reproduced faithfully in the transcription throughout the text, approximately one phrase or word per line, demonstrate the process of working that material into desired form.

Boswell has, of course, been accused of cleaning up Johnson, of making him presentable, and one can find instances of this in the manuscript. For example, it is interesting to find that Boswell originally wrote (as was reported from Hume through Garrick), in reference to the green room at Garrick's theater, that Johnson insisted that he would "come no more behind your scenes David; for the silk-stockings and white bubbies of your actresses excite my genitals." Boswell then changed "bubbies" to "bosoms" and "genitals" to "amorous propensities," but added the footnote: "Mr. Hume mentioned stronger words as used by Johnson. I give the meaning" (pp. 146-47; corresponding to Hill-Powell, I, 201). But this footnote was dropped before printing. Here we see Boswell moving step by step towards an acceptable Johnson. The anecdote was first recorded by Boswell before he met Johnson, and presumably was "cleaned up" and neutered while Boswell shaped his later portrait.

The evidence concerning Johnson's pension may be relevant to this aspect of Boswell's presentation. He first wrote "George III [changed to "His Majesty George III"] on his accession to the throne of Great Britain, having wisely and liberally resolved as much as possible to abolish distinctions of party and reward literary merit, Johnson though distinguished for his aversion to the two first Princes of the House of Hanover was now considered to be a most proper object of Royal Munificence, and the King was pleased to grant him a Pension of £300 a year." This was replaced in manuscript by "This year Johnson having been represented to the King as a very learned and good man without any certain provision his Majesty was pleased to grant him a pension of three hundred pounds a year." This paragraph is also deleted in favor of the following, which was eventually printed:

> The accession of George the Third to the throne of these kingdoms opened a new and brighter prospect to men of literary merit who had been honoured with no mark of Royal favour in the preceeding reign. His present Majesty's education in this country as well as his taste and beneficence prompted him to be the Patron of science and the arts and early in this year Johnson having been represented to him as a very learned and good man without any certain provision his Majesty was pleased to grant him a pension of three hundred pounds a year.

This is followed by Boswell's firm assertion that, despite reports to the contrary, the pension was given for literary merit and without political stipulations (pp. 258-59; Hill-Powell, I, 372-73).

The movement is obviously towards complete removal of any suspicion of political motives behind the offering, and the acceptance, of the pension. "That there was nothing inconsistent," Boswell continues, "or humiliating in Johnson's accepting of a pension offered to him as his was, I am fully satisfied. His definitions of Pension and Pensioner partly founded on the satirical verses of Pope which he quotes, may be generally true and yet every body must allow that there may be and have been instances of pensions given and received upon generous and honorable terms" (p. 260). Not satisfied with this formulation, especially, presumably, his personal protest "I am fully satisfied," Boswell replaces these sentences with a new formulation: first, the sentence beginning "His definitions," followed by, "Thus then it is clear that there was nothing [etc.]." In this way he avoids the rather lamely subjective "I." But Boswell clearly protested too much, and his urge to protest even more makes it clear how unconvinced he was of Johnson's ethical position in the matter. Waingrow's notes tell us that, in the printer's copy after the words "upon generous and honourable terms," "The bottom of the leaf has been cut at this point, obliterating perhaps three or four lines of writing [on p. 202 in the manuscript]. On the first line may be read the beginning of a revised sentence, 'It has been regretted by some'" (p. 260). Waingrow points out what we might have guessed: that in fact Boswell himself was one who had regretted that Johnson had accepted the pension. Whatever was written here was presumably intended to be deleted from the printer's copy.

Boswell continued to wrestle with this problem. Waingrow tells us in the next note (p. 261):

> In the margin of this page of the MS JB instructed the compositor, "N.B. Pages 203 & 204 are cancelled. So do not look for them." It is evident from the surviving fragment at the bottom of page 202 . . . that JB continued his discussion of the pension, but exactly to what purpose and length cannot be known. The first draft clearly proceeded with the page numbered 205—had JB cancelled pages 203 and 204 immediately on having written them he would have numbered the next page 203 again—and the decision to cancel was made later. The verso of page 202 is blank; whatever revisions were made in the text of page 203 were made on that page and none (as commonly) on the page opposite.

It would appear, then, that Boswell spent two further pages discussing Johnson's pension and the reactions to it, apparently trying to exonerate him from charges of opportunism and hypocrisy. Might there have been further evidence concerning the pension originally introduced here, which has since been lost? At the very least, the evidence of Boswell's continued treatment of this issue makes it clear that he finds himself caught justifying a position he is not comfortable with, and almost certainly regrets. Readers of the *Life* who may well have received the impression that Boswell approves without hesitation of Johnson's position in accepting the pension will be dissuaded from that view by this evidence. The pension represented a problem for many of Johnson's friends and admirers. For those who want to see in Johnson a figure with either anti-monarchical views or anti-Hanoverian, pro-Stuart sympathies (like Boswell himself), Johnson's pension poses particular problems. Those involved in the debate concerning the significance of Johnson's pension should profit from a consideration of this crux in Boswell's manuscript.

The examination of the manuscript allows Waingrow to clear up some confusions of attribution and reference in the printed *Life*. For example, it is now clear from Waingrow's discussion of the manuscript evidence (p. 316, n. 1) that it is Wilkes, not Burke, who is the subject of the conversation of Thursday, 28 July 1763 (Hill-Powell, I, 453): "Has not ______ a great deal of wit, Sir? [etc.]." And on 2 August 1763, in the discussion in which Boswell mentions "an imprudent publication" of "a certain friend of his" (Hill-Powell, I, 463), it would appear from Waingrow's analysis (pp. 325-26, n. 4) that Croker's identification, followed by Hill and Powell, of the publication and author as the *Vindication of Natural Society By a Late Noble Writer*, written by Burke, is incorrect. The actual references are (much more interestingly!) to the *Letters between The Honourable Andrew Erskine, and James Boswell, Esq.*, published by, yes, Boswell himself in 1763.

Many of the explanatory notes in this edition reach convincing conclusions, or construct plausible hypotheses, concerning the most confused textual matters and the most intricate relations between different pieces of archival and documentary evidence. Waingrow is often superb in his persistent working out of problems. He is also refreshingly open to other interpretations than his own, and is fully prepared to correct explicitly his previous conclusions published in his *Correspondence Relating to the Making of the Life of Johnson*. His notes will prove a valuable addition to Boswellian/Johnsonian annotated texts, like Hill-Powell and his own volume of Boswell's *Correspondence*. This volume is a great pleasure to read and use, and Marshall Waingrow has set an example of intelligence and perseverance for us all. The *Life* is a text whose history and making seem to live already on the surface of the text. And with this superb edition, the text is loosened, and deepened, its lines and levels of history illuminated and made articulate. As the editor insists, this is the *Life* as process, and an extraordinary one it was.

ALLEN REDDICK

BRUCE REDFORD, ed. *The Letters of Samuel Johnson*, volume IV: 1782-1784; volume V: Appendices and Comprehensive Index. Princeton: Princeton Univ. Press, 1994. Pp. xix + 462; xv + 174.

Here they are, the volumes which bring us Johnson's final three years, that "last" at which he felt such horror. Undatable letters are very properly banished to the final volume. Impossible, then, for this reader, at any rate, not to approach volume IV backwards. It concludes with a letter to William Strahan (not present in the previous collection, edited by R. W. Chapman) written three days before Johnson's death, which reads, in its entirety:

> Sir, I am very unwilling to take the pains of writing, and therefore make use of another hand to desire that I may have whatever portion of my pension you can spare me with prudence and propriety. I am, Sir, your humble servant, Sam. Johnson. (IV, 446)

A practical enquiry about money, performing the same function as pressing the "balance" button on a cash-dispenser at the bank; also a single-sentence arrangement of words with sure and perfect musical rhythm; also a moment of contact with another mind.

Johnson had enquired three days earlier how much of his pension was already due, and how much would be due at the Christmas which he did not live to see. The day before that he had written to John Nichols to convey through him to posterity a scrap of paper which disentangled the multiple authorship of "the ancient universal History" (IV, 445), printed by Richardson in 1755. Before that he had arranged an epitaph for his parents and brother, had reported to Lucy Porter the epitaph arranged for his wife, had attempted to trace a first cousin and some first cousins once removed, had prepared to publish his Latin poems, and had sent presentation copies of his English works to Pembroke College, Oxford. His family and his literary works divided his attention; in considering each, he was thinking about how the world would soon go on without him. Less than two months after his death, Boswell dreamed of Johnson as showing anxiety about leaving his library in disorder.

From all this the conclusion easily follows that Johnson would be delighted with the present edition of his letters. Editing of this kind aims, like Johnson's latest letters, at commemoration, perpetuation, order; it is to take prodigious pains about a man. The lavish though tasteful print and paper, handsome layout, and informative, non-intrusive notes have already been commented on at some length with reference to volumes one to three. The illustrations to volume four reproduce an unfamiliar, affecting sketch of

Johnson near the end of his life, by Thomas Trotter, and pages from four letters: three from his last year, and one written within a few hours of his stroke. Volume five reproduces the opening of Boswell's account of his dream. Then we have the appendices and comprehensive index.

Hill, Powell, and Chapman have set a tradition of appendices and indices which positively bury the reader in information. Here these things are handled with restraint, and contain not the encyclopaedic but the strictly relevant. Undated letters tend towards the small change and small beer of life ("Mr. Johnson will wait [on] dear Miss Reynolds in about two hours": V, 18), but they also include longer and richer specimens. Some letters not only of lasting interest but even of some importance refuse to form part of the chronological narrative. They include three to Charlotte Lennox (all advising how to keep friendships in repair), one to Joseph Simpson expressing generous indignation about his father's treatment of him, and one to Hester Thrale contrasting her (as "You that waste your lives over a book at home") with himself (who, like Odysseus, travels and sees strange things: V, 29).

More enigmatic still are the contents of Appendix II, "Letters Substantiated by Mention." This spoor of lost letters ranges from the barest record, through scraps of paper bearing address, or date, or signature, to descriptive detail that whets a hopeless appetite: Johnson's own account of his letters to his mother from school; Edmund Hector's account of letters on the tedium of schoolmastering; a letter to Catherine Talbot "so highly polite and complimental that it infinitely distressed" her (V, 38); one to the Thrales "expressive of anger" after a quarrel at Brighton (V, 39); one to Thomas Hervey about marital discord, which apparently (unlike Johnson's letters to John Taylor in a similar situation) sided with the wife. The loss of one letter was bitterly lamented, even by the correspondent who failed to preserve it: the one to Taylor written immediately on Tetty's death.

Letters in the Latin which Johnson used for academic and medical occasions are translated here in a manner which catches their dignity, symmetry, and powerful forward momentum. Supplementary annotation is supplied for the first three volumes. Those volumes have flushed out three additional letters, now safely in the Hyde Collection and that of Mr. Arthur Rippey, and printed here. Most tantalizingly, the Addenda conclude with the statement: "As of July 1993, three previously unknown letters have been recovered . . ." (V, 58). Two of these are addressed jointly to Eva Maria Garrick and Hannah More (whose count of surviving letters from Johnson is thus doubled); one is addressed to Mary Nollekens, who has not been hitherto known as a correspondent of his at all. Assuming that it is publishers' timetables, not owners' intransigence, that have kept the texts of these new letters from us, a moment's nostalgia may be indulged for an age of less technology but more flexibility. Cadell or Dilly, Strahan or Tonson, or the other booksellers whose names figure in the table of Johnson's correspondents, would surely have made it possible to incorporate the text of those new letters somewhere in these volumes.

The final component, the Comprehensive Index, reflects a shift in orthodox opinion about indexing. During the 1960s, the idea of listing general topics (reading, reason, religion) was a kind of heresy, which would cause a well-trained indexer to mutter darkly the word "thesaurus." The prescribed procedure was to include a section (duly present here) on "Johnson, Samuel—religion," and to index other people's religious opinion or practice under their personal names, but to omit anything not susceptible of being subsumed under some proper name. According to this system, George Biggin, balloonist, would never have found himself rubbing shoulders with *Bibliothèque raisonnée des Ouvrages des Savans*, nor Thomas Chatterton with the game of chess, nor Shakespeare (William) with *Shakespeare* (Johnson). Indexes of the old school imposed a higher degree of regularity and uniformity on the formless heterogeneity of life, and demanded too a busier exercise of the classifying faculty on the part of the reader, who would not have found "Royal Academy" or "Charterhouse" without knowing that they were in London, or "English Chapel" without knowing that it was in Edinburgh, or *Rasselas* without knowing that Johnson wrote it.

The small, separate indexes of volumes one to four, "compiled mainly by Phyllis L. Marchand with the assistance of Marcia Wagner Levinson and Judith A. Hancock," are old-style indexes listing proper names only. Samuel Johnson finds no place in them; nor do the Royal Academy, the Charterhouse, the English Chapel, or *Rasselas*. The Comprehensive Index includes many new features along with the contents of the four small indexes: substantial entries under "controversies and quarrels," "death," "education," "friendship," "health," "letter writing," and so forth, as well as under "Johnson, Samuel"; briefer entries under "idleness," "intellect," and many more; single references under unexpected topics like "neighbours" and "wigs and hair." These are the kinds of categories which classic indexes exclude; which R. W. Chapman sets aside in their own separate space; which, obviously, testify to the indexer's exercise of subjective or individual judgment.

A dozen people working independently would produce identical entries under a proper name; but under "friendship," for instance, they might produce a dozen somewhat different collections of references. This is an admirable and delightful index, although much of what follows will sound like criticism: to expect the real existence of the best of all possible indexes would be highly unJohnsonian. This index has the habit of quoting generously, which is the most effective method for reminding readers of a familiar passage. So the "friendship" entry is in part a tissue of Johnsonian maxims:

"the greatest benefit which one friend can confer upon another, is to guard and excite and elevate his virtues" . . . one of the "few states of which it is reasonable to wish the continuance through life" . . . "to take friends such as we can find them, not as we would make them" . . . "a friend at once cheerful and serious is a great acquisition."

To balance this lavishness, the "friendship" entry adopts the pinchpenny locution, "and J. Elphinston," "and R. and C. Congreve," "and H. Boothby," to signal passages where Johnson writes of the importance to him of friendship with such and such a person. The use of initials, which is general throughout the index except in headings, grates a little. H. and H. Thrale are exempted from this treatment, presumably on grounds of ambiguity; so is Boswell, who warrants the abbreviation JB. There is, however, an expansiveness about their references—"on value of JB's," "SJ hopes his friends (the Thrales) will never forget him," "on acquisition of the Thrales' friendship"—which suggests that some qualititative principle, some measure of importance, has dictated the choice of such detail instead of "and JB," "and Thrales."

Three passages are indexed under "on Henry Thrale's friendship" (referring to "the acquisition of such a friend . . ." "it is not likely that I should ever find such another friend," and "If I could have done any good, I had not delayed an hour to come to you"). Of Hester Thrale there is no mention under "friendship." This is surprising, but the reasons for it are not hard to divine. Take a particularly eloquent passage:

> This is the pleasure of corresponding with a friend, where doubt and distrust have no place, and every thing is said as it is thought. . . . These are the letters by which souls are united, and by which Minds naturally in unison move each other as they are moved themselves. I know, dearest Lady, that in the perusal of this such is the consanguinity of our intellects, you will be touched as I am touched. I have indeed concealed nothing from you, nor do I expect ever to repent of having thus opened my heart. (III, 90)

If this is not friendship, where is friendship to be found? Yet I have also argued, and am prepared to argue again, that the opening notes of this cadenza ("In a Man's Letters you know, Madam, his soul lies naked, his letters are only the mirrour of his breast": III, 89) are purely ironical. Johnson did not believe what he says here, that the epistolary mode guarantees sincerity. Yet this self-mocking irony, I believe, leaves the reader (ourselves as well as Hester Thrale) free to take the later sentences at face value. Logically, the pleasure of corresponding, the movement of minds in unison, is as effectively demonstrated by her sharing his ironic scepticism as by her sharing his faith, if that really existed. And tonally, I hear a shift, a change of key, between the impersonal "great truth" about "a Man" and the personal and particular passage about this pleasure, these letters, our intellects, and my heart. This declaration of friendship is as remarkably oblique as those about Henry Thrale are remarkably direct.

The exclusion of this passage from index listings under "friendship" prompts some further musings on the vexed and complicated issue of gender. Only Hill Boothby, of all Johnson's female friends, appears there; and indeed, it would be hard to argue that such expressions of affection as "my dearest dear," which dot his letters to Hester Thrale, Frances Reynolds, and others, can be precisely categorized as expressions of friendship. But then, if one is not to deny that these relationships *were* friendships, is one to postulate two separate categories of male friendship and friendship between the sexes, of which only one is namable or indexable?

Under her own name, Hester Thrale occupies almost three columns of the index, Henry rather over two. (Boswell has two and a half, and Johnson himself five and a half.) But the Thrales differ from each other less in their entries' length than in their proportions. Henry's entry uses just three lines to list letters to him; just over two pages are occupied with facts and events, including a largish section on health ("visited Lichfield," "presents *Thraliana*," "SJ borrows money from," "recovery and improvement of"). Hester's entry takes nearly a whole column to list the letters to her, but slightly *less* than her husband's for facts and events. There is no reference that reads either "receives *Thraliana*" or "writes *Thraliana*," and no separate paragraph on her health. The listing of references to her pregnancies is very incomplete. Again, this is perhaps not surprising. The index lists "SJ's concern over health of"; but how would it have classified "I am glad that you are sick" and, three weeks later, "I am glad you are so sick—and nobody to pity" (III, 80, 88)? Are these "SJ's affected unconcern over health of"? One sees why an indexer might regard such a description as over-interpretive, either with *or* without the softening adjective.

These two unindexed remarks convey the information that Hester is pregnant and that Samuel is glad of it; they express a complex of personal feelings and gender attitudes that one would hardly dare set out to unravel in an article, never mind an index. On the second occasion Johnson went on to express his hope that Hester would bear a girl—this of a pregnancy from which, distraught at their son's death, the Thrales were desperately hoping she would produce a boy. Who is to say whether Johnson is propitiating fate, or urging the moderation of vain wishes, or voicing a genuine preference?

A year after his letter about letters, sincerity, and unanimity of mind, Johnson addressed to Hester Thrale another passage of ironic self-mockery, in which, I believe, he again says what he does not believe: this time, that she does not value their correspondence.

> I know you will pretend to be angry if I do not write a letter which when you know the hand you will perhaps lay aside to be read when you are dressing to morrow, and which when you have read it, if that time ever comes you will throw away into the draw and say—Stuff! (III, 126)

Are we to read this at face value? Or if not, how else? Is Johnson, in imitating her alleged impatience, making a plea to her to value his letters, or is he, ironically, expressing confidence that she does value them? Whatever the individual reader's answer, the need for interpretation is again, it seems, too acute to permit indexing. These passages exemplify the way, as Bruce Redford writes in his Preface in volume one, Johnson's letters define and complicate the friendship they were at length to dissolve. One cannot expect an index to perform so delicate a critical task as a preface.

Johnson's friendships with women, it seems, were typically built and sustained through raillery and deliberate double messages, which include warm expressions of affection but not explicit statements about the value of this friendship to him, nor about the nature of friendship as such. The Johnson of male friendship is more aphoristic than the Johnson of friendships with women, less subtle, multivalent or perhaps ambivalent. It seems we are all lucky that his relations with women are now attracting so much more attention, since in complicating our view of him we must surely deepen and enrich it.

ISOBEL GRUNDY

STEVEN LYNN. *Samuel Johnson after Deconstruction: Rhetoric and the Rambler*. Carbondale and Edwardsville: Southern Illinois Univ. Press, 1992. Pp. 191.

By most accounts, deconstruction is dead. In a sense it was swallowed by one of the extra-textual realities it had denied: history. The exact date of its demise may be difficult to pin down, but by the end of 1992 an essay in the *PMLA* was conducting a postmortem on it and citing Barbara Johnson's remarks (made in the summer 1992 School of Theory and Criticism) on "the wake of deconstruction."[1] Thus Steven Lynn's splendid book, also published in 1992, with its deliberately ambivalent title, manages to be both prescient and timely. Lynn's double-meaning pivots on the word *after* in his title: it invokes both the sense of what we get after subjecting Johnson's writings to the critical process(es) of deconstruction and that of the canonical Johnson left standing after deconstruction has passed into obscurity.

In what is (according to the dustjacket précis) "the first book-length study of *The Rambler*," Lynn "uses deconstruction to illuminate Johnson and Johnson to illuminate deconstruction." The result is a provocative book full of radical and daring ideas, at least some of which will prove appealing if not persuasive to even the most traditionalist of Johnsonians. Certainly this reviewer's initial skepticism was overcome (although not always and everywhere) by Lynn's balance of theoretical sophistication with historical evidence, common sense, skillful close readings, and a logical and readable style. Every Johnsonian would benefit from reading this book, or even just one or two of the five relatively autonomous chapters in which it is laid out.

The first chapter, rather mystifyingly titled "Introduction: Johnson upon His Bedside, after Deconstruction," nonetheless sets out an array of Lynn's most intriguing ideas. Under a much more intelligible motto, sensibly chosen from W. J. Bate ("Whatever we experience, we find Johnson has been there before us, and is meeting and returning home with us"), Lynn pursues his overarching idea that in important ways Johnson's writings anticipate deconstruction and that deconstruction in turn has special value in interpreting them, particularly *The Rambler*. Lynn questions the tradition that Johnson composed *The Rambler* hastily and haphazardly and he argues quite persuasively that *The Rambler*, despite its apparently fragmentary and heterogeneous makeup, deserves and rewards reading as "a coherent whole, a single symphony" (p. 17). Using a commonsensical reader-centered approach, Lynn shows, through example after example, that individual *Rambler* essays cannot be read intelligently on their own but must be read with and against groups of others: numbers 170 and 171 (Misella the prostitute's story, to which he might have added 191 about Bellaria); numbers 113, 115, 119, and 167 (Hymenaeus and Tranquilla on marriage); numbers 86, 88, 90, 92, and 94 (all on Milton, to which Lynn connects 93, on criticism); and so forth, in a method that highlights the extensive interconnectedness of *The Rambler* and underlies most of Lynn's critical findings.

Lynn concedes that "my method is not fully deconstructive," explaining that:

> There would seem to be little point in dismantling a text that is already perceived to be an essentially arbitrary collection of fragments—fragments that are themselves already the trace of a contradictory, associational, rambling logos. My goal instead

> is to reconstitute Johnson's controlling evangelistic aim. (p. 17)

Here Lynn touches on his most radical idea: connecting Johnson with deconstruction on the basis that both are deeply skeptical about any system of thought and the claims made for it, and both are involved in an endless search for an elusive, ineffable truth beyond language, and thus perhaps beyond knowledge. In reading *The Rambler*, Lynn contends, "the reader experiences a series of repetitions, qualifications, reversals, and diversions" (p. 18) and these (with an implicit nod to one of Johnson's favorite books, *Pilgrim's Progress*) are analogous to the reader's struggles to find religious truth, to achieve salvation. Lynn wants to show, he says, that Johnson's "strategy in *The Rambler* . . . involves the reader in a process of education that is not by any means a direct and confident route to salvation (Johnson himself could not find such a way), but is instead a troubling, difficult, evasive, confusing, and especially rambling journey" (p. 20). In connecting this process with our state of mind in late modernity Lynn concludes: "Such forces of skepticism and uncertainty, exposing the absence of any ultimately ordering Word, have arguably come together in our time most acutely in the project of deconstruction" (p. 20).

But ultimately Lynn's book is less about skepticism than about faith. In the course of chapters 2 through 5 he takes us to a point where he can say that "deconstruction and belief are not mutually exclusive but may in fact be ultimate allies" and that in this book he is "striving to reverse the secularization of Johnson's thought . . . bringing to the forefront the religious nature of *The Rambler* that has been subordinated to other themes: in short, affirming the Other in Johnson's discourse" (p. 159). Each of his chapters contributes to that end, while also offering incidental nuggets of critical insight that will reward the serendipitous reader less interested in the overarching theme.

Thus in chapter two, "(Mis)Reading *The Spectator*: *The Rambler* in Bloom," Lynn employs Harold Bloom's model of "misreading"—"how strong writers make their own work possible by misreading their precursors"—to read *The Rambler* against its "most threatening and enabling forerunner, *The Spectator*" (p. 25). The results are fascinating and original insights into Johnson's method in *The Rambler* (whether conscious or unconscious) and into the intertextual presence of *The Spectator* in his writing. Along the way Lynn offers very suggestive ideas about such topics as the misogynist implications of *The Spectator* and the contrast with Johnson's attitudes towards women in *The Rambler*; the sustained distinction between author and persona, and the acute sense of "writing as absence" that pervades *The Rambler*; the links between Bate's "accommodating and inclusive burden of the past" and

Bloom's "more competitive and disruptive anxiety of influence;" and the anxiety of influence that some might see in Bloom's own treatment of Johnson (Lynn might also have explored Bloom's possible anxiety vis-à-vis Bate, his *Anxiety of Influence* [1973] having followed by three years Bate's seminal *Burden of the Past and the English Poet* [1970]).

Chapter three, "A Difference in Nothing: Johnson and Derrida," boldly poses the central pairing on which the whole book is based. Starting from Jean Hagstrum's provocative observation that Johnson "may be regarded as one of the most original and persistent deconstructionists in our language," Lynn argues that Johnson and Derrida are radically connected in their thinking about language and meaning. In Lynn's view, "Johnson's philosophy of language and his ideas of freedom and meaning . . . [are] based on or derived from our perception of life's vacuity, and that we get to [his underlying theory] by thinking through, in several senses, something very much like deconstruction" (p. 70). In an ingenious close reading of *Rambler* 41, Lynn quotes key phrases to demonstrate Johnson's deconstructionist impulse:

> If we are "forced" by our lack of presence, our life of gaps, to turn "every moment" to the past and future, then we cannot be said to have experienced life itself in any full or direct sense as we live it. As Johnson says shortly in this essay, the present is "in perpetual motion," and it "leaves us as soon as it arrives, ceases to be present as soon as its presence is well perceived, and is only known to have existed by the effects it leaves behind" (3:223-4). How is this statement, aside from its clarity, different from Derrida's . . . description of consciousness as the trace of an effect, a trace that is caught up in the past and future, which are also characterized by difference and deferral? (p. 74)

In Lynn's presentation, Johnson's misgivings about the instability of language and his view of humankind's insatiable "hunger of the imagination" flow quite readily into deconstructionist ideas about "the space between language and reality as well as between desire and its object" (p. 73). More ambitious, and much more difficult to grasp, are Lynn's efforts to show that deconstruction can also acccommodate religious belief. Noting that "a growing number of thinkers are asserting that deconstruction clears a space for faith" (p. 96), Lynn demonstrates that "deconstruction . . . does not require atheism" and then suggests that, like a deconstructionist facing a space cleared by his own skeptical argumentation, Johnson develops "a faith based finally on nothing, on the absence of something (everything) for which we yearn" (pp. 98-99). In other words, the irrepressible sense of nothingness at the heart of things makes faith necessary, even irresistible. All this is interwoven with close readings of *Ramblers* 5 to 8, in what is the book's most difficult and challenging chapter.

Close readings also inform Chapter four, "'The Order in Which They Stand': (Re)Writing Johnson (Re)Writing," but they are of scattered essays (numbers 24, 122, 134, 155, 166, 172, 174, and 184) read both in themselves and against other groups or subsets of *Rambler* essays. Lynn's rhetorical analysis is subtle and effective. His main point in this chapter is to prove that Johnson did not dash off *Rambler* essays haphazardly and randomly, but with a pattern and a purpose. In Lynn's words, "Johnson's essays rely on certain paradigms or structural formulas, and the formulas are holistic, not connective; rhetorical and not investigative" (p. 143). Along the way, Lynn succeeds in debunking some old myths in a style worthy of Donald Greene, such as the story that Johnson composed *Rambler* 134, on procrastination, in Joshua Reynolds' parlour while the printer's boy waited. (Not only was Reynolds away in Italy throughout the months of *The Rambler*'s composition, he did not even meet Johnson until seven months after he had published the last essay.) Most importantly, after reading Lynn's careful critical explications, few readers could ever again read individual *Rambler* essays as hasty and unmethodical in themselves, or as unconnected with other essays, whether nearby or far separated, in the overall series of *The Rambler*.

In chapter five ("Conclusion: 'The Only Proper and Adequate Relief'"), as he reiterates his vision of Johnson's overriding purpose and pattern in *The Rambler*, Lynn also becomes personal, almost confessional:

> Motivated by my own desire to salvage and justify some sort of belief in a Transcendent Other, I have read that interest back into Johnson's *Rambler* to the exclusion of other concerns. I have read *The Rambler*, which appears to be concerned with many things, as a single-minded evangelical document, casting Johnson as a lay preacher out to win our souls for God.

In this brave self-examination, Lynn recognizes that he also risks deconstructing his own utterance and defeating himself:

> In this reductive misreading deconstruction has been conscripted as my accomplice, disarming the normative forces of genre and history that would otherwise restrict my movements. Able to violate these conventions, reading *The Rambler* differently, I have disagreed at/on one point or another with virtually every Johnsonian living and dead including at times

> the original Johnsonian himself. In positing such a unity and center, I have violated the very method I invoke. (p. 157)

Lynn needn't worry. His final chapter offers still more new insights, including the passage in which Derrida acknowledged a truth and a person so great they stand "outside deconstruction" (I will not give away the identities here) and some further instances of exemplary close reading. And it closes as gracefully and evocatively as any of the others, with Lynn quietly affirming that "faith is quite simply the only 'proper and adequate' choice we ramblers have" and then modestly leaving the final word to Johnson himself, quoting a very moving passage from *Rambler* 69 about piety as humankind's only relief in the face of mortality (p. 161).

In a book with so much to admire, it seems grudging to suggest faults. Some readers might welcome chapter titles that seem more immediately intelligible and convey more directly what the chapter is actually about. Others may flinch a little at the apparently unavoidable diction of poststructuralist criticism—the frequent *moves* and *manuveurs* and, above all, *strategies*, which make criticism sound more like war games than anything else. (Writing as we do, after Swift, can such language be anything but derivative? What might psychoanalysts tell us about the mentalities behind the proliferation of the vocabulary of warfare in recent critical discourse?)

The final complaint is really a compliment: the book is too short. One would welcome more of Lynn's brilliant readings of other series and subsets of *Rambler* essays. He has made the possible permutations and recombinations seem an open-ended field of prospective critical discovery. Moreover, his book manifests an open and healthy—even healing—desire to draw together the best from seemingly incompatible schools of critical thought, at a time when criticism seems more factionalized and belligerent than ever. Finally, Lynn has given us a genuinely original and valuable reading of one of Johnson's greatests texts and, whether one agrees with his particular conclusions or not, he renews one's faith in the power of literature to connect in meaningful ways with real issues in our lives.

JAMES G. BASKER

1. Jeffrey T. Nealon, "The Discipline of Deconstruction," *PMLA* (Oct. 1992), p. 1266. Nealon opens by observing that "Deconstruction, it seems, is dead in literature departments today" and goes on to analyze the competing theories about its cause of death, whether by "suicide" or "murder" (pp. 1266-67).

HOWARD D. WEINBROT. *Britannia's Issue: The Rise of British Literature from Dryden to Ossian*. Cambridge: Cambridge Univ. Press, 1993. Pp. xvii + 625.

Perhaps it is no hyperbole to suggest that between the cultural reordering that took place in 1066 in an invasion of England that reshaped its language and a failed invasion in 1940 that insured the persistence of its culture and language, there is no more important date than 1660. With the return of Charles II in May, expectations of rebirth and renewal were so high that poets and others were drawn to a grand historical parallel—the imperial establishment after a period of civil wars of Octavius Augustus Caesar in Rome in 31 B.C. Dryden in *Astraea Redux* celebrated the occasion by linking Rome's glorious past with England's potentially glorious future:

> Oh Happy Age! Oh Times like those alone,
> By Fate reserv'd for great *Augustus*' Throne!
> When the joint growth of Arms and Arts forshew
> The World a Monarch, and that Monarch you.

Dryden alone did not have the power to join the classical to the neoclassical—to suggest connection between England and the civilization defined by Cicero, Aristotle, Longinus, Quintilian; with Homer and Virgil; with Horace, Juvenal and Pindar. He was merely part of a larger chorus sensing a parallel between the Roman Peace and a modern British Peace in the making. The post-1660 world has been forever endowed by the terms "neoclassical" and "Augustan" as it has been similarly defined by such labels as The Enlightenment and The Age of Reason.

While it would be easy to compile a substantial bibliography of scholarly works suggesting such linkage, a more practical review of texts might be more instructive. If scholarly argument reaches thousands, commercial college textbooks reach tens of thousands. Their influence in defining a cultural period, such as the Restoration and Eighteenth Century, can hardly be underestimated. In their *Eighteenth Century Poetry & Prose* (1939), Louis I. Bredvold, Alan D. McKillop, and Lois Whitney review a wide range of influences that shaped the post-1660 world, neoclassicism prominently among them. The latter as a theory, they argue, "sought to mediate between nature and art, imagination and reason, delight and instruction. Boileau, Dryden, and Pope all tried to steer some such middle course. We may speak of an Augustan compromise just as easily as we speak of a Victorian compromise" ("Introduction," p. xix). In their widely used *English Prose and Poetry*

1660-1800 : A Selection (1961), Frank Brady and Martin Price suggest that some of Rome's many roads led to England:

> The desire for order which marks the search for agreement and compromise, for disinterested or practical inquiry rather than acrimonious speculation, shows itself in the new literary forms of the day. There are survivals of the wit of the Metaphysicals in the new style; but, typically, it turns back to the model of Rome for a greater simplicity, urbanity, and dignity. This look backward is another assertion of man's intrinsic powers; for it is the literary virtues that antedate Christianity that are sought, with a humanistic confidence in the greatness of classical antiquity. . . . The turning to classical models and examples was part of a new sense, moreover, of cultural stability. The forms of English politics were now closer to those of Republican Rome. ("Introduction," pp. xiii-xiv).

While commentary such as the above could easily be multiplied, one must avoid the simplification that neoclassicism dominated a period so complicated that a variety of labels seems essential to define it. Although it is difficult to move in any direction without bumping into the Augustan heritage, whole zones are relatively free from its influence, especially as one moves through the eighteenth century. Fielding may attempt to shape *Joseph Andrews* as some version of epic, but it is more to be referred to the modern novel genre defined by Defoe and Richardson. The celebration of the contemporary rather than the classical is nowhere better seen than in the periodical paper—from Addison and Steele to Johnson and beyond. The practical concerns of the latter are magnified in such utilitarian compendia as the *Gentleman's Magazine* and the *Encyclopaedia Britannica* where only faint classical echoes are heard. The period is festooned with literary originals sprung from powerful individual imaginations: Swift's *A Tale of a Tub* is *sui generis*; so, too, is Gay's *Beggar's Opera*. Samuel Johnson, who is among the last great English writers to have intuited classical greatness in grammar and text, is more to be remembered for his *English Dictionary* than for the Juvenalian resonances heard in *London* and *The Vanity of Human Wishes*. These and other exceptions aside, however, the classical heritage remains a major prop in illuminating this age, as pervasive in literature as it is in the other arts and architecture itself. It remains as tangible as Wren's St. Paul's cathedral; less obviously in the extant portions in the Adams brothers Adelphi scheme off the Strand modelled after the Palace of Diocletian at Spalatro; or it is seen in an entire city, Washington, D.C. Have we, even at this late date, really achieved a proper perspective on the impact of the classical order on Britain of the post-1660 period? Professor Howard Weinbrot thinks not. "Neither wooden stake nor silver bullet can keep Augustanism in its tomb," he writes in *Britannia's Issue: The Rise of British Literature from Dryden to Ossian* (p. 19).

While one may legitimately doubt whether creatures of the night can be permanently put to rest, Professor Weinbrot provides an intellectual best effort to amend views of a neoclassicism too easily received on authority and too often repeated without serious examination of its power to shape 1660 and beyond. *Britannia's Issue* provides just such a redefinition in a study that compares favorably in scope and originality to the best that have assessed a complicated cultural period. Michael McKeon's *The Origins of the English Novel: 1600-1740* (1987) and David Spadadora's *The Idea of Progress in Eighteenth-Century Britain* (1990) come to mind. Compared to studies merely polemical or theoretical, *Britannia's Issue* is clearly historical in its insistence that the best definition of the era examined will be found in what it had to say about itself, not in later superimpositions of what it might have thought about itself. Students familiar with Professor Weinbrot's valuable previous studies will see much the same scholarly method at work. The latter include *Studies in Augustan Imitation and Satire* (1969), *Augustus Caesar in "Augustan" England: The Decline of a Classical Norm* (1978), *Alexander Pope and the Traditions of Formal Verse Satire* (1982), and *Eighteenth-Century Satire: Essays on Texts and Contexts from Dryden to Peter Pindar* (1988). While *Britannia's Issue* is linked to these studies, it is much more than a mere amalgam of them. There is a quality of *summa*, a term that can refer simultaneously to Professor Weinbrot's scholarship and the sweeping intellectual landscape of his work. The latter includes nothing less than how between 1660 and the 1760s England achieved its modern British identity, chiefly in the way its artists and intellectuals negotiated a new relationship with the classical heritage that so profoundly influenced them. To achieve an understanding of this cultural shift required, as Professor Weinbrot notes, "excursions into areas like biblical history, economics, ethnography, mythology, philosophy, reception, religion, and of course varied hills and dales of intellectual, literary, and political history" (p. 11).

While the book as a whole is defined throughout by historical data, the world revealed as a factual entity capable of objective evaluation, its title draws on a shaping myth of culture seen in Thomas Gray's "The Bard" (1757). In surveying England's past, the poet predicts its future. The Bard commits suicide to prevent Edward I from killing him, but not before he sees woven a tapestry reflecting future national glories: Welsh Elizabeth Tudor assumes the throne; a larger union is formed joining disparate parties into a free nation capable of producing a great literature. "No more our long-lost Arthur we bewail," Gray writes, "All hail, ye genuine Kings, Britannia's Issue, hail," Of the larger architecture of his study suggested by Gray's poem, Professor Weinbrot-writes:

> This statement is by an eighteenth-century English poet, about a thirteenth-century fancied event, and British national pride and history resumed under a sixteenth-century Welsh-British queen, predicted by a Celtic Bard carrying on the wisdom of first century Druids, in turn often thought the outgrowth of itinerant Hebrew-speaking perhaps Jewish Phoenicians long before the birth of Christ, and expressed in a fifth-century B.C. rigidly regular, but thought irregular, Greek poetic form. So amiable and complex an amalgam exemplifies what *Britannia's Issue* is about. (p. 14)

Britannia's Issue is indeed a formidable intellectual amalgam, but one carefully planned: its five major sections, each of which is prefaced by an overview, provide a chapter-by-chapter account of the emergence of modern Britain between 1660 and the 1760s. During this time the nation's creative and intellectual population worked to declare its independence from a classical past. This act of emancipation took place in the broadest time-space continuum imaginable: in real estate it involved a movement from south to north—from Athens and Rome to London and Edinburgh; in time it involved more than the two millennia necessary to see the world of classical paganism replaced by a pervasive Christianity, which itself was subsumed by a culturally diverse nationalism in a greater Britain setting out to become the first world culture. Mere emancipation from Rome, however, was not enough to produce the latter, which was made possible through inclusion of vital native elements. The latter included Celtic myths and peoples as well as an indigenous Jewish population whose Hebraic values greatly enriched a kingdom. Admittedly, this is a grand hypothesis, one that might elicit admiration from those who enjoy seeing an active imagination at work. There is more than imagination in evidence in *Britannia's Issue*, which is a demonstration of scholarship in the classic sense of the term. Without proclaiming it as such, Professor Weinbrot's methodological focus can be seen in his citing of George Bright's remark in his 1684 edition of John Lightfoot's *Works*: "I am not to set down my own but my Authors sense, be it what it will, or what others can make of it. . . . It is not my business to make my Author, but to give him made; not to tell what the Author should say, but what he hath said; every one may take it or leave it as he pleaseth" (pp. 12-13).

The approach suggested above is everywhere evident in *Britannia's Issue* as Professor Weinbrot concentrates on text and context, on the exegesis of particular literary works in the larger universe of thought that contained them. The latter is suggested in particularized documentation that characterizes this volume. The subtext of this scholarly text should be taken, not left, for it reveals the forensic, evidentiary, and judicial power of the library to stabilize the meaning of literary texts years after the fact of their issuance. The parts as well as the whole of *Britannia's Issue* speak to the possibility of coherency.

Part I, which is entitled "Contexts: Intellectual, Psychological, and National," is divided into four chapters: "Moderns, Ancients, And The Secular: The Limits Of Southern Hegemony"; The Spiritual: Truth Was Not The Inclination Of The First Ages"; "An Ambition To Excel"; and "The Making Of A Modern Canon." In the last chapter particularly, one sees enunciated the argument developed in previous ones—that contrary to the widely-held notion that the literature of the post-1660 world often represented *incorporation* of the classical, the situation was more complex. Incorporation there was but also the distancing needed as a modern nation tried to define itself: abridging a southern literary and cultural hegemony was crucial in establishing an independent northern nation.

Part II moves to a more specific and local ratification of the broader movements detailed in Part I. Its larger argument, entitled "Texts Within Contexts, Essaying England: Our Genius, Our Clime," includes four chapters: "Dryden's 'Essay Of Dramatick Poesie': The Poetics Of Nationalism"; "Homeric Wars"; "The 'Pax Romana' and the 'Pax Britannica': The Ethics of War and the Ethics of Trade"; "'Windsor Forest' And 'The Rape of the Lock.'" The titles here suggest what the various chapters demonstrate—Professor Weinbrot not only provides broad historical perspectives (context) but also the exegetical review (of text) that give to his broad thesis a local habitation and home. Many will agree that Dryden's "Essay" has never been given a finer historical domestication; the conclusion to this study is too stunning to be given away in a review. It is hardly enough to note in passing the originality of Professor Weinbrot's commentary on the ethics of war and the ethics of trade—that the new economic configurations emerging in the eighteenth century could be joined to cultural, aesthetic, and literary domains, not as they were later and more famously to class struggle. Pope's best known poem requires additional commentary later.

Part III is the most generically specific in its examination of the ode in a section entitled "Growing One's Own. The British Ode From Cowley To Gray." In two chapters—"Greek Jockeys And British Heroes: The Rise And Fall Of The Pindaric Ode" and "Odes To The Nation And The North: Dryden, Collins, And Gray"—Professor Weinbrot reveals how one of the most pervasive genres of the time can be seen not as suggesting ligature between the classic and the modern but as illustrative of the effort to sever present from past. "During some 150 years of experiment," he writes, "poets invent, vary, exhaust, and redefine the form that becomes so British that even Greek relicts take the native hue" (p. 334). In traveling north, then, Pindar is forced to accommodate himself to local use—in Young's *The Merchant: An Ode On The British Trade And Navigation*; in Dryden's evocation of the Stuart Court in his ode to Anne Killigrew; or as Collins populates a classical form with native mythologies.

Part IV, entitled "Expanding The Borders. Jews And Jesus: This Israel, This England," is shaped by two chapters—"The House Of David And The House of St. George: Philosemitism, Hebrews, And Handel" and "Beyond The Hebrew Leaven: Smart And The God In Christ." Here Professor Weinbrot shows that while numerically insignificant, indigenous Jews provided mechanisms, both financial and cultural, to help construct a modern state. The latter were more important, particularly as the Old Testament offered a powerful monotheistic counterweight to polytheistic Greece and Rome. Literary Hebraism, the colorful and evocative voice instantly recognizable to generations of readers of holy texts, directly endowed such works as Handel's *Israel In Egypt* and Smart's *Song To David*. From this legacy developed an eighteenth-century "philosemitism" rooted in shared cultural values and elaborated through political ones: a Protestant nation, as it looked back to its Catholic past and across the channel to a continent dominated by Catholic France, felt as something of a minority itself kinship with its Hebraic community. Finally, both were defined by a mutual interest in the civilizing value of trade. While the eighteenth century saw an expression of international Hebraism in the special context of British culture, it also witnessed the beginning of an internationalized British culture. Before the latter could take place, however, the incorporation of vital elements was necessary.

In Part V, "Celts, Germans, And Scots: Towards A United Kingdom," Professor Weinbrot completes his amalgam of the modern British nation in two chapters—"Celtic Scotland," and "Ossian In Scotland; Great Britain, And Modern Europe: Joining Britannia's Issue." Each chapter involves a paradox: while clans and with them an entire people went with the wind of Culloden and the catastrophe of '45, and Ossian proved to be a fake, both had the effect of making a lesser England into a greater Britain. The latter needed both the physical territory of the north and the rich myth generated by Ossian that endowed it to fully distance itself from its classical past. In his "Conclusion: Synthesizing All The Nations Under Heaven" Professor Weinbrot writes: "In this attractive eighteenth-century British paradigm a myth of race includes heterogeneity. It also includes the approximate shape of the century-long development I have chronicled: national self-esteem that once came through association with southern classicism, becomes national self-esteem through association with many native voices in one British chorus" (p. 558). The consequences of the changes brought about as the dynamics of the post-1660 world caused the world to refocus on an island once peripheralized off the coast of Europe will continue to be debated. What cannot be debated is how many have become, willingly or not, issues of Britannia, whether in language or as part of the cultural and commercial values it carried to the world.

While Professor Weinbrot's text provides valuable material for cultural historians to ponder, the way he ratifies literary texts through cultural history deserves special note. Among the varied merits of *Britannia's Issue* is that it allows us to reconsider the ways we read works, even one so well established as Pope's *The Rape Of The Lock.* Even in an age of a rapidly changing canon, it remains central. A scholarly Rip Van Winkle waking up to reclaim his place in the classroom would be in for some surprises as he attempted to reclaim this work in the larger context of the Restoration and Eighteenth Century. Although general agreement about the period's beginning still exists, there is new doubt when it ended. Familiar signs are down, the more personal "Age of Johnson" (this journal excepted) is replaced with a more generic "later eighteenth century." Advertising and scholarship seemed to have converged, both sharing the mantra "new," a term descriptive of individuals as well as the period as a whole. The French, with theory, seem to have achieved a cultural occupation they failed to effect with land and naval forces in the eighteenth century. The latter, he would note, has been increasingly reconstituted in terms of race, class, and gender. With some minimal instruction he could identify in bibliographic cyberspace essays that would bring him up to date, particularly about a poem he once knew so well. Among these Professor Brijraj Singh's "Pope's Belinda: A Feminist Rereading," (*CLA Journal*, 34 [1991], 467-85) would serve well. A brief digression summarizing this reading of Pope for later comparison with Professor Weinbrot's analysis would seem to be appropriate.

Professor Singh notes initially the distinction Elaine Showalter makes between *feminist criticism* and *gynocriticism*, the former still endowed by "white fathers" like Jacques Lacan, while the latter seeks to identify that "wild zone" or "female space" that remains special and unknowable to men. Gynocriticism, Professor Singh argues, can lead to permanent gender separation, something mitigated by androgyny, which is defined as something extant in males capable of sensing within "a female inner self" that allows for something other than a single-gender reading of life. Androgyny, then, is a tool allowing for a reading of *The Rape* at its deepest level, its "repressed subtext," which is defined by "strategies of concealment, repression, appropriation, and denial," within which Professor Peggy Kamuf argues "phallocentrism hides its fictions" (p. 469). "The stated meaning of the poem,"

Professor Singh writes, "is that Belinda is beautiful but also trivial, vain, empty-headed, and lacking in a discriminating moral sense" (p. 474). Clarissa "expresses not a female but a phallogocentric view of life" (p. 475). Although a spokesman for a repressive bourgeois and phallocentric society, Pope is capable, chiefly because he was short and Catholic (two minority conditions) to appreciate the excluded female other and hence achieves an androgynous appreciation of Belinda. The cutting of her lock is balanced by its stellification. Gynocritics, Professor Singh notes, have exacerbated a gender split capable of some healing through androgyny. The meaning of *The Rape*, though, is gender-linked, however that phenomenon is defined.

The lately awakened Rip could also turn to Professor Weinbrot's exegesis of the same poem and enter a different critical universe. Here the poem, while consistently referred to gender, is recentered in the context that generated its original composition, or at least a good portion of it—epic. Professor Weinbrot's contention is not, however, that Pope invoked epic essentially to contextualize the folly of the Baron or the vanity of Belinda. Rather, he suggests, the poem is an integral part of a larger process by which classical values were replaced by those more contemporary and closer to home. Pope "insists that Homer is not his contemporary, and thus must be replaced by the superior human values of his own age and its preference for a culture united by the bonds of an at least tentatively rational culture. Homer's Sarpedon is the noble exemplar of the *Iliad*'s ideals; Pope's Clarissa will be even better" (p. 303). Others are referred to epic, Thalestris chief among them as she is connected with Plutarch's original who is reputed to have visited Alexander. She is best defined by a vast literature concerning Amazonia, to which even Samuel Johnson contributed in his translation of the Abbé Guyon's *Histoire des Amazones* in 1741. The amazon, however, is no model for a contemporary eighteenth-century femininity, which while it might exult in triumphing at Ombre, would hardly desire to meet the amazons' test of killing three men before mating with a detestable gender. Clarissa's famed and controversial words of advice to Belinda probably transcend simple sourcing, though they echo Sarpedon's speech to Glaucus in the *Iliad*, 12.371-96. Here, Professor Weinbrot suggests, the brevity of youth is seen as a sufficient cause to yield to "a paradigm of aristocratic responsibility" (p. 320) in the classical context as Pope suggests one must yield to the earthly imperfections of marriage in his modern eighteenth century.

Arguably, there is no settling the debate over which reading of Pope's poem is more adequate, though most would agree the two approaches are tactically different. Professor Singh produces essentially two facts concerning Pope's stature and religion. These are used in a reductio that suggests they were instrumental in generating in Pope an androgyny making him sympathetic to Belinda. This could be the case but an act of faith is required to subscribe to it. While Professor Singh honors the scholarly convention of sourcing her argument, it is done so through references chronologically distant from Pope's poem. Professor Weinbrot's contextualizing and sourcing of *The Rape* has hardly been suggested in the brief summary here. He virtually drowns his reader in the world that most immediately pertained to this text. It was not his business to make Pope but to give him made by the richly complex world that generated his most famous work; the latter was particularly defined by the fascinating story he tells in *Britannia's Issue* of Britain's march to its modern identity as it separated itself from its classical heritage and redefined itself through an ecumenical incorporation of invigorating native elements. Professor Weinbrot's book has no rival past or present and is a cause for scholarly celebration.

JOHN L. ABBOTT

Boswell: Citizen of the World, Man of Letters. Edited by Irma S. Lustig. Lexington: The University Press of Kentucky, 1995. Pp. xvii + 270.

During the second half of the twentieth century, literary scholarship has been generous in its examination of James Boswell. The 1950s opened with the resounding success of *Boswell's London Journal*, the first volume of the Yale trade edition; companions arrived periodically (if to less popular acclaim) throughout the next four decades, with the final journal published in 1989. Several volumes of Boswell's correspondence, well-researched and carefully prepared, also appeared as part of the ongoing research series of the Yale Boswell Editions. The revision of the authoritative Hill-Powell edition of the *Life of Johnson* was completed in 1964. The quality of editing that Boswell's writings have received, with no obvious exceptions, is exemplary.

Scholarship seeking to explain and illuminate these texts has also been plentiful. Anthony Brown's recently updated bibliography, *Boswellian Studies*, is no short handlist. This surprising amount of attention derives, in part, from the abundance of available source material: the rich Boswell archive makes research an attractive undertaking. Boswell's complex personality has also intrigued many scholars. But most investigation has revolved around the absorbing questions raised by the *Life*.

Boswellian studies have maintained such a high level of interest in the past decade that, alongside the steady stream of articles in the periodical journals, three volumes devoted to reexamination of his life and work have appeared. *Boswell's "Life of Johnson": New Questions, New Answers*, edited by John A.

Vance, is the first (1985) and arguably the most important. *New Light on Boswell: Critical and Historical Essays on the Occasion of the Bicentenary of The Life of Johnson*, edited by Greg Clingham, appeared in 1991. *Boswell: Citizen of the World, Man of Letters*, edited by Irma S. Lustig, is the most recent (1995) and the volume under review here.

The volume consists of an introduction by Lustig and eleven essays divided into two sections: "Boswell and the Enlightenment" and "The Life of Johnson." It closes with a new count of Boswell's meetings with Johnson. The scholarship is mature and penetrating, if not always immediately recognizable as significant.

In "Boswell's Travels through the German, Swiss, and French Enlightenment," Marlies K. Danziger delineates the various scholarly, religious, and political luminaries that Boswell met during his studies and travels in Northern Europe. She demonstrates that Rousseau, Voltaire, and Paoli were not alone in challenging and shaping Boswell's thinking, an important point. I am tempted to suggest, however, that a close reading of the trade journal would serve much the same purpose. Nonetheless, the essay is clear and focused, providing a glimpse of what scholars can expect from Danziger's work on the Yale research edition of Boswell's journals in Germany and Switzerland.

Peter F. Perreten's essay, "Boswell's Response to the European Landscape," makes the most out of Boswell's limited interest in this area. The study, based on the continental journals, categorizes Boswell's reactions to landscape:

> . . . a direct response, based on a deep affection for what Boswell called the "romantic beauties" of his native Scotland; a consciously developed response to "improved" landscape learned from his friend and surrogate father, Henry Home, Lord Kames; and a complex, indirect response shaped by classical authors and guidebooks. (p. 37)

Perreten suggests that Boswell's reaction to landscape "is as sophisticated as that of most young gentlemen on the tour" (p. 37) but stops short of demonstrating that Boswell's aesthetic reactions to nature had profound effect on his life or writings.

Boswell's indebtedness to Lord Kames is more profitably studied by Richard B. Sher in "'Something that Put Me in Mind of My Father': Boswell and Lord Kames." In one of the strongest essays in the collection, Sher develops a familiar theme—Boswell and surrogate father figures—delivering new and productive insights. Sher argues persuasively that Boswell's biographical projects, whether focused on Kames, Sir Alexander Dick, Alexander Lockhart, or Johnson "constituted in part a literary pretense for intimacy" (p. 79). All of these projects allowed "Boswell to spend more time with his subject, and establish a closer relationship with him, than would otherwise have been possible" (p. 79).

Michael Fry traces another of Boswell's Scottish acquaintanceships from beginning to end in "James Boswell, Henry Dundas, and Enlightened Politics." Immediately following is Thomas Crawford's related essay "Politics in the Boswell-Temple Correspondence." Both essays develop the familiar thesis that Boswell was not temperamentally suited for politics. Crawford's clear delineation of Boswell's movement from conservatism to a more liberal stance, measured against Temple's contrary movement from early "Commonwealthman" to vigorous conservatism, is fascinating reading.

The final essay in the "Enlightenment" section is John Strawhorn's "Master of Ulabrae: Boswell as Enlightened Laird." Strawhorn provides a succinct examination of Boswell's accomplishments as heir presumptive and as Laird of Auchinleck. Boswell's concern for the day-to-day aspects of estate affairs, with his signature mix of innovation and conservatism, demonstrates that even when removed to London, he took is duties as Laird quite seriously.

The first half of *Boswell: Citizen of the World, Man of Letters* attempts, with varied success, to observe Boswell within diverse contexts of the European Enlightenment, yet Boswell's complex personality is perhaps resistant to such examinations. The second half of the volume achieves greater success by exploring various aspects of Boswell's acknowledged strength: his writing of the *Life of Johnson*. Carey McIntosh opens with a close reading of prose style in "Rhetoric and Runts: Boswell's Artistry." He describes Boswell's manipulation of high diction in order to achieve "prose of authentic grandeur in major celebrations of Johnson's character" and low diction to convey "a degree of intimacy with the formidable subject of this biography" (p. 152). On the subject of whether Boswell presents Johnson's conversation with authenticity, McIntosh is convinced that he does not. Comparison of Boswell's Johnsonian dialogue with Johnson's written work, especially his *Rambler* essays, and with passages of Johnson's dialogue recorded from dictation, proves otherwise. The examination, according to McIntosh,

> . . . confirms the notion of Boswell's Johnson as an interpretation, an independent creation, not a replica of Samuel Johnson. . . . (p. 153)

William Paul Yarrow makes no distinction between Boswell's record of Johnson's speech and Johnson's authentic conversation. Taking the two to be roughly equivalent, he sifts the *Life*, finding surprisingly extensive and varied use of metaphor. In an engaging and perceptive essay, Yarrow identifies Boswell's lifelong fascination with metaphor and illustrates his increasing skill in its use. Metaphor becomes central to Boswell's depiction of Johnson.

> Johnson needed to be crystallized metaphorically because the picture of Johnson for most readers, Boswell recognized, was an abstract one. (p. 167)

Yarrow sees Boswell's preoccupation with metaphor working at several levels, not only within the *Life* to more clearly present Johnson's character and thought, but in Boswell's day-to-day conversation, where he cajoles Johnson into more metaphorical statements than he might otherwise have spoken.

> *Johnson*. "No, Sir. I never heard Burke make a good joke in my life."—*Boswell*. "But, Sir, you will allow he is a hawk."—Dr. Johnson, thinking that I meant this of his joking, said, "No, sir, he is not the hawk there. He is the beetle in the mire."—I still adhered to my metaphor,—"But he *soars* as the hawk."—*Johnson*. "Yes, sir, but he catches nothing." (p. 178; quoted from *Life*, V, 213)

John Radner, in another strong essay entitled "Pilgrimage and Autonomy: The Visit to Ashbourne," also finds Boswell and Johnson collaborating, if unwittingly. Radner argues that the extended visit to John Taylor's house, in September 1777, served to bring about a fundamental shift in the relationship between Johnson and Boswell. Especially significant was their heated debate over England's American policies. Boswell began to see himself as Johnson's peer (if not his equal). Radner has carefully reviewed the chronology behind the writing of Boswell's Ashbourne journal, the subsequent letters between the two men, and Boswell's eventual revisions of journal and letters for inclusion in the *Life*. Radner argues that by reflecting on the visit, and journalizing the experience, Boswell attains the power of objective judgment that he would later use as biographer of his friend. Particularly compelling is the discussion of Boswell's collaborative reflections in the journal from this period. According to Radner:

> Instead of endorsing Johnson's opinions while perhaps adding an example or an illustration, or rejecting Johnson's claims with arguments to support his opposition, Boswell incorporated but substantially augmented Johnson's positions. (p. 211)

Isobel Grundy investigates Boswell's tolerance for uncertainty within his portrait of Johnson, contrasting Boswell's efforts to create a permanent monument to Johnson with Virginia Woolf's literary experiments demonstrating the impossibility of fixing character. Grundy remarks on some passages where Boswell attributes thoughts and motivation to Johnson but identifies many others, especially in the latter portion of the *Life*, where Boswell resists the temptation to speculate.

At times Grundy lapses into obscure prose:

> We may suspect the biographer, who has given hostages to fortune undreamed of by the novelist, of being ready to claim a factitious certainty of interpretation, so that any unsolved mysteries or dilemmas of understanding are to be resolved not by his text but by his exclusions, his silences. (p. 188)

On the whole, though, "'Over Him We Hang Vibrating': Uncertainty in the *Life of Johnson*" is a thoughtful essay that has located a new facet of discussion about Boswell's biographical methodology.

Irma Lustig, in the final essay of the volume, "'My Dear Enemy': Margaret Montgomerie Boswell in the *Life of Johnson*," contributes a perceptive analysis of Johnson's epistolary campaign to win favor with Margaret Boswell. What might be viewed as a gratuitous inclusion of personal detail within the *Life* is convincingly shown to be yet another example of Boswell's subtle and richly complex art of characterization. Lustig, in addition to this essay, has done a fine job as editor of this volume. The introduction, presentation, and indexing all show the work of a practiced hand.

The Appendix deserves notice: Hitoshi Suwabe has recounted the meetings between Boswell and Johnson, basing his count on the trade edition of Boswell's journals, published correspondence, as well as the *Life*. Donald Greene's count of 327 days (his calculation was based on the *Life* and *Journal of a Tour to the Hebrides*) has been revised to 400 days (or 404 if you include four meetings noted in the *Life* but unsubstantiated elsewhere). Of what importance is a recount of this type? Or for that matter, of what relevance is another volume of essays establishing esoteric historical contexts and offering yet more analysis of Boswell's well-analyzed masterpiece? In some instances, the details of Boswell's life and times uncovered in this volume may strike one as insignificant. But when compounded in a volume of this sort, and when joined with the whole of the generally intelligent scholarship pertaining to Boswell, they add yet more strokes to the "Flemish portrait" of man and author. Certainly Boswell would have understood the significance of the undertaking.

THOMAS E. KINSELLA

DUSTIN GRIFFIN. *Satire: A Critical Reintroduction*. Lexington: University Press of Kentucky, 1994. Pp. x + 245.

CLAUDE RAWSON. *Satire and Sentiment, 1660-1830*. Cambridge, New York, and Melbourne: Cambridge University Press, 1994. Pp. xviii + 309.

Dustin Griffin begins *Satire: A Critical Reintroduction* with a review of satire criticism over the past thirty years which is particularly useful in two regards. First, it makes clear how often satire has appeared to be directed towards all the "wrong things," given the tastes and temperament of the past critical generation. The traditions of satire are conservative, often classical, and overwhelmingly male; satire and satirists are generally impatient with philosophical refinements and abstract theorizing about culture, and most satires are grounded in local situations and particular histories; satire is aggressively evaluative, often narrow and intolerant and ungenerous and tendentious; satire features exaggeration and indirection and fictionality rather than the apparent straightforwardness of lived experience deeply considered. The second thing that Griffin's survey makes clear is that satire criticism has been developing along two different tracks and at significantly different speeds: the practical criticism of particular satires and satirists has been moving forward purposefully, variously, fruitfully, while more general theory and critical models for satire have mostly "stood still" for a generation—or perhaps even since the time of John Dryden.

Griffin turns from recent criticism back to Dryden's "Discourse concerning the Original and Progress of Satire" (1693)—an essay given significant extra weight by the poet's success as a practicing satirist and the translator of Juvenal and Persius—in order to question the premises that Dryden passed on to many of his critical successors. In Griffin's view the "Discourse" defined and exemplified satire too narrowly and selectively, and thereby curtailed and misrepresented the rich variety of the tradition it pretended to encapsulate. More specifically Dryden played up the moral and formal assurance of satire, its historical pertinence and power, its reaching out after conclusive meanings; he also shaped his arguments to imply that "real" satire was formal verse satire and the many other kinds were merely undignified intruders upon an essentially Roman form of art. Griffin would "reintroduce" satire by questioning Dryden's influential assumptions—his eighteenth-century successors tended to codify satire into a neat and widely shared set of moral, intellectual, and artistic polarities—and by reconstructing a picture of satire that is larger, more problematic and doubtful, more essayistic than Dryden's. Satirists "try out" a variety of ideas in their works, readers "try out" various responses, and satire works more to unsettle comfortably assured visions of the world than to confirm preexisting notions of approved and disapproved values.

Griffin does not aim to rewrite the history of satire—his approach is more topical and problem-oriented—but his arguments are carefully situated in a broad comparative context and informed by a wealth of literary examples. He reminds his reader of several perspectives that converge to support a broader vision of the traditions and theory of satire: the breadth of Menippean practices in classical satire (Lucian, Apuleius, Petronius), the reimagining of Menippean strategies in the Renaissance (More, Erasmus, Rabelais) and the eighteenth century (Fontenelle, Blake), the richness of English satiric traditions before and aside from Dryden (Donne, Marston, Hall); the ways in which Horace's, Pope's, and even Dryden's own satiric practices did not quite conform to his theories. As Griffin reconfigures satiric traditions, formal verse satire seems mostly the exception (an honorable one, to be sure) and freeform Menippean experimentation is the general rule and a very fruitful one.

Griffin also draws attention to the fact that Menippus was an orator and urges that the classical understanding of rhetoric included more than simple persuasion. He proposes a satiric model based upon open-ended exploration rather than systematic progress toward a conclusion, an exploration conducted through multiple and mutually supportive rhetorics of inquiry, provocation, display, and witty play. Satirists write to expose the objects of their scorn, of course, but they also write to display their own ingenuity; satirists play for the applause of the audience at least as much as for the reform of the times. It follows from Griffin's celebration of the playful and open-ended qualities of satiric rhetoric that he should also notice that satire resists neat formal closure. In his view satire is naturally inclined "rather to inquire, explore, or unsettle than to declare, sum up, or conclude" (p. 95), and in the Menippean tradition in particular, satire regularly borrows from other literary kinds that are themselves significantly open-ended—journey literature, digressive narratives, catalogue literature, dreams or tall tales or fantasies. Even moments of apparent stasis in satire, when approved and disapproved attitudes seem firmly and clearly delineated, should be understood as pauses in a more dynamic process of ongoing evaluation. "The point is to keep moving, and the satirist is supple enough to do so" (p. 113). Sterne's Tristram Shandy, himself the proprietor of an open-ended, playfully provocative, and continually evolving literary form, would agree enthusiastically: ". . . so much of motion, is so much of life, and so much of joy—and . . . to stand still, or get on but slowly, is death and the devil—" (Book VII, Chap. xiii).

After celebrating satire as a potentially inexhaustible reservoir of evaluative meanings, Griffin reverses field to explore one of the limiting factors, specifically, the place of the fictionalized worlds of satire in relation to the world of narrow purposes and historical particulars from which satire draws materials. As in earlier sections, he begins with a review of traditional critical issues—What is the "power" of satire? Is satire best conceived as a kind of history, or a rival to it, or perhaps a rejoinder? And what does satire *accomplish* in the non-satiric world?—and he offers a clear summary of traditional answers, although he does not have much that is distinctively new to add. In the end, he sees satire as radically indebted to the historical world it criticizes, not just for materials or the freedom to speak out, but more generally for the systems of education, dissemination, and cultural evaluation which allow satire to achieve whatever power it finally possesses. Satire may have its moments of sublime independent-mindedness (or sublime perversity), but it continues to be deeply implicated in the world as it exists: "Satire remains a radically impure art, both contaminated and energized by the world outside itself" (p. 123).

Griffin draws toward a conclusion with a strong chapter on the mixed pleasures of satire, including the guilty one often ascribed to it, that it takes pleasure from inflicting pain (or seeing it inflicted) upon others. He begins with psychological theories of humor—Hobbes's "sudden glory" theory of laughter, Freud's influential notion of the coexistence of humor and hostility—and then moves on to more specifically literary kinds of pleasure. Satire offers the pleasure of skillful and economical and witty representation of its objects; satire offers the pleasure of decoding oblique and ironical expression; satire offers pleasure in its uncompromising reductiveness, conveying at least a temporary sense of mastery of the world; satire offers a kind of demotic lexicography, name-calling and low abuse raised into admiration. Satire offers mental stimulation certainly, erotic stimulation possibly, the free play of indignation and a festival of publicly-inflicted punishments, and perhaps a mechanism for assuaging secret guilts. Undoubtedly satire gives readers a chance to indulge the fearful human fascination with all kinds of deformity. Griffin closes by urging more interdisciplinary approaches to satire and trying out an extended quasi-anthropological meditation on satire's associations with foods and occasions for eating: perhaps satire is a necessary and sometimes violent force-feeding inflicted upon disordered constitutions that require strong measures.

The final value of this study will vary with the needs of the reader. Griffin suggests at the outset his hopes that the discussion can be useful to beginning students, literary generalists, and to satire specialists. Certainly all three groups can profit from his clear, even-handed, and undoctrinaire presentation of where satire criticism has been moving in the past thirty years; it is intrinsically valuable to see difficult and divided issues articulated so clearly—and doubly so when the interpreter does not feel compelled to reduce all complexities to a single pattern. Beyond this, the different kinds of reader will find different strengths and limitations: beginning students will find Griffin's considerable range of literary reference daunting—but perhaps inviting as well; literary generalists are likely to wish he had nodded a few more times toward satire in the novel, a subject he set aside at the outset as beyond his current project. Specialists are in Griffin's debt for researching and annotating this study so carefully and for reasserting the pertinence of Menippean models for satire (we *are* influenced by whatever seems "normal" and, thanks largely to a gifted group of Pope's interpreters, formal verse satire has seemed normal for more than a generation). But the same specialists are likely to wish that Griffin had spent less time with past critical arguments and more time seeking to step beyond them.

My own most significant reservations about Griffin's approach have to do with his seemingly unlimited commitment to satiric tentativeness, his preference for a version of satire that is "problematic, open-ended, essayistic, ambiguous in its relationship to history, uncertain in its political effect, resistant to formal closure, more inclined to ask questions than to provide answers, and ambivalent about the pleasures it offers" (pp. 4-5). Granted, some very important satires fit this description (Swift's *Tale of a Tub* is the example to which Griffin most often returns). And insofar as this characterization of satire is offered as a necessary corrective to Dryden and those who followed in his footsteps, overplaying their moral, intellectual, and artistic assurance, Griffin has a good point to make and his approach does lead back to Menippean satire in a way that makes it appear both stimulating and inviting. Still there ought to be some middle critical way between total assurance and none at all, or alternatively, some way to describe the limits of satire's newly ascendant tentativeness. Even Griffin concedes that much of the time satirists are simply guiding readers through new materials, toward new attitudes. But at crucial moments a special doubtfulness takes over: ". . . we may suspect that satire gets out of hand, that the satirist has raised so many questions as to lose control of the inquiry . . . we may sense that satire, once set in motion, acquires a momentum of its own" (p. 64). That motion is sustained not by calculation but rather by the instabilities inherent in satiric irony: ". . . as the satire gains momentum, one suspects that the satire draws the satirist on, soliciting a libelous imagination to make claims and charges not part of any premeditated plan . . . let the satiric inquiry go where it will" (p. 69). What Griffin is characterizing as the unpremeditated *nature* of satire in its finest moments, I would try to recharacterize as yet another well-chosen *strategy* of satire, the calculated (and therefore limited) use of doubtfulness as a teaching tool.

One question that faces all critics who write about satire is, by what analogies shall the satirist's role be described? Is he an orator, a guide, a friend, a performer, a master of ceremonies, or even a cook preparing a mixed stew (*lanx satura*)? Of course, each of these ascribed roles goes with a different approach and tonality of satire and most specialists have at one time or another used all of them. But there is another analogue that should be used more often than it is: the satirist-as-teacher. The obvious reason for making this connection is that teachers so often play the same roles suggested above: they too (we too?) are friends, guides, performers, makers of mixed stews, and so on, and they too use various approaches and tones to do their work.

But there are subtler reasons to assert the notion of the satirist-as-teacher in response to Griffin's very open-ended idea of satire. Teachers are strategists as satirists are, and also dealers in doubt. More specifically, teachers working their way through difficult materials are pedagogical pragmatists, trying one approach and then another, turning their materials this way and that, trying to provoke students first into interest, then into forming some opinions of their own—and so it is with satirists. An experienced teacher would most likely feel neither total assurance nor total puzzlement concerning the difficult materials to be examined, although he or she might pretend either attitude. But that teacher is more likely to offer students along with new materials a controlled and progressive set of strategic doubts and interpretive options which are neither foregone conclusions nor wholly open-ended choices—and so too with satirists. Teachers and satirists alike seek to empower those they would lead by putting them into situations in which they must think for themselves, follow the various consequences of different values and different interpretive choices, and eventually learn the difficulty and rewards of good judgment by experiencing both. To put the matter more bluntly still, both teachers and satirists have a vested interest in showing their respective audiences that thinking for oneself actually *works* and helps one to understand the world better, and asserting too many doubts and open-endednesses along the way does not conduce to those ends. Paying more attention to the pedagogies of satire has the additional advantage of making it easier to ask after the shape of the whole satiric curriculum: after all the doubtful choices and witty amusements along the way, what values are finally recommended and what ones condemned?

Griffin has written a good book, substantial and persuasive enough to be admired *and* to be argued over.

Claude Rawson's *Satire and Sentiment, 1660-1830* is a collection of eleven separate essays, loosely brought together by his aspiration to use diverse pieces of literary history to describe and analyze "the energies of a patrician culture in decline" (p. xi). A few of the writers he touches on had direct patrician connections (Rochester, Fielding, Boswell, Byron, Shelley), but most were non-patricians who nonetheless adopted tastes, loyalties, and cultural aspirations that were quasi-patrician or Augustan: Oldham, Dryden, Swift, Pope, Johnson, Reynolds, Burke, Austen. Rawson also nods repeatedly in the direction of a few writers (notably Defoe and Richardson) who staunchly resisted patrician attitudes and uses their gestures of resistance to dramatize the contests of cultural values then being played out in literature and elsewhere. What Rawson is in effect offering then is a series of glimpses into the divided attitudes and experiences of writers whose works reflect a defensive and somewhat nostalgic commitment to Augustan values, even as they increasingly reflect the pressure of more modern and middle-class tastes and attitudes.

The different essays vary considerably in their weight, resonance, and methodology. At least six of them derive in whole or in part from journalistic originals and bear the marks of their origins—a certain informality of tone and explanatory style, a lack of scholarly annotation, a tendency to "survey" materials rather than to analyze them closely. Two other essays derive in significant part from Introductions written for new student editions (Boswell's *Life of Johnson* and Austen's *Persuasion*), so again the emphasis is more introductory and appreciative than it might otherwise be. There is nothing intrinsically wrong with bringing the interests of a specialist to a broader audience—the common reader is an important reader—but both substantively and methodologically these popular essays seem oddly matched with the two longer, carefully wrought scholarly essays that are original to this volume and make up nearly half its length. Those two are substantial enough to deserve individual treatment here.

The first of these, which appears as chapter 3 (pp. 29-97), is an extended and thoughtful account of the literary styles and strategies used to convey human violence, particularly the violence of war, from classical times down through the Renaissance to the eighteenth century. Rawson's aim is, first, to trace the cultural impact of improved technology and the consequent increasing scale and horror of organized combat over the centuries, and second, to assess the changing moral impact of warfare and the various oblique ways in which anti-war sentiments, suspicions of cruelty and ethical ambivalence, crept into accounts of events that another time might have deemed simply "heroic." In more down-to-earth terms Rawson sets up three useful discussions which eventually come together: first, exploring the genealogy of mock-heroic style; second, proposing that there are several alternative styles and strategies of mock-heroic, and when one won't suit, another might; and third, looking for ways in which English Augustan satirists, notably Swift and Pope, worked to contrive mock-heroic strategies without on one hand subverting the classical heroic poets whom they admired, or on the other, endorsing classical constructions of heroic morality which now seemed brutal. Even as he writes within a broadly comparative literary perspective,

Rawson is also a persuasive close-reader of particular texts and he conveys very effectively what fine control Swift and Pope used in the *selective* evocation (or the selective neglect) of available heroic backgrounds; beyond that, he argues persuasively the notion that stylistic borrowing of the mock-heroic sort has ethical implications, not just strategic ones.

In his second previously unpublished essay, chapter 5, entitled "Revolution in the Moral Wardrobe: Mutations of an Image from Dryden to Burke" (pp. 133-96), Rawson brings to bear his broad learning in eighteenth-century studies to reconstruct an important conversation that eighteenth-century Britons had among themselves concerning the arts of civilization. He begins and ends with Burke's famous indictment in his *Reflections on the Revolution in France* of the general tendency of events there—and specifically of the enforced removal of the royal family from Versailles to Paris in October of 1789: "All the decent drapery of life is to be rudely torn off. All the superadded ideas, furnished from the wardrobe of a moral imagination, which the heart owns, and the understanding ratifies, as necessary to cover the defects of our naked shivering nature, and to raise it to dignity in our own estimation, are to be exploded as a ridiculous, absurd, and antiquated fashion." Rawson asks, why "decent drapery," and proceeds to unpack the many meanings and associations of the metaphor over the previous century: one art leads to another, from dress to painting to architecture to music and dance, and the aesthetic ideals of the fine arts apply as well to personal, social, and civic life—but then the countervailing notions of drapery as artifice or as concealment also come into play. And what would it signify to strip the drapery away? The other half of his discourse is to explore the meanings of the shivering nakedness figure, which the eighteenth century turned into a sustained debate concerning what is "natural" to man and might therefore be expected of him or proposed to him in a civic setting. Rawson's argument moves from writer to writer, text to text, one small increment to another, so it is not always possible to see exactly where he is headed. But he brings in a wealth of interesting materials along the way, and in the end it is clear that he has constructed a carefully documented and nuanced analysis of eighteenth-century understandings of civility and the arts that sustain it—and that is a complex and important composition worth having.

This book has less to say—and particularly, less *good* to say—about the literature of sentiment than the title might lead one to expect.

PETER M. BRIGGS

BARBARA M. BENEDICT. *Framing Feeling: Sentiment and Style in English Prose Fiction, 1745-1800.* AMS Studies in the Eighteenth Century, Number 26. New York: AMS Press, 1994. Pp. 261.

In his rigorous and important study, *Sentiment and Sociability* (1988), John Mullan maintains that, while sentiment played a critical role in the developing arguments about social bonds and sociability during the eighteenth century, systematic philosophy (Hume's, that is) proved to be the wrong medium to convey the language of feeling or to sustain its social import—because, in general, philosophical inquiry lacked a wide readership. Unlike the philosopher, the novelist "did not have to find forms of public address" to communicate or value sentiment; nor was the novelist restricted "to any exact specification of the world."[1] These reasons, coupled with the formal properties of the genre itself, made the novel, according to Mullan, the best suited vehicle to ensure the cultural viability of the language of sentiment.

Even though Mullan makes a strong case for his position, we do not have to rank the mediums of sentimental exchange in order to appreciate the impact that sentiment had upon the age. Students of the eighteenth century are well aware of how sentiment infiltrated almost every conceivable cell and stage of life: from the physiology of the body, to the psychology of the being, to the culture that helped to define and shape social existence during the period. Recent works by Ann Van Sant and J. Barker-Benfield, among others, have further elaborated and reaffirmed these points. Had this recent vintage of scholarship on sentiment never been written, however, there is still nothing new about the study of sentiment as such. Whatever novelty emerges about the topic does so from amid congestion and as a function of approach. In this respect, and following Mullan's lead about the importance of the novel as vehicle, Barbara Benedict's contribution stands as a noteworthy example of critical resourcefulness and an often probing investigation into the inner workings and structural makeup of sentiment in the eighteenth-century novel.

Like many scholarly books these days, *Framing Feeling* is relatively short, with the body of the text running a bit over two hundred pages, but Benedict is ambitious in her scope. She begins with late Haywood, concludes with Austen, and makes expected and unanticipated stops between. We expect to hear from Sterne and Mackenzie—and are not disappointed. We do not anticipate the likes of Henry or Frances Brooke—neither of whom has attracted a large body of scholarly pursuers, though Mullan himself comments on both—but are pleased by their appearance. Equally appealing is Benedict's argument. She maintains that we must approach sentimental displays cautiously, certainly not as some inflated or overdone endorsement of female values, democratic fervor, or bourgeois social advancement.

Through her close inspection of the works before her and by sparing us many of the afflictions of programmatic criticism, she makes the strong case that readers of eighteenth-century fiction ought to recognize the ways in which novelists harnessed sentimental representations "within conventional discourses" in order to "placate contemporary moralists and please contemporary readers" (p. 6).

By staking this claim, Benedict departs from Mullan who restricts his analysis to feeling that, in her words, "reflect[s] a philosophical skepticism about sociability"; and, earlier still, Brissenden (*Virtue in Distress* [1974]), whose stress falls on sentimentalism as the agent of individualism. Benedict announces instead that sentimental literature "controls both skepticism about sociability and individual alienation by adapting sentimental values to conventional rhetorical and structural forms. These discourses and structures signal social feeling and traditional values to the reader despite the portraits of isolated or revolutionary feeling that the characters present" (p. 4). Sentimental literature does so, according to her terminology, via the frame. Implicit in this critical apparatus is Benedict's firm persuasion that if, somehow, literary sentiment has raised the specter of anomie, then the potential threat of this bogy has also been misunderstood. The division between story and the manner of its telling, between spectacle and its often moral or satirical rendering, establishes the distance readers would have needed not to have been infected by the mystery, mysticism, or dark contagion that sentiment seemed to scatter indiscriminately.

One might suspect, however, that since it has taken till the last decade of the twentieth century to come to this realization, its truth may have eluded many an eighteenth-century reader, especially since "reader" evokes the impression of a flesh-and-blood human being with book propped up and open. Audience, in this regard, may have been the safer term, unless Benedict is convinced that all period readers may be subsumed by one. I, for one, lack such conviction and believe there is sufficient evidence available about eighteenth-century reading habits to ask that we be careful about the terminology we choose.

I raise this issue not as a quibble but to point out a concern I have about Benedict's effort; let me use her own words to help illustrate what I mean. In chapter 1 ("Gender and Narrative Authority"), she states: "Among the favorite topics of the eighteenth-century periodical is social behavior, a topic that pulls together diverse readers and diverse reactions" (p. 21). While it is true that any group of readers all reading the same source of printed material will be pulled together, as it were, by what they share in common, that does not mean that their diversity somehow dissolves. But Benedict wants to emphasize the pulling together more than the diversity, even though gender issues and the friction they occasioned, for example, remain a strong subtext throughout her book. When she remarks that "sentimental novels placate moralists and please contemporary readers," should we assume that when moralists move their eyes across the page of a sentimental tale they are moralizing and that everyone else without these ethical qualifications is, as a group apart, merely reading?

The impression she leaves us with over and again is that, faced with the cultural and moral authority of the frame, readers had no choice but to yield to its persuasive powers and distancing effects. Hence, from the first chapter: "Like Haywood's *Female Spectator*, Fielding's *Amelia* separates social authority from female narrative; the private story of women's experience becomes a topic in a tale of self-control, and sentimental narrative is told from a masculine perspective. Both works warn women against indulging their feelings and laud the power of spectatorial distance" (p. 46). At the end of the century the lesson is pretty much the same: "In *The Mysteries of Udolpho*, Radcliffe centers her plot and her style on the importance of fighting fear and of knowing how to see the truth. *The Romance of the Forest* dramatizes the hazards of unobserved behavior. In both novels, Radcliffe pictorializes sentiment to control feeling by means of artistic convention. Her Gothic fictions portray the hazards of the indulgence of private feeling" (pp. 194-95). I do not wish to dispute the existence of the cues in the stories that lead to formulations of this sort. Benedict is in fact remarkably adept at bringing them to the surface, but I do seriously wonder whether readers were nearly so impressionable as she represents them to be.

Benedict's quite formidable skill is as a vertical rather than a horizontal thinker. In her seven numbered chapters one invariably samples her enviable talent at reading bits and pieces of sentimental tableaux. Missing from this labor, however, is any real sense of what sentiment and umbrella ideology or frame mean for the novel as a whole, except in the frequently unconvincing and often awkwardly imposed moral coda she supplies in the last sentence or two of a chapter. She does not read the novels in any systematic way or horizontally plotted fashion; as a consequence, the contents of chapters generally comprise what appears to me as some kind of classificatory scheme, such that one could move around a part here and a part there without unbalancing the whole. In addition, though I recognize that she covers more than fifty years of British literary history in her work, I find it enormously difficult to detect any real pattern of development in what she offers. Even though chapter headings change key words as they proceed, there is a pervasive sense about *Framing Fictions* that when you have read one chapter you have read them all. It is thus with relief that one arrives at the last ten pages of the Conclusion where Benedict starts answering all the questions that have been brewing over the course of the previous seven chapters. I should mention too that the second half of the Conclusion has almost nothing to do with *Sense and Sensibility*, which is the titular focus of the chapter ("Conclusion: Watching Feeling in *Sense and Sensibility*").

The classificatory tendency and vertical structuring that I see in Benedict's study are akin to the kid-glove approach she grants narrative authority. About *Sense and Sensibility* she comments: "While the sisters strive to identify the particular figure on horseback, the narrative presents a conventional picture of figures meeting in a landscape," which is most perceptive on her part, but then in an apposition she writes, "a picture accessible only to the structurally detached narrator and reader" (p. 203), as if both are perfectly devoid of self-interest or, in the case of the former, inaccessible to examination. Similarly she wants to give the narrator of *Amelia* an "Olympian perspective that epitomizes authority" (p. 32), which *may* serve for *Tom Jones* but suggests inattention to this narrator and almost no consideration to his relative weight among Fielding's earlier narrators. In the Conclusion Benedict does characterize Fielding's last fictive narrator as being "detached but not impartial" (p. 207), while calling attention to the narrator's masculinized parti pris in the chapter that deals with Fielding. Yet somehow we are left with the impression that if these biases affect readers, especially women readers, they do so in ways that ultimately make them submit to authority.

Similarly, Tristram Shandy's relationship with audience is "primarily authoritarian, indeed schoolmasterly" (p. 79); while *The Mysteries of Udolpho* discusses Madame Montoni "in the very language of neoclassical, antifeminist satire," consistent with "Pope's maxim that 'most Women have no Characters at all'" (p. 175). It is perhaps salutary at this point to recall Mary Wollstonecraft's words toward the end of *A Vindication of the Rights of Woman*. There she writes:

> If women are to be made virtuous by authority, which is a contradiction in terms, let them be immured in seraglios and watched with a jealous eye.—Fear not that the iron will enter into their souls—for the souls that can bear such treatment are made of yielding materials, just animated enough to give life to the body.
>
> 'Matter too soft a lasting mark to bear,
> 'And best distinguish'd by black, brown, or fair.'[2]

I raise these objections not to be invidious but to express the disappointment and frustration I feel as a reader of *Framing Feeling*. This book begins with great promise and is consistently shrewd and insightful throughout; these virtues, however, are not matched in the broad execution of the study.

Perhaps some explanation for my response to Benedict's work can be reconstructed from her Acknowledgments page, where she tells us that, exclusive of the Introduction, five of her seven chapters have already appeared in different forms as published articles. It is one thing to write an article and an entirely different enterprise to write a book. No doubt, any number of unnamed factors could help to explain the final composition of *Framing Feeling*, but for me it cries out for a greater distancing from its original and inspiring materials.

This sense is never clearer than in the longest chapter in the work, Benedict's chapter 6 ("Selling Sentiment: Pictures of Feeling in Sentimental Miscellanies"), which first appeared in *ELH* "in slightly altered form." I had first read this chapter in its unaltered form some years ago when it appeared in *ELH* and, frankly, was a bit perplexed then about the point that Benedict wanted to affirm. This time she offered what I thought of as more explicit instruction about how to interpret the sentimental tableaux in the miscellanies. Their popularity, as she urges in the Introduction, indicates "that some novel-readers desired particular effects that could be provided by parts of sentimental prose fictions severed from" their original sources (p. 7). Benedict goes on: "As a hybrid form that includes prose, poetry, and fiction, sentimental miscellanies seem to transgress the generic restriction of [*Framing Feeling*]. In fact, they are at the heart of this book because they exemplify the ambiguities of sentimental literature and demonstrate the link between culture and form" (p. 7). This assertion reminded me of Terry Castle's treatment of *A Simple Story* in *Masquerade and Civilization*, in which Inchbald's story becomes the most carnivalesque novel of the century, despite the fact that Inchbald gives over only a sentence or two to the actual masquerade—the what-you-get-is-not-what-you-see approach.

Critical paradox is perfectly acceptable and, with writerly bravura, can be rendered plausible. Unfortunately, Benedict seems to forget her introductory promise once she arrives at the chapter. Sentimental vignettes, at once unframed and reframed, become framed according to a process that I cannot easily decipher. In addition, many of the vignettes she describes, often at great length, strike me as having precious little to do with sentiment. Even Benedict catches herself. After discussing a sermon included in a miscellany, she remarks that "This is not the technique of sentimentalism, but it is the traditional rhetoric of the fathers of religion" (p. 143). Later, when forced to explain why Johnson's famous letter to Chesterfield qualifies as sentimental, she notes that "nothing beyond Boswell's explanation of the circumstances of the letter . . . frames it. Why is this 'sentimental'? It is sentimental partly because it snubs a middle-class nose at the nobs and proves success outside the patronage system—the triumph of true values. . . . Primarily, however, it is included because it was written by Johnson, the exemplar of private charity" (p. 151). Even if Johnson were the exemplar of charity during the eighteenth century, should his altruism suggest that anything or everything he wrote is sentimental? A brief letter, *whole* unto itself and from the pen of a worthy, may instead explain its appearance in a miscellany.

Lost in this welter of details and detailing is any real justification for the existence of the chapter. Throughout, Benedict goes to sometimes painful lengths to try to fashion a frame, but the results are unwieldy and confusing—a frame that needs a frame. When discussing optics, spectatorship, and visual stimulation—a few additional elements of the frame—she recalls the jigsaw puzzle and defines it this way: as "an activity which requires visual skills to replicate the pattern of the external world from a pile of fragmented pictorial details rather like the patchwork fields newly created by enclosure" (p. 158). Try that out on your seven-year-old. One also gathers that enclosures from the sixteenth or seventeenth century did not factor in the jigsaw wars. Were I Benedict's editor I would recommend that she simply cut out chapter 6, with a jigsaw if need be, because it does not fulfill the pivotal role she accords it. Chapter 6 is more akin to the appendix than it is to the "heart."

Despite my criticisms, which, for better or worse, reflect both my biases and expectations as a reader, *Framing Feeling* is not only a good book but also a noteworthy book for scholars who follow the debates over sentiment. It could have been a better book, however, not so much in conception, but by virtue of its execution. Benedict is both an impressive and insightful reader herself, and others are bound to learn, as I did, more about the import of sentiment from having read her pages. Had I to do it over again, I would read her Introduction and the last half of her Conclusion first, before beginning the body of the text. The physical book itself is generally well produced and has a series of illustrations that are referred to. I must confess to my irritation, however, with the abundance of conspicuously uneven word spacings that occur throughout. That may be part of the price of publishing in the electronic age; the other, $45, is considerable, but not out of keeping these days for relevant acquisitions to our libraries.

JAMES CRUISE

Notes

1. John Mullan, *Sentiment and Sociability: The Language of Feeling in the Eighteenth Century* (Oxford: Clarendon Press, 1988), pp. 13, 14.
2. Mary Wollstonecraft, *A Vindication of the Rights of Woman*, ed. Carol H. Poston (New York: Norton, 1975), p. 188.

JON THOMAS ROWLAND. *Faint Praise and Civil Leer: The "Decline" of Eighteenth-Century Panegyric*. Newark: Univ. of Delaware Press, 1994. Pp. 190.

"The complexity of panegyric in the late seventeenth and early eighteenth century was not a result of its decadence; it was always complex" (p. 22). This opening statement demonstrates one of the strengths of Jon Rowland's rhetorical study of panegyric; he tries not to simplify his subject, and, at times, refutes those whose categorizations on the subject have been useful. For instance, Rowland refutes Leon Guilhamet's argument that satire is demonstrative rhetoric and always suggests its opposite, panegyric, by demonstrating that panegyric has a "deliberative capacity" (p. 19). In general, Rowland's introductory survey of the historical definitions of the term "panegyric" is nicely set out, although in light of his focus on praise, his passing mention of blame is a bit of a simplification in itself. The complexity of Rowland's discussion of what he calls the "instability" of panegyrical writing and his analysis of the effect of overhearing or overreading on the reception of panegyric is a welcome addition to a more complete description of the genre.

The scope of Rowland's study is very limited, as he relies on examples that he finds representative. The problem with this is that he does not clearly relate his criteria for identifying representatives of the genre. In discussing Samuel Daniel's "Panegyrike Congratulatorie," his first example of the genre, Rowland convincingly demonstrates that the topicality of panegyric "almost inevitably clashes with its topoi, its conventions, and commonplaces . . . [and] that such conflict may be instructive, a warning to the king about how self-mythologizing can turn into satire when it fails to take reality into account" (p. 35). Although Rowland asserts that this is "the first panegyric in English," he provides no support for the claim, saying merely that he "feel[s] it contains nearly everything of interest that can be found in panegyrics of the period" (p. 35).

The choice of Marvell's *Horatian Ode* and *The First Anniversary* as poems which "belong to the panegyric tradition simply by challenging it . . . [and] by fulfilling its potential, by realizing its capacity for reversal and conversion" (p. 36) adds some needed complexity to this discussion of the genre, at the further expense of defining it. One of Rowland's most astonishing claims in this section is that Marvell is capable of controlling panegyric's potential to flip over into vituperation: "what this alternation does is alert us to the role of the poet, himself controlling and guiding as power ought to do but does not" (p. 38). While there may be some truth to the theory that good panegyric and panegyrical satire can be defined by authorial skill, it is a circular sort of claim. Rowland makes it again in his section on *The Rehearsal*

Transpros'd: "Marvell's language is itself 'transpositional,' but *intentionally* so, and with Marvell conspicuously in control" (p. 94). The "conspicuous" evidence offered consists of the standard prefatory and digressive comments traditional in satiric types of panegyric, and, as such, is not very convincing. A related claim is even more dismissively stated: "it is inevitable that when [satire and panegyric] are juxtaposed they will change roles more than once. No doubt the frequency of such alternation was controlled for contemporaries by their awareness of certain facts in conjunction with what the poet was advising" (p. 84).

The temptation to treat the complexities of panegyric as a result of development of the genre is not one that can properly be resisted; Rowland duly notes that with the advent of Marvell's panegyrical efforts "panegyric, standing for the means of presentation if not *signification* itself, becomes itself a metaphor for obfuscation and obstruction" (p. 51). The tracing of various topoi throughout the series of "Painter-Poems" is well done, culminating at one point with the observation that both Waller and Marvell "manipulate events in time, placing them in or out of chronological sequence, to score a point. The goodness or badness of events is entirely a matter of presentation; or certainly it must have seemed so to contemporary readers" (p. 66). Rowland's detailed readings of individual works are the most valuable part of his book.

Rowland's continued focus on the tradition of panegyric, as distinct from its parallel satiric tradition, seems unnecessarily narrow when he begins to discuss panegyrical aspects of *A Tale of a Tub*. His narrow focus leads him to the unsatisfying conclusions that "the bad that is the target of the satire is inseparable from the good that is not" (p. 115) and "the point may be that one *ought* to be uncertain" (p. 122). Rowland describes the Hack's predicament as that of a man "who must praise satire because the conventions of panegyric have been exhausted" (p. 144). His narrow focus leaves him with nothing more to say than that, with *A Tale of a Tub*, panegyric has, indeed, become decadent. But especially in light of Rowland's continuing attention to the manifestation of self-praise in pieces that present themselves as panegyric, he would have done well to widen his scope enough to describe the characteristics of a parallel, satiric genre (with self-praise as one of its main characteristics).

A section entitled "Postmodern Panegyric," focusing on Swift's "Stella" poems and Charles Churchill's *Gotham*, further develops the idea of increasing generic complexity. With one of the "Stella" poems, Rowland says, we see "the final stage in the apotheosis—not of a king into a god, but of a poem into nothing" (p. 153). Similarly, in *Gotham*, Rowland agrees with Thomas Lockwood about the problem of abstractions in Churchill's poem, saying that "as abstractions, 'Plain Truth,' etc., are obviously as empty as the expressions they are meant to replace" (p. 157). What panegyric has become, Rowland suggests, is "a tradition of self-consciously analogical signification" (p. 145).

In his Afterword, Rowland attempts to differentiate Swift's "mock-panegyrics of the 1730s" from "mere mock-encomia" (p. 166), claiming that mock-encomia can only mock the exaggeration of praise itself, but that mock-panegyric "mocks abusive relationships, not just of patrons and clients, but also of tenors and vehicles" (p. 166). While the point may be valid, quibbling over the terms is curious, especially in light of the fact that the introductory historical survey of terms does not make a clear distinction between encomia and panegyric (or blame and satire, for that matter).

That panegyric does not degenerate but becomes more complicated as the eighteenth century progresses is well demonstrated in this study; as Rowland puts it, "what develops in the panegyric tradition, I believe, is a sophisticated understanding of the problems of signification" (p. 167). Rowland also points out that the panegyric tradition was connected "intimately" to the Stuarts, and that this could be one of the reasons for its continued development throughout the century. If Rowland's use of the term "satire" was as careful as his use of panegyrical terms, perhaps the differences that he sees would be clarified. As it is, though, his conclusion, that "Eighteenth-century satire ridicules the conventions of a discredited genre, panegyric, while exploiting them to expose current abuses" (p. 169) sounds curiously close to previous analyses of panegyric as a degenerate genre in the late eighteenth century.

JEANNE GRIGGS

Writers, Books, and Trade: An Eighteenth-Century English Miscellany for William B. Todd, ed. O M Brack. New York: AMS Press, 1994. Pp. xii + 412.

Resurgent interest in the history of the book is reaching a crucial moment. Scholars of bibliography, initially pleased with the validation of their too often unacknowledged labors that cultural materialism provided, are now asserting the primacy of their knowledge. Many of the most prominent critics who enthusiastically gush about the history of the book and its importance for both literary formal criticism and a new sociology of knowledge seem to know little about the printing, selling, marketing, or distribution of books. References to "print culture" are rewarded with solemn nods when the composition of such a "print culture" is often misunderstood. The present reviewer must count himself among those who profess an interest in material aspects of literature without possessing more intimate knowledge of historical bibliography than can be gleaned from a reading of Philip Gaskell's *New Introduction to Bibliography*.

The pioneering work of D. F. McKenzie, Roger Chartier, and others, though, has energized both traditional bibliography and the new history of the book, bringing together histories of the spread of print, such as Martin and Lefebvre's *Apparition du Livre* and Eisenstein's *Print Revolution in Early Modern Europe*, and more detailed examinations of the people, customs, and technologies that comprise the literary marketplace of the hand-press era. O M Brack, Jr., has edited a festschrift (or, as the volume is appropriately subtitled, "an eighteenth-century English miscellany") for William B. Todd that illustrates some of the current movements and controversies in this field of cultural studies. *Writers, Books, and Trade*—McKenzie, we are told, provided the suggestive title—includes an array of essays that covers many aspects of bibliography and literary history. And while not all of the contributors to the volume appear to appreciate the implications of their arguments and research for those concerned with broader cultural issues, even the reader annoyed with technical detail will appreciate many of the quirky facts unearthed here.

Brack summarizes William Todd's life work in a preface that hints at the concerns of many of the volume's contributors: "Among his numerous scholarly publications, however, it is his analyses of the eighteenth-century book—press figures, concealed editions, concurrent and recurrent printing, the use of advertisements—culminating in his *Bibliography of Edmund Burke*, that have had the greatest impact." Todd's work on press figures—those seemingly mysterious symbols appearing at the bottom of the pages of some eighteenth-century books—inspired two contributions to the present volume, G. Thomas Tanselle's "Press Figures in America: The Shop of Thomas Dobson" and B. J. McMullin's "Further Observations on the Incidence and Interpretation of Press Figures." Tanselle's work here describes the occurrence of press figures in books printed by Dobson from 1790 to 1799. Tanselle shies away from the active interpretation of press figures as reliable guides to distinguish particular impressions or presses or pressmen. While hoping that his work might spur further research in the field, Tanselle avoids making broad or particular claims for the meaning of his research. B. J. McMullin's more speculative work reluctantly concludes that "on their own, press figures may often enough not be capable of translation into verifiable information about the production of a volume" (p. 193), but that they might be used as corroborating evidence in complex bibliographic cases. Both the mildness with which McMullin treats over-eager bibliographers who wish press figures to say too much—and his refusal to draw absolute conclusions of his own—might well be a model for literary historians who can sometimes confuse speculation with fact.

Indeed, it is not only the bibliographers in this volume who avoid the trap of wild enthusiasm. Many of the articles included approach the material elements of book production and publishing in ways that even non-bibliographers could use. James Raven's essay, "Selling One's Life: James Lackington, Eighteenth-Century Booksellers and the Design of Autobiography," looks at the published memoirs of Lackington, William Hutton, and George Miller in the context of the peculiar genre of the bookseller's self-writing, which may be said to begin with the remarkable writings of John Dunton at the end of the seventeenth and beginning of the eighteenth centuries. The bookseller was a public figure in the eighteenth century. Jacob Tonson was at the center of controversy over his Kit-Kat Club, Edmund Curll was as well known as a "personality" as for the content of his publications, John Almon became famous for his own published apology, and Joseph Johnson's role as a radical bookseller in the 1790s was related to the importance of the authors he published. Lackington was both an innovative retailer of books and a self-promoter who, in the manner of a Lee Iaccoca or Donald Trump, drew attention to himself as an entrepreneurial exemplum in a published autobiography. Lackington was most well-known for his "Temple of the Muses" in Finsbury Square: there Lackington stocked and sold, reports James Raven, over a million mostly remaindered and used books. He allowed his customers no credit: "By refusing to extend credit Lackington made certain of full payment, gained the interest on the income, was free of debt collection operations, and was able to simplify his accounting system" (p. 3). Raven later points out that Lackington does not necessarily deserve credit for the "originality" of his business practices: he was successful, however, in creating a monopoly of scale for himself. Raven does not suggest why readers would be interested in the autobiographies of men like Lackington, Miller, and Hutton. One key might be their shared origins in traditionally non-literate classes. As Raven argues, "All stressed the late age by which they learnt to read, crucial as it was to their new trades" (p. 17). And he quotes the introduction to an edition of Lackington's book published in 1827: "Although the rambling Memoirs of this fortunate bookseller belong to a class which principally exhibit the importance of the writers themselves, it is not without interest as a record of the progress of natural sagacity, industry, and frugality, to riches and independence" (p. 16). The publisher's autobiography, then, might be a genre uniquely geared to take advantage of both narrative patterns explored, for example, by Bunyan and Defoe and of an economic system that was beginning to open up to manufacturers and retailers serving an incipient consumer revolution.

Patricia Hernlund's important article, "Three Bankruptcies in the London Book Trade, 1746-61: Rivington, Knapton, and Osborn," counterbalances the triumphant pats on the back issued by Raven's booksellers to themselves. Hernlund suggests that these three eventual failures were weighted down by "out-moded business practices" and the section on Rivington, in particular, shows rather that innovation does not always succeed. Rivington attempted to make spectacular earnings by underselling all competing booksellers in the American colonies with books to which he did not hold a copyright. He was, as Hernlund argues, guilty of practicing piracy and a man with "no discernible code of honor" who broke "ethical, moral, and legal codes" in his dealings with the colonies (p. 82). It is unfortunate that the volume does not include more information on trade with the American colonies, since notions of copyright, transatlantic trade, and the capacity for the production of print all come together in cases like Rivington's. David L. Vander Meulen's interesting piece on "unauthorized" editions of Pope's *Dunciad* makes the important argument that what we have been calling "piracies" for two hundred years were not necessarily illegal or unethical. Like the used and remaindered market pioneered by Lackington, the reprint trade, argues Vander Meulen, is more important than is usually credited. Again, intellectual and historical issues of copyright, colonialism, and international trade and law come into conjunction with concrete bibliography in order to demonstrate how the literary marketplace met the demands of readers and shaped possibilities for new authors while preserving recent works that came to be known as "classics."

As many have observed in the past, the book trade provided unique possibilities for class mobility in a society obsessed with the adventures of a Pamela Andrews but still deeply invested in traditional structures. Hugh Amory's contribution to the book brings some of these issues into focus. On the surface "Master Arthur and the Boys: Author, Editor, and Compositor in Fielding's *Amelia*" is a slightly ill-tempered attack on some of the choices Martin Battestin made while editing the Wesleyan edition of Fielding's last novel. What emerges from Amory's analysis of Battestin's choices, which hinged upon his interpretation of Fielding's intentions, is an understanding of print culture that accepts the diversity of printed variations and the multiple agents—writers, publishers, editors, compositors—whose efforts might be seen as cooperative rather than antagonistic. Amory suggests that Battestin set up an opposition between an "Olympian" authorship and the wretched "functionaries," the compositors, who have been wrongly thought to be mere "corruptors" of texts. He argues for an important influence that Arthur Murphy as editor had in the 1762 *Works* upon which Battestin relies. Amory's piece is oddly written with elliptical pronouncements, but it does raise serious questions about any simple interpretation of print culture. Issues of class not only have impact upon the kinds of authors who published and whom we read now, but can affect our understandings of textual authority, intention, and meaning.

As C. J. Mitchell's exhaustive research shows, print culture also became an economic arena open to women. "Women in the Eighteenth-Century Book Trades" provides a summary of the various paths followed by women who, mostly, took over printing and bookselling operations after their husbands died. The article does provide information on apprenticeships served by women and therefore suggests that it was not only widows who printed, published, and sold books. "It is clear," she writes, "that women were involved in the day-to-day work of all the book trades" (p. 34). This article might act as an essential introduction to the variety and importance of women in aspects of production in eighteenth-century print culture.

Beverly Schneller provides more intensive analysis of one female bookseller in the eighteenth century. "Using Newspaper Advertisements to Study the Book Trade: A Year in the Life of Mary Cooper" is interested in showing how newspaper advertisements can be used by bibliographers to establish facts about a particular publisher's work. Advertisements have often been overlooked in studies of early print culture, perhaps as too vulgar and commercial an aspect of writing and communication, but they are worthy of the kind of analysis that Schneller's piece begins to attempt. Through looking at the advertisements placed particularly in *The Daily Advertiser* and *The Public Advertiser* (and it is not insignificant that these newspapers emphasized the profit motivation in their titles) by Mary Cooper in 1753, Schneller reaches a couple of useful conclusions: first, that profit rather than ideology motivated most publication, as Cooper published works both for and against the "Jew Bill" that was introduced in the spring session of Parliament that year; and second, that the burgeoning appetite for "news" motivated many of the publications in the middle of the eighteenth century. As Schneller argues, the advertisements she analyzed indicate that "Cooper's publishing was largely derived from news items" (p. 140). If further work can expand this individual case as a general rule, our understanding of the relationship between "the news" and other aspects of print culture will increase. What Lennard Davis called the "news/novel matrix" might be shown to have bearing on genres and forms other than the novel.

O M Brack, Jr.'s own contribution to the volume also concerns a kind of advertising: the "puffs" Tobias Smollett wrote for his *Complete History of England*. Between the "Plan" used to promote the quarto first edition, advertising in newspapers for the seven-volume octavo edition, and proposals and advertising for the quarto edition sold in numbers, the publishers Rivington and Fletcher spent nearly £800. Smollett himself wrote many of these proposals, which Brack quotes extensively. Smollett's clever promotional material for his *Complete History* stresses, as Brack argues, not a comprehensive philosophy of history (which Smollett was in other places happy to provide) but an understanding of his readers' demands. Smollett, that is, says his own *History* "will be more easy in the purchase, more agreeable in the perusal, and less burthensome to the memory, than any work of the same nature, produced in these kingdoms" (p. 268).

Betty Rizzo's "Bonnell Thornton, Reviewer: Evolution of a Technique" looks at an aspect of print culture that arose with advertisements as a paraliterary discourse. Book reviews are neither fully independent essays nor simple advertisements for books. Rizzo's important essay illuminates both the gradual changes in the ways Thornton prepared his reviews in the *Public Advertiser* and the *St. James Chronicle* and more generally sheds light on the overlap between puffery and literary criticism. Book reviews at this point consisted mainly of summaries and extracts of work with a few comments. Thornton was part of a literary and political circle inhabited also by John Wilkes, Robert Lloyd, and Charles Churchill. Indeed, the reviews examined by Rizzo were of poems by Churchill, and Rizzo notes that Thornton's "primary purpose" was "to exalt Churchill and his works" (p. 340). He did so by marking out passages for extraction and writing his prefatory and concluding remarks in the margins of the printed poem. Thornton eventually came to cross out passages from the poems under consideration not to be extracted and so the compositor for the reviews merely had to set type for Thornton's remarks and all the text readable on the page. Rizzo's analysis is interspersed with the narrative of Thornton's relations with other writers in his circle, and while she does not reach major conclusions about the importance of reviews in establishing and breaking the important literary "reputation" in the eighteenth century, her essay goes some way towards drawing the outline of such work.

Jan Fergus and Ruth Portner, on the other hand, reach some very interesting conclusions about the importance of book reviewing in the second half of the eighteenth century, after the *Monthly Review* and *Critical Review* became established almost as national journals. Fergus and Portner examined the records of two provincial booksellers, the Clay family with operations in Daventry, Rugby, Lutterworth, and Warwick and Timothy Stevens of Cirencester to see if a direct correlation between favorable reviews and the sale of books could be established. Interestingly, from this limited sample the answer would appear to be "no." While the purchase of books by clubs of readers seemed to be somewhat influenced by favorable reviews, individual purchasers showed a surprising reluctance to follow the recommendations of reviewers. (And with the extensive logrolling done by interested reviewers in purportedly disinterested reviews, one can understand why the stubborn consumer refused to be led.) The data analyzed by Fergus and Portner is limited, but both the statistics and their interpretation of them will be of interest to scholars concerned with ideas of consumption and literary reputation, literary criticism, and the commercial nature of publishing in the eighteenth century. Fergus and Portner correctly note that with the establishment of the *Critical* and *Monthly Reviews* the book review becomes a semi-autonomous genre: "Clearly, however, individual subscribers did not need to use the Reviews as guides to purchasing in order to get their money's worth from them. The Monthly and Critical certainly offered instruction and entertainment, perhaps combining the qualities of our own *Reader's Digest* and *Times Literary Supplement*" (p. 166).

J. D. Fleeman's "Johnson's Shakespeare (1765): The Progress of a Subscription" gives a decidedly non-triumphal view of developments in the literary marketplace of the eighteenth century. Publishing by subscription had been a popular way for a well-known author to capitalize on his or her fame, or for booksellers and authors to bankroll an especially expensive production. Pope's editions of Homer are the best-known examples of subscription used to increase remuneration and establish control over the copyright, production, and distribution of literature. Johnson's own authorial independence perhaps rebelled at the element of charity implied by publishing by subscription. In any case, the publication of Johnson's edition of Shakespeare through subscription was relatively unsuccessful in monetary terms, as Johnson's indolence and poor record-keeping seemed to express a disdain for the procedure that resulted in fewer subscriptions than might have been possible. Fleeman not only recounts the story, but he attempts to resolve a puzzle over the number of subscribers to the edition, coming up with a probable figure of 400 or 500 total, fewer than the 750 estimated by Johnson's contemporary Thomas Birch.

While in general the eighteenth century saw the distinctions between booksellers, publishers, and printers widened, K. I. D. Maslen and Nicolas Barker discuss two prominent eighteenth-century printers who were able to expand their trades and become prominent in the world of letters. Maslen's essay, "Slaves or Freemen—The Case of William Bowyer, Father and Son" looks at the Bowyers, who are now remembered mainly for their preserved ledgers, which have provided invaluable information on the book trades. Barker's essay concerns the relationship between the Bowyers' former journeyman-compositor, William Strahan, who in his own words, "taught [printers] to emancipate themselves from the Slavery in which the Booksellers held them" (p. 145). Maslen's essay is a valuable exploration of some of the opportunities and difficulties faced by printers in the eighteenth century: dependent upon booksellers for much of their income, they also interrupted larger projects for "advertisements, bills, labels, tickets, hymn sheets, and so on" (p. 150). Without making the error of confusing the process of printing various forms of reading material and their meaning and reception, one might argue again that print culture must be seen to comprise non- or para-literary elements as well as the literature traditionally studied. In this instance, bibliographers have been ahead of literary scholars in understanding the reliance of modern literature upon the means of its production, not only in simple relationships between "editors," "authors," and "readers" (all terms that have recently come under scrutiny) but in the wide world of advertising, retailing, bookmaking, and packaging. Indeed, William Strahan emerges as a pivotal figure in eighteenth-century print culture in *Writers, Books, and Trade*. A friend of Benjamin Franklin, Strahan pursued in England a diverse and ambitious career that encompassed printing, publishing, and writing. Barker's short piece shows how an ambitious printer like Strahan could use the debts often run up by booksellers employing them to muscle their way into portions of the copyright, thereby retaining an interest in the capital of what the material they produced. Strahan was able to buy part of the copyright to works like *A Sentimental Journey* and *Tristram Shandy* from Thomas Becket, the third bookseller for Sterne's works after Sterne himself (who had the first two volumes "printed for the author" in York after their rejection by the Dodsleys), the Dodsleys (who quickly realized their mistake), and Thomas Becket, who became Sterne's publisher with volumes three and four of *Tristram Shandy*. Barker's essay traces in meticulous fashion Strahan's printing of Sterne's works with the concurrent increase in Strahan's permanent interest in Sterne's works, and he notes in conclusion that Strahan's own standing in literary London became equal to Sterne's.

John Horden's essay on "John Freeth: The Birmingham Poet" is an entertaining foray into sub-literary provincial writing and publishing. Freeth was a Birmingham publican who apparently had quite a knack for composing topical "verse" which he would set to well-known tunes and sing in his pub, the Leicester Arms or "Freeth's Coffee-House." During a kind of career that spanned the years 1766-1803 he published collections of his songs, many under the title of *The Political Songster*, ostensibly for the maintenance of his family:

> . . . each day, when children nine,
> In perfect health sit down to dine—
> Think not, the whole can be maintain'd
> By what is from the ale-score gain'd. (p. 310)

Horden is evidently working on a full biography of Freeth. Even if occasionally Horden makes too great a claim for Freeth's significance, this work will be interesting to those who wish to recapture the feel for the everyday life of a print culture that overlaps with performance and, not insignificantly, drinking.

Writers, Books, and Trade also includes a couple of contributions of strictly bibliographical interest, Donald D. Eddy's "Bibliography of John Brown: A Supplement" and Thomas B. Adams's "Mount and Page Imprints, 1684-1800." Indeed, the variety contained in the collection is its strength. Even if none of these essays fundamentally changes the way scholars will approach the history of the book in the eighteenth century, their eclecticism enacts a broad approach to the history of print that will surely form the basis of future work that attempts to combine material aspects of the book and other printed matter with more "literary" or "theoretical" approaches. More and more books are being published that address material aspects of writing, reading, and publishing in order to understand more fully the impact that print culture had, and continues to have, on people's lives. Even if many of these essays are perhaps overcautious in their refusal to explore the implications of bibliography for cultural study, the volume will be welcomed on library shelves by bibliographers and literary critics alike.

GEORGE JUSTICE

INDEX